FOOD PROCESSING
PRINCIPLES AND APPLICATIONS

FOOD PROCESSING
PRINCIPLES AND APPLICATIONS

K. PRABHAKARAN

KNOWLEDGE BAKERS

ISBN: 978-93-49964-26-6

Published by : **Knowledge Bakers**
Pune, India
info@knowledgebakers.com

PREFACE

Food Processing: Principles and Applications offers a comprehensive examination of the science, technology, and methodologies involved in transforming raw agricultural products into consumable food products. As the global population continues to grow, the need for efficient, safe, and sustainable food production becomes increasingly critical. Food processing plays an essential role in meeting this demand, ensuring food safety, enhancing nutritional value, and extending shelf life. This book aims to provide readers with a solid understanding of the principles behind food processing and their real-world applications in the food industry.

Designed for students, professionals, and researchers in food science, agricultural engineering, and related fields, this book serves as an essential resource for anyone interested in understanding the science and techniques that contribute to food preservation, innovation, and quality. It covers a wide range of food processing methods, from traditional techniques to modern, cutting-edge technologies, focusing on how these processes impact food quality, safety, and sustainability.

In the first part of the book, we explore the core principles of food processing, including the chemical, physical, and biological transformations that occur during processing. We examine the fundamentals of heat transfer, mass transfer, and preservation methods such as canning, freezing, drying, and fermentation. Additionally, we discuss the importance of food microbiology and how food processing contributes to the elimination of harmful pathogens and spoilage microorganisms.

The second part of the book delves into the wide array of food processing applications across different food categories, including fruits and vegetables, dairy products, meat and poultry, grains, and beverages. Each section provides a detailed examination of the specific methods and technologies used to process these foods, along with the considerations for maintaining their nutritional integrity, taste, and safety. We also address the technological advancements in food packaging, automation, and product development, as well as the growing importance of sustainable and eco-friendly practices in the food processing industry.

Furthermore, this book highlights the emerging trends in food processing, such as the use of biotechnology, nanotechnology, and high-pressure processing, which are revolutionizing the industry. We also explore the growing demand for functional foods, plant-based alternatives, and processed foods with enhanced nutritional profiles, addressing the role of food processing in responding to modern consumer needs and health concerns.

We hope that Food Processing: Principles and Applications will serve as a valuable guide for professionals and students alike, providing both the theoretical knowledge and practical insights required to navigate the complex and ever-evolving field of food processing. By offering a blend of traditional methods and cutting-edge innovations, this book encourages readers to consider the broader implications of food processing on public health, sustainability, and food security.

K. Prabhakaran

CONTENTS

Preface v

Section-I: Introduction

1. **Introductory Chapter: Food Processing** 3
   Introduction 3
   References 6

Section-II: Processing

2. **Effects of the Incorporation of Arabinoxylans Derived from Selected Cereals (Rice Bran and Corn Fibre) and Sugarcane Bagasse on the Quality of Baked Foods: A Systematic Review** 9
   Introduction 9
   Methodology 11
   Results 15
   Discussion 41
   Conclusion 49
   Abstract 50
   References 51
3. **Honey Production Process** 59
   Introduction 59
   Other beekeeping products 63
   Microbiological risks in honey and honey products 64
   HACCP in honey production 66
   Conclusions 67
   Abstract 68
   References 68

4. **Chocolate: Health, Processing, and Food Safety** 71

Introduction 71

Health benefits 72

Side effects associated with chocolate consumption 73

Production of chocolate 74

Processing of cocoa 77

Food safety of chocolate 81

Conclusions 85

Abstract 86

References 86

Section III: Packaging

5. **Pectin and Its Applicability in Food Packaging** 95

Introduction 95

Pectin 97

Pectin in food industry and food packaging 100

Conclusion 106

Abstract 106

References 106

6. **Utilization of Agro-Industrial Wastes as Edible Coating and Films for Food Packaging Materials** 115

Introduction 115

Agro-waste based renewable sources used in development of edible coatings and films 116

Preparation of edible coatings and films 118

Classification of edible coatings and films 120

Incorporation of bioactive compounds into edible coatings and films 120

Properties of edible coatings and films from agro-waste based polymers 122

Physiochemical and morphological characterization of edible coatings and films 124

Application of edible coatings and films 125

Conclusions and future prospects 126

Abstract 127

References 127

Section IV: Computational Applications

7. Computational Applications for the Evaluation and Simulation of the Thermal Treatment of Canned Foods 139

Introduction 139

General and technical aspects of the heat treatment of canned foods 141

Heat penetration studies of canned foods 143

4. Heat treatment evaluation general method 144

Calculation and prediction of heat treatment by Ball's method 145

Heat treatment modeling and simulation 145

Optimization of heat treatment 148

Conclusions 149

Abstract 149

References 149

8. Enhancement of Heat Transfer Using Taylor Vortices in Thermal Processing for Food Process Intensification..163

Introduction 163

Taylor–Couette flow 164

Food process intensification using Taylor–Couette flow 166

Conclusions 179

Abstract 179

References 180

9. Thoughts for Foods: Imaging Technology Opportunities for Monitoring and Measuring Food Quality 185

Introduction 185

Image analysis procedures 188

Software available 190

Sources of digital images 191

Gamma rays imaging 192

Conclusions 222

Abstract 224

References 224

Index 245

Section-I

INTRODUCTION

Chapter 1

INTRODUCTORY CHAPTER: FOOD PROCESSING

INTRODUCTION

Evidence suggests that the earliest food processing was undertaken in human history, using the heating process to make raw foodstuffs more palatable. When agricultural activities are started, the need for storage and preservation of raw food materials arose by 3000–1500 BC. Drying under sunlight, fermenting plant-based foods, grinding cereals, and baking bread in oven were the earliest attempts of food processing techniques. Exchanging of foods in trade and explorers-oriented technologies resulted in the change of food processing techniques in distinct food products (e.g., dairies, bakery, fermented foods). All the efforts brought today's reached point.

In today's world, food processing is a part of manufacturing industry. What we consume as foods are produced from plant-based or animal-based raw materials, in the meantime called as agricultural sources. Apart from directly consumed foods (e.g., raw eaten fruits) at postharvest term, the raw materials are required to be processed for a healthy consumption, as intermediate or finished value-added food products. Energy, equipment, labor-ship, science are used for a step (unit operation) or a series of steps (process) to produce a marketable product. For example, exposing milk to a heat source could be given as a simple example for a unit operation in terms of heat treatment (pasteurization or sterilization) in the case of milk processing at dairy industry. Therefore, from starting commercial sterilization the way what we observed up to now shaped the food industry.

To meet the needs (e.g., shelf life stabile, nutritive, and variety in convenience with diet) of the global market, postharvested raw materials are processed in different ways. To serve a marketable food product, there are a number of intrinsic and extrinsic parameters that determine specific processing design of each product. Raw materials as sources of foods to be used for processing are complex substances. As intrinsic parameters the content of foods, carbohydrates, proteins, fats, vitamins, minerals along with water are first accounted. However, biological, chemical, and physical properties might be classified both in intrinsic and extrinsic parameters.

The content and surrounding atmosphere (e.g., in-pack conditions, storage room conditions) of foods vary for almost each food product. Biological (bacteria, yeast and molds, viruses, parasites), chemical (pesticide, fungicide, allergens, mycotoxins), and physical (stones, dirt, metal, glass, insect fragments, hair) hazards should be controlled for a safe, nutritious, and wholesome consumption.

The basic processing steps consist of raw material harvesting, pretreatment (e.g., washing, separation,), basic unit operations (e.g., heat treatment, freezing, drying), and packaging. There are currently a number of different ways of processing technologies being applied in different ways at industrial scale. It is clear that almost every food types need different processing conditions to be an optimal product in its distinctive package. An overall view to food processing is presented in Table 1.

Table 1. An overview to food processing. Adapted from Park et al.

Unit operations in food processing	Thermophysical properties, microbial aspects, other considerations	Common food preservation/ processing technologies	Other food processing/ preservation technologies
Heat transfer	Raw material handling	Processes using addition or removal of heat (pasteurization and blanching, thermal sterilization, aseptic processing, sous-vide cooking, microwave heating, ohmic heating, drying, refrigeration and freezing)	Fermentation
Fluid flow	Cleaning and sanitation	Nonthermal food processing and preservation (irradiation, high-pressure processing, pulsed electric field processing, ultrasound)	Extrusion

Unit operations in food processing	Thermophysical properties, microbial aspects, other considerations	Common food preservation/ processing technologies	Other food processing/ preservation technologies
Mass transfer	Engineering properties of food, biological, and packaging material		Baking
Mixing	Microbiological considerations		Hurdle technology
Size adjustment	Role of acidity and water activity in food safety and quality		Packaging
Separation	Reaction kinetics		

However, both thermal and nonthermal food processing technologies are up-to-date widely used, alternative food processing technologies are now involved in the field. Some examples for these thermal technologies, microwave, radiofrequency, infrared heating, pressure-assisted thermal sterilization, and sous-vide processing, could be given. Whereas, nonthermal technologies in generally apply high hydrostatic pressure, irradiation, ultrasound, pulsed electric field, pulsed light technologies, and 3D printing. Nanotechnology (e.g., bio-sensing, packaging, agro-chemical production) is being started to implement besides these traditional processing techniques. The novel nonthermal or cold-pasteurization technologies provide less energy and water usage than traditional processing technologies to sustain the scarce resources in globally. These technologies in sum serve for better quality, more healthful, minimally processed, traceable, and safer foods. Beyond product performance, the expectation of consumers is to consume safe foods that are produced under hygienic and sanitary conditions.

The food industry ensures food safety with the rise of pasteurization and sterilization techniques and food security with modern agricultural practices at reasonable prices. Meanwhile, the main problem still seems to be the issue of hunger and insufficient food, and the demand for food for the growing world population is still under discussion. Moreover, to meet the dynamic conditions and consumeroriented needs, synchronized innovative solutions in food processing technologies should be found and applied.

So far all the time, the urgent need was to feed the world, by growing sufficient raw material and producing food products (e.g., wheat flour, dairies, meat products, aqua culture products, etc.) at reasonable prices. However, in today's world, the climate crisis threatens sustainable food production, whereas sustainability concern has now become a real endangering issue. The United Nations defined "*the needs of the present without compromising the ability of future generations to meet their own needs*" for sustainable development [6]. From starting this end, a food production should ensure secure, environmentally sustainable, and healthy supply. The food industry produces more accessible and convenient foods, in terms of having reasonable price and remaining stable in predetermined shelf life periods. However, the quality of the foods, mainly snacks or on-the-go food or beverages, should be questioned for quality aspects. These types of foods and beverages containing carbohydrates, lipids, sugar, and food additives, which are increasing day by day and are purchased relatively cheaply, should be discussed for their unhealthy conditions. For the case of this concern in food industry, a high amount of negative perception on processed foods has become among consumers. The rate of literate people, social media, TV programs, magazines have all contributed to aforementioned negative concern on foods in pack.

To overcome current negative conditions, efficient and economic use of scarce resources (water, land, energy, air), environmental protection, and waste (food and non-food) management are requiring innovative attempts, unless carrying no suspicion on health concern of consumers.

REFERENCES

1. Fellows PJ. Food Processing Technology. Cambridge, UK: Woodhead Publishing; 2009
2. Park SH, Buddhi PL, Balasubramaniam VM. Principles of food processing. In: Clark S, Jung S, Lamsal B, editors. Food Processing: Principles and Applications. West Sussex, UK: John Wiley & Sons Publishing; 2014. pp. 1-15
3. Neetoo H, Chen H. Alternative food processing technologies. In: Clark S, Jung S, Lamsal B, editors. Food Processing: Principles and Applications. West Sussex, UK: John Wiley & Sons Publishing; 2014. pp. 137-169
4. Gunasekaran S. Nanotechnology for food: Principles and selected applications. In: Clark S, Jung S, Lamsal B, editors. Food Processing: Principles and Applications. West Sussex, UK: John Wiley & Sons Publishing; 2014. pp. 171-205
5. Schoenfuss T, Lillemo JH. Food safety and quality assurance. In: Clark S, Jung S, Lamsal B, editors. Food Processing: Principles and Applications. West Sussex, UK: John Wiley & Sons Publishing; 2014. pp. 233-247
6. Murphy F, McDonnell K, Fagan CC. Sustainability and environmental issues in food processing. In: Clark S, Jung S, Lamsal B, editors. Food Processing: Principles and Applications. West Sussex, UK: John Wiley & Sons Publishing; 2014. pp. 207-232

Section 2

PROCESSING

Chapter 2

EFFECTS OF THE INCORPORATION OF ARABINOXYLANS DERIVED FROM SELECTED CEREALS (RICE BRAN AND CORN FIBRE) AND SUGARCANE BAGASSE ON THE QUALITY OF BAKED FOODS: A SYSTEMATIC REVIEW

INTRODUCTION

There are numerous dietary fibre enriched food products developed in the food industry during the last decade encompassing various popular and widely consumed foods such as bread, cakes, cookies or biscuits, yoghurts among others [1–4]. It is well established in the literature, that dietary fibre intake at levels greater than 25 g per day, tend to be associated with numerous health benefits, namely the reduced risk of coronary heart disease, type 2 diabetes, enhanced physiological functions of the human body, improved weight maintenance and other positive effects on various disease risk factors and the alleviation of certain types of cancers [5–8].

In recent years, several drivers such as consumer awareness of the nutritional value of dietary enriched foods, and governmental policies promoting healthy lifestyle behaviours have contributed to the continual increase in the use of dietary fibres in foods [3, 9–11]. Consequently, the value of the dietary fibres global market is expected to experience an astonishing annual rise, with the latest estimates projected growth of about 9.74 billion U.S. dollars by 2025 [12]. Dietary fibres can be considered as non-digestible carbohydrates which are inclusive of lignin, resistant oligosaccharides, resistant starch, non-starch polysaccharides (NSP) such as cellulose, pectins, hydrocolloids and hemicelluloses of which can be eaten and are not prone to enzymatic digestion and absorption within the small intestines, but can undergo complete or partial fermentation in the large intestine of the human body [1, 3, 6]. In previous studies, Foschia, Peressini, Sensidoni, and Brennan [13] indicated that the major dietary fibres (DFs) are consist of arabinoxylans, β-glucans, resistant starch and inulin. In the non-starch polysaccharides (NSP) fraction, arabinoxylan is the main polysaccharide, additionally, arabinoxylan structure is comprised of a framework of β-(1–4) connected xylose residues to which α-Larabinose tend to linked unto the second or third carbon positions [14, 15]. Agricultural by-products from milling industries namely fibres from selected cereal crops rice, corn, and energy crop sugarcane and plant parts from other fruits and vegetables can be regarded as dietary fibres and subsequently tend to contain arabinoxylans in varying amounts [16–18]. The addition of fibres to food products tends to influence the consistency, texture, rheological tendencies and sensory characteristics of the finished food products [3]. The incorporation of fibres in breakfast cereals, bread, cookies, cakes, pasta, yogurt, beverages and meat products among others have been widely reported with desirable results [2, 13, 19–23].

Arabinoxylans have been considered as an important dietary fibre of selected cereals (rice bran and corn fibre) and sugarcane bagasse and it has been suggested that when incorporated in the optimum proportions into food products they are capable of improving its quality, which includes, but not limited to only changes in the physical, rheological, and sensorial characteristics of food products [24–26]. Moreover, various technological functions of food products are also enhanced by dietary fibre incorporation into foods such as its nutritional value, functional properties and improvement of other chemical properties [27–29]. It has also been reported in several studies that dietary fibre arabinoxylans from agricultural byproducts such as rice bran, corn fibre: brans and other corn parts and sugarcane bagasse can be safely utilised in the baking industry and consequently be used whole or as extracts of soluble dietary arabinoxylans to facilitate the production of functional and health-promoting food products through the supplementation of bread, cakes, cookies and other food products at varying incorporation rates [2, 17, 30–34].

However, there is a lack of consensus in the body of literature relating to the beneficial effects the incorporation at various levels of dietary arabinoxylans derived or originating from sources such as selected rice bran, corn fibre (bran and other parts) and sugarcane bagasse have on the sensory and physical qualities of baked foods including cookies/biscuits, bread and cakes [23, 35–40]. This systematic review, therefore, endeavours firstly to review the role of dietary fibre derived from selected cereals (rice bran and corn fibre) and sugarcane bagasse in baked foods, namely cookies/biscuits, bread and cakes; secondly to evaluate the effects of the incorporation of dietary fibre derived from selected cereals (rice bran and corn fibre) and sugarcane bagasse on the physical and sensory qualities of these baked food products and finally to critically assess the beneficial impacts of dietary fibre incorporation derived from selected cereals (rice bran and corn fibre) and sugarcane bagasse in baked foods, including bread, cakes and cookies.

METHODOLOGY

The methodology follows applicable aspects of the Preferred Reporting Items for Systematic Reviews and Meta-Analyses (PRISMA) guidelines [41] and includes details of the search strategy, studies selection, inclusion and extraction criteria, data extraction and assessment of study validity, risk of bias assessment and data analysis of the relevant studies utilised in this systematic review.

Research strategy

A systematic literature search using a Keyword-based concept was conducted for articles catalogued in ScienceDirect, Web of Science, PubMed, Medline, CINAHL and Google Scholar that were published between January 1, 2010 and August 1, 2020 using the following search strategy. The keyword-based searches included Boolean operators and were constructed using words from the research question along with the combination of truncations and wildcards to access as much primary research material as possible. The searches were limited to Scholarly and Peer-reviewed and English language. Grey areas of the literature such as government reports, conference reports and food magazines were also searched. Further, the reference lists of recent systematic reviews were searched for additional references.

Study selection, inclusion and exclusion criteria

The eligibility criteria utilised for the selection of relevant research studies followed the population, intervention, comparison, outcome and study settings (PICOS) research question framework. Further details are summarised in **Table 1**.

For studies to be included in the systematic review the following criteria were established, namely:

- The keywords and phrases e.g. arabinoxylans incorporation, rice and corn brans/fibre, sugarcane bagasse, baked food products among others relating to the research question should be included in the particular article title,

Table 1.: *PICOS criteria for study selection.*

Population	Human subjects and quantitative parameters of nutritional and physical properties of baked foods including bread, cakes and cookies.
Intervention	Rice bran, corn fibre and sugarcane bagasse
Comparison	Nutrient profile, volume, texture, weight, height, colour and sensory evaluation metrics.
Outcome	Improve quality
Study Setting	All research studies investigating the effects of the incorporation of dietary fibres on the nutritional, physical and sensory qualities of baked food products.

- Human subjects including adults/children,
- Food vehicles included bread, cakes and cookies,
- Source of arabinoxylans rice bran, corn bran and sugarcane bagasse,
- Quantity of arabinoxylans from the particular sources that was incorporated,
- Sensory evaluation,
- Quantitative parameters of nutritional profile e.g., fibre, fat, carbohydrates and protein content and physical profile e.g., texture, volume, weight, height and colour of baked foods including bread, cakes and cookies,
- Full text of articles (abstract only of articles would not be considered),
- Articles should be written in English language,
- The published dates of articles should be between January 1, 2010 and August 1, 2020,
- Controlling bias by using randomisation and mean measurements of food products from triplicates instead of a single measurement and
- Articles should be able to be placed into sub-groups to facilitate statistical analysis.

Studies under consideration for inclusion in the review which did not contain the required information as outlined previously were not selected. All articles chosen were sent to the EndNote reference database, which facilitated the identification of duplicate articles, which were also excluded from the review.

Articles consisted of a sensory evaluation of a particular baked food product, including bread, cakes and cookies; a nutritional and physical profiles of bread, cakes and cookies were deemed fundamental to the review since they provided data relating to the nutritional, physical and sensory qualities of baked foods being studied. The use of articles investigating the different dietary fibre sources of arabinoxylans (rice and corn brans and sugarcane bagasse) incorporation into foods facilitated the assessment of the effects of arabinoxylans incorporation into foods, its possible implications, roles and the possible optimum inclusion proportions of arabinoxylans, which may result in improved nutritional, physical and sensory qualities of different food products for consumer consumption. Priority was given to studies examining arabinoxylans derived from agricultural by-products of selected cereals (rice and corn) and sugarcane bagasse. Articles produced earlier than 2010 were excluded, mainly to reflect the recent advancement of public health policy guidelines, modern processing techniques and equipment within the food industry.

Data extraction and assessment of study validity

The data extracted from the studies were performed independently by the researcher and it included the pertinent characteristics of the studies relating to the research question of the systematic review. The characteristics used for the data extraction sheet consisted of the following headings:

- Authors' name
- Title
- Location of study and funding source
- The objective of the study
- Source of arabinoxylans and amount used
- Food vehicle (bread, cakes and cookies)
- Control of bias – Randomisation usage in design, triplicate measurements, statistical analysis and sensory evaluation
- Sensory evaluation metrics
- The nutrient profile of food product
- Physical parameters of food product (volume, texture, weight, height, colour)
- Statistical analysis

- Results
- Conclusion

The data from studies were checked for errors and any unavailable data was denoted as not determined (ND) in the particular table.

The assessment of study validity was conducted using the Downs and Black checklist [42]. This checklist is comprised of 27 questions and can be utilised in the assessment of the methodological quality of both randomised and non-randomised studies [42]. However, for this review only fourteen (14) questions were found to be applicable, the other questions were denoted as not applicable. The score ranges and corresponding quality levels used for the Downs and Black [42] were as follows: excellent (26–28); good (20–25); fair (15–19) and poor (≤14) [43]. Some of the included criteria relevant in the assessment of articles for this type of review comprised of the following areas:

1. Reporting
2. External validity
3. Bias
4. Confounding factors
5. Statistical power

Risk of bias assessment

The risk of bias assessment of the articles used in this review was evaluated using the Cochrane collaboration's tool for assessing the risk of bias. The statistical information presented in the articles was assessed for its appropriateness. The outcomes reported in articles used were verified for accuracy and the section of the study it was first reported was noted. Moreover, the credentials and attachments of the respective authors of the articles utilised in the review were also checked. Importantly, references to the disclosure of interest were keenly examined at the end of the articles.

Data handling

In this review, only eleven (11) research articles out of fourteen (14) studies were considered suitable for statistical data analysis. There were five (5) studies that focused on rice bran as the arabinoxylan source and both cookies/biscuits (3 studies) and bread (2 studies) as the food vehicle; four (4) studies used corn fibre as the arabinoxylan source and both cakes (2 studies) and cookies/biscuits (2 studies) and finally two (2) studies conducted research using sugarcane bagasse as the arabinoxylan source and cookies/biscuits as the food vehicle. In the area of sensory evaluation, various scales were used in the assessment of particular sensory attributes such as hedonic scales (5-point, 7-point and 9-point) and10 cm unstructured line scale. Thus, sensory assessment scores were standardised by dividing individual scores given by the panel by the maxima of the particular scale used, then multiply by 100 to convert to a percentage. Data of panellists were extracted and grouped into categories of trained, semi-trained and untrained; gender and nationality and utilised to conduct a descriptive statistical analysis. Students were grouped as semitrained and in studies that did not specify the training of panellists, they were classed as untrained. Studies were grouped into sub-groups according to the food vehicle and source of arabinoxylan that was incorporated into the food product. The weighted means of the papers was calculated by using the Statistical Package for Social Sciences (SPSS), version 26 software. Moreover, the percentage of fibre incorporation was also similar for each sub-group utilised for data analysis. The metrics extracted from the included studies relating to the physical analysis of the particular food vehicle were also standardised namely for bread: volume (ml), mass (g) and specific volume (g/ml); cookies/biscuits: width (mm), thickness (mm), spread factor (%) and colour: L*, a* and b* and finally cakes: crumb colour: L*, a* and b*, crust colour: L*, a* and b* and texture (Kg F).

Data outputs

The fundamental characteristics extracted from the articles were conducted using applicable aspects from the Preferred Reporting Items for Systematic Reviews and Meta-Analyses (PRISMA) guidelines [41]. Statistical analysis of data extracted from the various studies included in the review was performed using a Simple (oneway) analysis of variance (ANOVA) using Statistical Package for Social Sciences (SPSS), version 26. Post Hoc analysis was conducted using Tukey's test to identify where the difference lies between each group, (see Appendix F). The significant difference between the means was identified where ($p \leq 0.05$).

RESULTS

Study selection

In this systematic review, a total of fourteen (14) studies satisfied the inclusion criteria established, the process that guided the selection of these studies is illustrated in the PRISMA diagram (see **Figure 1**). However, only eleven (11) studies facilitated categorisation into a sub-group, namely bread, cakes and cookies/biscuits and then further sub-divided into fibre type and food vehicle and were thus considered suitable to be utilised for statistical analysis, (see **Table 2**). In the fourteen (14) studies included, all conducted a sensory evaluation study, a nutritional analysis/profile and a physical analysis of the particular food produced being studied, except the research studies completed by [17, 23], there was no nutritional analysis/ profile and the study carried out by [44], which did not perform a physical analysis of the chapatti, a type of fermented bread. The sensory studies consisted of 362 participants from eleven (11) countries, see **Figure 2.** Panel members used in the sensory studies ranged from 5 to 60 participants and the mode were 30 and 10 panellists. The panellists were grouped into trained, semi-trained and untrained, see Table 2. The panel members were also made up of both gender (male and female), age groups (which ranged from 18 to 50 years old) and occupations (staff members, post-graduate students and others not mentioned). In addition, eight (8) of the 14 studies were conducted in Asia, namely Sri Lanka-1, India – 3, Iran – 1, Pakistan – 2 and Bangladesh – 1. Two (2) studies were conducted in North America, namely United States – 1 and Mexico – 1 and two (2) studies were carried out in Africa, namely Cameroon 1 and Nigeria-1. Finally, one study was conducted in Brazil, South America. Studies examined various baked food products, namely bread-four (4) studies), cakes-three (3) studies, cookies/ biscuits studies-seven (7) studies.

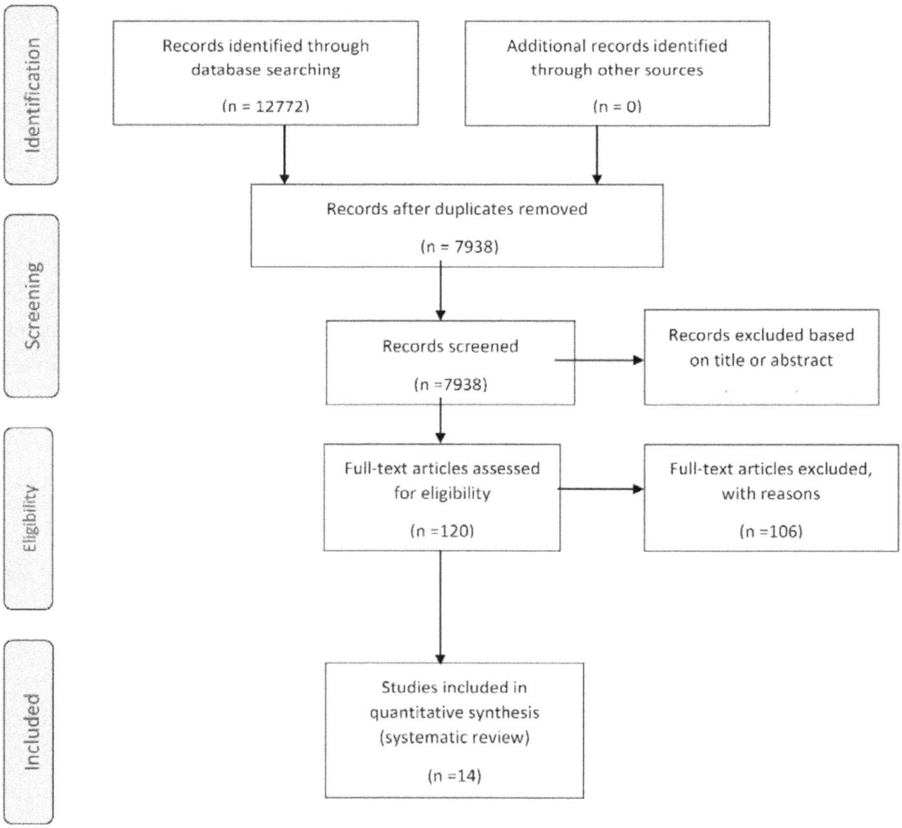

Figure 1. *Study selection process based on PRISMA guidelines. Adapted from [41].*

Table 2. *Sensory quality of sugarcane bagasse enriched cookies/biscuits.*

Detail	Appearance	Taste	Texture	Overall acceptance
Control (0%)	85.6 ± 8.6a	86.9 ± 9.8c	85.2 ± 9.2a	86.1 ± 9.2a
5% SCB	84.2 ± 7.5b	82.1 ± 7.4b	82.1 ± 7.4b	84.2 ± 7.5a
10% SCB	68.9 ± 18.9c	69.5 ± 21.0c	72.3 ± 13.4c	72.3 ± 13.4b

Values are means ± standard deviation of sub-groups. Means in the same column with different superscripts are significantly different ($p < 0.05$). Key: SCB – sugarcane bagasse.

The metrics utilised in the sensory studies of this review encompassed colour of crumb and colour of crust; aroma; flavour; appearance; taste; texture firmness or hardness and overall acceptability, see **Table 2.** Meanwhile, for the physicochemical and physical characteristics aspects of the studies, the parameters examined included moisture %, protein %, ash %, fat %, carbohydrates %, crude fibre %, mineral contents and phenolic contents; texture for example firmness or hardness; colour or luminosity of crust and crumb (for example lightness/brightness, redness/ greenness and yellowness/blueness); loaf weight, bread height and bread volume; diameter, thickness, spread ratio and texture, see **Table 2**. The intervention data extracted were the various dietary fibres derived from arabinoxylans selected sources, namely rice bran, corn fibre and sugarcane bagasse and which was supplemented at different rates ranging from 0–30% into baked foods bread, cakes and cookies/biscuits.

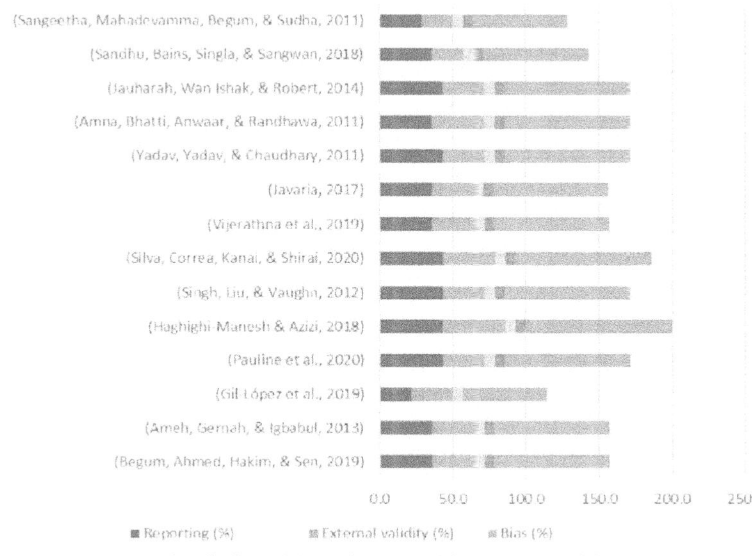

Figure 2. *Quality assessment results of the various studies used in the review, presented in percentages per category and overall total using the Downs and Black checklist.*

Data quality

The quality assessment of the fourteen (14) research studies included in the review was conducted using the Downs and Black [42] checklists, importantly only fourteen (14) questions were considered applicable, the other questions were denoted as not applicable to this review and consequently were not utilised to assess the included studies. The highest score was received by the research paper [17], namely 14/14, while the lowest score was obtained by the research paper [44], namely 8/14. Five (5) research papers received 11/14; four (4) research papers received 12/14 and one (1) research paper each received scores of 13/14, 8/14, 9/14 and 10/14 respectively, (see **Figure 2**). Thirteen (13) out of the fourteen (14) studies excelled in the category of reporting, only [44] scored poorly. In the category of bias or internal validity all fourteen (14) papers received satisfactory scores, contrasting in the category of external validity only two (2) studies excelled, the other studies failed to show that the panel members were chosen from a representative sample. Moreover, in areas of confounding factors and statistical power, the majority of studies received high scores. Based on the scoring scale of the Downs and Black [42] checklists the quality of the fourteen (14) studies would be considered in the range of fair to excellent [43].

Data analyses

This section entails the review and analysis of the findings of the primary research articles included in the review which were categorised into sub-groups to reveal common effects and enhance statistical power, except for those papers which did not enable statistical analysis. The studies were grouped as follows: the types of food or food vehicle namely, cookies, biscuits, bread and cakes and fibre; the type of fibre supplemented into the food product, namely rice bran, corn fibre and sugarcane bagasse and the particular outcomes: effects or no effects were outlined or highlighted, (see **Table 2**). The included studies were further sub-divided into specific food and fibre types; the same rates of fibre incorporation into the particular food product as the food vehicle and their respective similar sensory metrics and physical parameters were extracted from the eleven (11) studies to facilitate statistical analysis. The findings will be presented under three (3) main food type headings, namely cookies/biscuits, bread and cakes to facilitate a logical presentation.

Cookies/biscuits: sugarcane enriched

The incorporation of sugarcane bagasse up to the level of 10% showed significantly undesirable overall acceptance ($p < 0.05$). Moreover, based on the sensory evaluation results, as the level of sugarcane bagasse increased, the overall acceptance of enriched cookies/biscuits reduced significantly ($p < 0.05$), see **Table 2**. Cookies/ biscuits incorporated with sugarcane bagasse at 5% were similar to control.

Sugarcane bagasse incorporation up to 10% resulted in no significant differences ($p > 0.05$) in the physical quality of the cookies/biscuits. In **Table 3**, the thickness of fibre supplemented cookies was marginally less than the control sample and although the enriched cookies/biscuits were slightly wider, enrichment produced a reduction in the parameter of spread factor and was more 2 times harder than control cookies/biscuits.

Cookies/biscuits: rice bran enriched

Rice bran fibre supplemented cookies/biscuits up to 15% showed significantly undesirable overall acceptance. The sensory scores were all lower than the control sample according to sensory evaluation. Rice bran enriched samples obtained, namely 15% incorporation obtained the low scores for most of the sensory attributes assessed, see **Table 4**. Overall acceptance of cookies/biscuits significantly reduced ($p < 0.05$) in comparison to the control sample of cookies/biscuits. Incorporation of rice bran into cookies/biscuits at 10% performed only slightly better than 5% and 15% levels of incorporation. However, the control sample obtained the best overall acceptance based on sensory evaluation.

The incorporation of rice bran into cookies/biscuits significantly ($p < 0.05$) enhanced the thickness of the food product. The width and degree of spread factor showed no significant differences in comparison to the control. Further details are outlined in **Table 5**.

Cookies/biscuits: corn fibre enriched

Corn fibre supplemented cookies up to 20% obtained a significant ($P < 0.05$) desirable overall acceptance based on sensory panel evaluation. There were significant differences ($p < 0.05$) in colour between enriched corn fibre cookies/biscuits and the control. Interestingly, 20% supplemented obtained the best score for overall acceptance. Moreover, both 10% and 20% enriched cookies/biscuits were found to be statistically similar, see Table 6.

Table 3. *Physical quality of sugarcane bagasse enriched cookies/biscuits.*

Detail	Thickness (mm)	Width (mm)	Spread factor (SF%)	Texture (Kg force)
Control (0%)	5.3 ± 0.3a	48.9 ± 10.3b	100.0 ± 0.0c	5.7 ± 5.4d
5%	4.9 ± 0.2a	51.0 ± 6.6b	87.3 ± 3.0c	7.6 ± 6.6d
10%	4.9 ± 0.4a	51.9 ± 5.0b	83.4 ± 9.5c	13.0 ± 3.9d

Values are means ± standard deviation of sub-groups. Means in the same column with different superscripts are significantly different ($p < 0.05$). Key: SCB – sugarcane bagasse.

Table 4. *Sensory quality of rice bran enriched cookies/biscuits.*

Detail	Flavour	Colour	Taste	Texture	Overall acceptance
Control (0%)	79.8 ± 14.5a	80.5 ± 15.1a	76.5 ± 13.0a	72.9 ± 11.7a	78.4 ± 12.3a
5% RB	70.9 ± 10.8ab 73.3 ± 11.7ab 69.6 ± 14.2ab 67.3 ± 10.7b	72.1 ± 11.0b			
10% RB	72.9 ± 8.5ab	72.4 ± 4.7ab	71.3 ± 7.1ab	71.4 ± 9.7a	72.3 ± 5.4b
15% RB	71.7 ± 5.2ab	71.4 ± 7.2ab	67.4 ± 8.0ab[1]	71.5 ± 1.0a	71.0 ± 3.1b

Values are means ± standard deviation of sub-groups. Means in the same column with different superscripts are significantly different ($p < 0.05$). Key: RB – rice bran, ab[1] is significantly different from control and 10% RB samples.

Table 5. *Physical quality of rice bran enriched cookies/biscuits.*

Detail	Thickness (mm)	Width (mm)	Spread factor (SF%)
Control (0%)	8.6 ± 1.0a	42.7 ± 13.6b	51.7 ± 5.2c
5% RB	9.5 ± 0.5ab[1]	40.5 ± 13.2b	49.4 ± 5.6c
10% RB	9.6 ± 0.1ab	38.8 ± 12.9b	46.5 ± 6.6c
15% RB	9.9 ± 0.3ab	36.9 ± 13.8b	44.5 ± 6.9c

Values are means ± standard deviation of sub-groups. Means in the same column with different superscripts are significantly different ($p < 0.05$). Key: RB – rice bran and ab[1] – no significant difference between 5% and 10% RB supplementation.

Table 6. *Sensory quality of corn fibre enriched cookies/biscuits.*

Detail	Flavour	Colour	Appearance	Texture	Overall acceptance
Control (0%)	81.5 ± 10.4a	81.2 ± 10.9a	83.2 ± 10.9a	85.3 ± 6.3a	83.2 ± 10.9a
10% CF	83.8 ± 7.9a	82.6 ± 10.4ab	83.2 ± 10.9a	85.3 ± 6.3a	84.4 ± 8.4ab
20% CF	79.9 ± 13.4ab	79.3 ± 12.8ab	79.9 ± 13.4a	80.8 ± 13.6b	86.7 ± 5.3ab
30% CF	67.8 ± 14.5b	70.0 ± 19.8 ac	67.2 ± 17.7b	67.9 ± 18.2c	67.2 ± 17.7c

Values are means ± standard deviation of sub-groups. Means in the same column with different superscripts are significantly different ($p < 0.05$). Key: CF – corn fibre.

In **Figure 3**, both 10% and 20% enriched cookies/biscuits obtained the best scores from panellists. Meanwhile, the incorporation of cookies/biscuits at 30% was not well accepted during sensory evaluation.

The physical qualities of up to 30% corn fibre supplemented cookies/biscuits were statistically similar to the control sample. However, cookies/biscuits incorporated with corn fibre up to 20%, showed increased thickness in comparison to the control sample, see **Table 7**. Incorporation of corn fibre at 30% gave cookies/ biscuits the highest firmness 2.1 ± 0.2 (Kg Force).

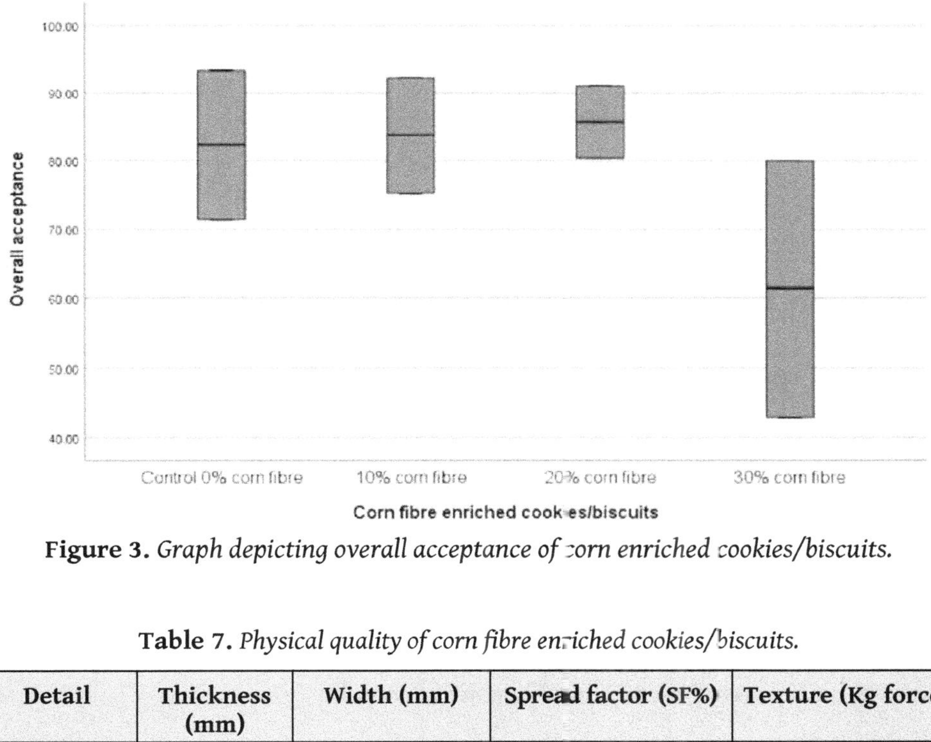

Figure 3. *Graph depicting overall acceptance of corn enriched cookies/biscuits.*

Table 7. *Physical quality of corn fibre enriched cookies/biscuits.*

Detail	Thickness (mm)	Width (mm)	Spread factor (SF%)	Texture (Kg force)
Control (0%)	5.9 ± 0.4a	49.1 ± 24.5b	75.9 ± 32.7c	1.6 ± 0.7d
10% CF	5.9 ± 0.4a	48.3 ± 19.9b	75.5 ± 25.6c	1.8 ± 0.5d
20% CF	6.1 ± 0.4a	48.4 ± 18.8b	74.0 ± 23.6c	1.9 ± 0.4d
30% CF	5.9 ± 0.7a	51.9 ± 5.0b	75.6 ± 19.5c	2.1 ± 0.2d

Values are means ± standard deviation of sub-groups. Means in the same column with different superscripts are significantly different (p < 0.05). Key: CF – corn fibre.

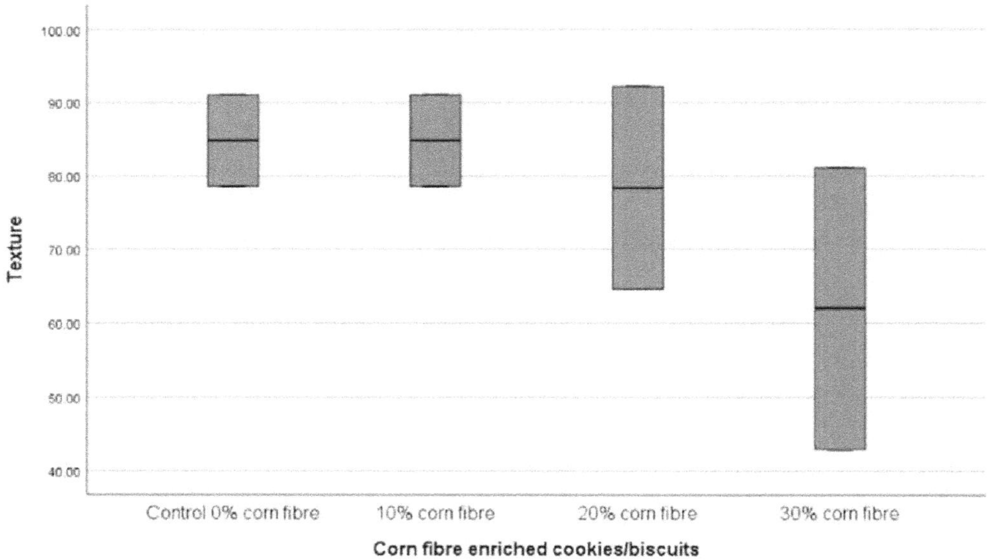

Figure 4. *Graph depicting the sensory attribute – the texture of corn fibre incorporated cookies/biscuits.*

In **Figure 4**, the texture of corn fibre enriched cookies/biscuits incrementally reduced with the incorporation of increased fibre up to 30%.

Rice bran enriched bread

In **Table 8**, incorporation of rice bran into up to 15% produced significantly ($p < 0.05$) undesirable overall acceptance. There was a significant difference ($p < 0.05$) in the aroma of the fibre enriched bread and control bread. All the rice bran fibre enriched bread was similar to each other but all significantly different ($p < 0.05$) from the control. There was no significant difference in the overall acceptance of 5% and 10% rice bran incorporated bread from the control sample.

In **Table 9**, there were no significant differences between rice bran supplemented bread up to 10%.

Sensory quality of corn bran enriched cakes

Based on sensory evaluation corn bran supplemented cakes up to 20% were found to be significantly desirable ($p < 0.05$) in comparison to control cake samples for the attributes of crust colour, taste, texture and consequently obtained the overall acceptance, see **Table 10**.

Table 8. *Sensory quality of rice bran enriched bread.*

Detail	Aroma	Crust colour	Crumb colour	Taste	Texture	Overall acceptance
Control (0%)	86.1 ± 3.9a	90.5 ± 2.4a	87.2 ± 3.9a	83.3 ± 1.6a	88.4 ± 0.8a	88.4 ± 0.8a
5% RB	73.9 ± 3.2ab	80.0 ± 7.7a	77.8 ± 6.4a	78.4 ± 3.2a	79.4 ± 5.8a	80.6 ± 0.6a
10% RB	71.1 ± 8.5ab	71.7 ± 11.6a	72.2 ± 8.5a	75.0 ± 6.7a	80.8 ± 13.6a	72.8 ± 10.4a
15% RB	68.9 ± 7.2ab	71.7 ± 11.3a	68.3 ± 6.5a*	68.9 ± 5.3a*	80.8 ± 8.3a*	68.4 ± 7.8a*

*Values are means ± standard deviation of sub-groups. Means in the same column with different superscripts are significantly different ($p < 0.05$). Key: RB – rice bran and * means 15% is significant in comparison to control only.*

Table 9. *Physical quality of rice bran enriched bread.*

Detail	Specific volume (ml/g)
Control (0%)	10.7 ± 9.1a
5% RB	10.1 ± 6.9a
10% RB	9.8 ± 6.6a

Values are means ± standard deviation of sub-groups. Means in the same column with different superscripts are significantly different ($p < 0.05$). Key: RB – rice bran.

Table 10. *Sensory quality of corn bran enriched cakes.*

Detail	Crust colour	Taste	Texture	Overall acceptance
Control (0%)	87.6 ± 6.3a	81.2 ± 10.9a	85.3 ± 6.3a	83.2 ± 10.9a
10% CB	86.1 ± 2.3a	82.6 ± 10.4ab	85.3 ± 6.3a	84.4 ± 8.4ab
20% CB	80.6 ± 0.6b	79.3 ± 12.8ab	80.8 ± 13.6b	86.7 ± 5.3ab

Values are means ± standard deviation of sub-groups. Means in the same column with different superscripts are significantly different ($p < 0.05$). Key: CB – corn bran.

Physical quality of corn bran enriched cakes

Corn bran incorporated cakes significantly ($p < 0.05$) impacted the crust luminosity of cakes at both 25% and 30% supplementation levels. All the other physical parameters were similar to the control sample. The texture, namely firmness increased as the level of corn bran increased in the cakes, see **Table 11**.

Initially, the texture of corn bran enriched cakes increased with the addition of corn bran, then reduced at 10% level of incorporation and thereafter incrementally increased as the supplementation levels were elevated in cakes. Not surprisingly, the highest degree of firmness in cakes is at the 30% level of corn bran incorporation in cakes, see **Figure 5**.

Table 11. *Physical quality of corn bran enriched cakes.*

Detail	Crust L*	Crust a*	Crust b*	Crumb L*	Crumb a*	Crumb b*	Texture (Kg Force)
Control (0%)	84.4 ± 4.2a	-3.9 ± 6.9a	28.6 ± 11.6a	51.1 ± 17.9a	11.7 ± 3.7a	34.5 ± 14.2a	6.7 ± 1.5a
5% CB	80.8 ± 2.9a	-3.6 ± 6.6a	31.7 ± 6.4a	50.7 ± 15.3a	11.1 ± 3.2a	35.2 ± 11.7a	6.9 ± 1.2a
10% CB	78.1 ± 1.4a	-5.1 ± 6.2a	31.8 ± 7.6a	48.8 ± 17.5a	10.3 ± 2.6a	36.1 ± 9.8a	6.9 ± 0.9a
15% CB	73.9 ± 3.6a	-5.0 ± 6.6a	33.2 ± 6.5a	47.1 ± 16.8a	9.8 ± 4.1a	36.3 ± 9.2a	7.1 ± 1.1a
20% CB	71.9 ± 3.5a	-5.4 ± 7.5a	34.6 ± 5.6a	46.9 ± 17.7a	8.0 ± 3.8a	36.9 ± 7.4a	7.4 ± 1.2a
25% CB	68.2 ± 6.7b	-5.8 ± 8.5a	35.5 ± 5.2a	46.5 ± 18.4a	6.8 ± 4.6a	37.5 ± 6.6a	7.7 ± 1.2a
30% CB	64.7 ± 7.0b	36.6 ± 5.0a	36.6 ± 5.0a	45.7 ± 19.3a	5.4 ± 4.4a	37.5 ± 4.9a	7.7 ± 1.1a

Values are means ± standard deviation of sub-groups. Means in the same column with different superscripts are significantly different ($p < 0.05$). Key: CB – corn bran.

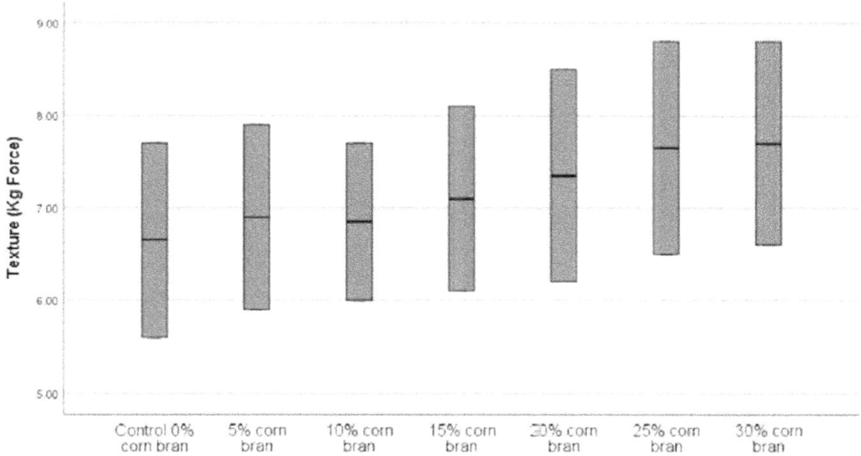

Figure 5. *Graph depicting the texture qualities of corn bran enriched cakes.*

Data Characteristics and features

Several baked food products, namely cookies/biscuits, cakes and bread illustrated enhanced levels of dietary fibre, moisture, ash, minerals, vitamins contents and on the other hand reduction in proteins, fats and carbohydrates, see **Table 12.**

The incorporation of rice bran up to the level of 15% into biscuits significantly

($p < 0.05$) enhanced its nutritional value [20]. In previous studies Yadav, Yadav, and Chaudhary [39] also found that the supplementation of biscuits up to 15% significantly ($p < 0.05$) enhanced its nutritional value in comparison to the control sample, namely significant increase in protein content from 7.3% to 15.4%; there was a non-significant increase in ash and fibre contents and a reduction in carbohydrates content. Several studies of cookies and biscuits using both higher rates of corn fibre (up to 40%) and rice bran (up to 20%) also reported a significant increase in protein, fibre, minerals and moisture contents, on the contrary, fat contents and total carbohydrates showed a reduction in cookies and biscuits [45, 46].

Table 12. *General study features and characteristics of cookies/biscuits, bread and cakes.*

Reference	Downs and Black quality assessment	Food vehicle	Fibre type and % incorporation	Control	n	Analysis parameters	Panela	Outcomes
[45]	11/14	Cookies	Rice bran – acid stabilised and heat stabilised (0–20%)	Wheat flour	5	Physical parameters: • Width (mm) • Thickness (mm) • Spread factor (%) • Texture (Kg F) Sensory attributes: flavour, colour, appearance, taste, texture and overall acceptability.	Postgraduate students – Food Technology Department (T)	No significant difference was detected in the chemical and physical properties of cookies incorporated with acid stabilised rice bran (ASRB) and heat stabilised rice bran (HSRB). The incorporation of brans in cookies resulted in a significant incremental increase of moisture, crude protein, fat and mineral contents. The parameters of average width, thickness and spread factor of cookies were also increased as rice brans were added. The scores for colour of cookies decreased significantly as the level of rice bran increased, but not was non-significant at the 10 percent level of substitution.

Reference	Downs and Black quality assessment	Food vehicle	Fibre type and % incorporation	Control	n	Analysis parameters	Panela	Outcomes
[20]	1/14	Biscuits	Rice bran (0–10%)	Wheat flour	26	Physical parameters: • Width (mm) • Thickness (mm) • Spread factor (%) • Texture (Kg F) Sensory attributes: flavour, colour, appearance, taste, texture and overall acceptability.	Males and females (50% males and 50% females), age ranged 20–50 years, untrained panellists (UT)	Biscuits supplemented with 5–10% heat stabilised rice bran using flour containing (rice flour 35% + maize flour 35% + pea flour 30%) resulted in desirable overall acceptance by the sensory panel. Incremental addition of stabilised rice bran significantly increased moisture, ash, crude protein, fibre and thickness of the biscuits. Width and spread factor of biscuits reduced with increasing levels of stabilised rice bran.

Reference	Downs and Black quality assessment	Food vehicle	Fibre type and % incorporation	Control	n	Analysis parameters	Panela	Outcomes
[39]	12/14	Biscuit	Rice bran protein concentrate (RBPC) (0–15%)	Refined wheat flour	10	Physical parameters: • Width (mm) • Thickness (mm) • Spread factor (%) • Texture (Kg F) Sensory attributes: flavour, colour, appearance, taste, texture and overall acceptability.	Semi-trained members (ST)	The substitution of refined wheat flours up to 10 per cent using RBPC resulted in the formulation of protein-enriched biscuits with favourable overall acceptability. There was a significant increase in the protein content of biscuits produced from 7.3% in control biscuits to 15.4% in the biscuits with 15% rice bran supplementation. The fracture strength was also significantly higher than that of the control biscuits ($p < 0.05$).

Reference	Downs and Black quality assessment	Food vehicle	Fibre type and % incorporation	Control	n	Analysis parameters	Panela	Outcomes
[46]	12/14	Biscuits	Corn fibre (0–30%)	Wheat flour	60	• Texture • Diameter (W):mm • Thickness (T): mm • Spread ratio • % spread factor • Sensory attributes of appearance, colour, texture, taste, flavour and overall acceptability.	Untrained panellists	Significant enhancement of protein and dietary fibre contents of the cookies. Hardness of the cookies also increased as the level of dried young corn increased in the cookies. Supplementation of cookies with 10% dietary obtained the best scores in comparison to the control.

Reference	Downs and Black quality assessment	Food vehicle	Fibre type and % incorporation	Control	n	Analysis parameters	Panela	Outcomes
[36]	10/14	Biscuits	Corn fibre (10–40%)	Refined wheat flour	10	• Texture • Diameter (W):mm • Thickness (T): mm • Spread ratio • % spread factor • Sensory attributes of appearance, colour, texture, taste, flavour and overall acceptability.	Semi-trained panellists (ST)	Corn fibre replacement in refined wheat flour resulted in enhanced hardness, thickness, moisture, total dietary fibre (TDF) and moisture in biscuits and simultaneously a significant reduction in carbohydrates and energy. Reduced diameter was observed and a reduced spread ratio was derived. Sensory evaluation results of the attributes of colour, appearance, texture, taste, flavour and overall acceptability of biscuits were similar to the control up to the level of 20% addition of corn fibre.

Reference	Downs and Black quality assessment	Food vehicle	Fibre type and % incorporation	Control	n	Analysis parameters	Panela	Outcomes
[38]	11/14	Cookies	Sugarcane bagasse – with peel (0–10%)	Wheat flour	30	Physical parameters: • Width (mm) • Thickness (mm) • Spread factor (%) • Colour: L*, a* and b* • Texture (Kg F) Sensory attributes: colour, appearance, taste, texture and overall acceptability.	Trained panel members (T)	The incorporation of sugarcane bagasse with a peel at 5% to enrich cookies displayed the highest overall acceptability. Enriched cookies with 5% peeled sugarcane contained increased phenolic content, moisture content, fat and ash content, but slightly lower protein content in comparison to the control. 5% enriched sugarcane bagasse (with peel) possessed a golden-brown colour, along with increased hardness, diameter and spread ratio but reduced thickness in comparison to the control cookies. Both cookies were of similar weight.

Reference	Downs and Black quality assessment	Food vehicle	Fibre type and % incorporation	Control	n	Analysis parameters	Panela	Outcomes
[37]	9/14	Biscuits	Sugarcane bagasse (0–15%)	Wheat flour	6	Physical parameters: • Width (mm) • Thickness (mm) • Spread factor (%) • Colour: L*, a* and b* • Texture (Kg F) Sensory attributes: colour, appearance, taste, texture and overall acceptability.	Panel members (UT)	Biscuits with the incorporation of 10% steamed sugarcane bagasse and additives were highly acceptable based on the results of sensory evaluation. Hardness and moisture of fibre enriched biscuits increased as the percentage of fibre added increased, contrastingly the surface features decreased. At 10% level of addition with additives, the texture of biscuits was enhanced and surface features improved. The colour of the crumb at 10% fibre addition with additives was similar to the control – pale to yellow-brown, also its spread value and width increased, but thickness decreased. Dietary fibre content increased in fibre enriched biscuits but protein and fat contents were reduced in comparison to the control. Micro-bacterial load of bagasse was reduced through the steaming process, and rheological features such as dough development, stability and viscosity were all negatively affected with increasing levels of fibre addition

Reference	Downs and Black quality assessment	Food vehicle	Fibre type and % incorporation	Control	n	Analysis parameters	Panel[a]	Outcomes
[47]	11/14	Bread	Full fatted rice bran (0–15%)	Wheat flour	20	Physical parameters: • Volume (ml) • Mass (g) • Specific volume (g/ml) Sensory attributes: aroma, crust colour, crumb colour, taste firmness, taste and overall acceptability.	Staff members and students (T)	There were significant percentage increases in the nutrient contents of the rice bran supplemented bread, most notably the increase in protein, crude fat, carbohydrates, moisture, minerals and vitamins. Importantly, there was a reduction in the sodium content of the bread samples. As the level of rice bran incorporation increased in the bread samples from 5–15% it resulted in enhanced bread weight. Contrastingly, both the volume and specific volumes of the bread samples were reduced. There were significant differences illustrated between the physical properties of the control bread samples and the fibre incorporated bread samples: in both cases of dough and bread loaves.

Reference	Downs and Black quality assessment	Food vehicle	Fibre type and % incorporation	Control	n	Analysis parameters	Panela	Outcomes
[31]	11/14	Bread	Full fatted & defatted rice bran (0–15%)	Wheat flour	10	Physical parameters: • Volume (ml) • Mass (g) • Specific volume (g/ml) Sensory attributes: aroma, crust colour, crumb colour, taste firmness, taste and overall acceptability.	Staff members and students (T)	The hardness of composite bread increased with elevated supplementation of rice bran in flour. The specific volume of the composite bread reduced with the addition of rice bran from 5–15%. Rice bran incorporation in wheat flour enhanced the protein and crude fibre in comparison to the control wheat flour bread. Supplementation of wheat flour with 10% of full fatted and defatted rice bran in the preparation of composite bread was found to be desirable overall acceptability when compared to the control composite bread.

Reference	Downs and Black quality assessment	Food vehicle	Fibre type and % incorporation	Control	n	Analysis parameters	Panela	Outcomes
[48]	12/14	Bread	Maize bran (0%&30%)	Wheat flour	30	Physical parameters: • Volume (ml) • Mass (g) • Specific volume (g/ml) • Colour: L*, a* and b* Sensory attributes: flavour, crust colour, crumb colour, colour, taste, texture, and overall acceptability.	Untrained tasters (UT)	The incorporation of brans in bread formula resulted in reduced bread volumes, and enhanced water, ash, lipids, proteins, fibres and phytates contents. Maize bran supplemented bread recorded the highest phytate value (0.53 g/100 g DM) although lower than the control (2.5 g/100 g DM). The addition of maize significantly ($p < 0.05$) enhanced mineral contents such as magnesium, potassium, zinc, manganese and iron in comparison to the control bread but had no significant effect on calcium and copper contents.

Reference	Downs and Black quality assessment	Food vehicle	Fibre type and % incorporation	Control	n	Analysis parameters	Panela	Outcomes
[44]	8/14	Bread (cha-patti/ fer-mented brad)	Sugarcane bagasse (0, 5 & 8%)	Wheat flour	50	Physical parameters: ND Sensory attributes: flavour, aroma, crust colour, and texture. Overall acceptability: ND	Students, between the ages of 18 and 36 years old (ST)	Hedonic responses suggested that the addition of 8 wt.% of treated fibres during the production of chapatti-type fermented bread (with SCB) is suitable for human consumption. Incorporation of food with sugarcane bagasse enhanced its nutritional value. Increased in total fibre 7.4 ± 0.5–11.7 ± 0.6 g/100 g. Similarity in the inhibition activity of the 2–2 diphenyl-1 picrylhydrazyl (DPPH) free radical values between control and sugarcane bagasse enriched bread. Lower levels of crude fat, protein and ash content in sugarcane bagasse enriched chapatti bread in comparison to control samples and the other sugarcane tops enriched samples.

Reference	Downs and Black quality assessment	Food vehicle	Fibre type and % incorporation	Control	n	Analysis parameters	Panela	Outcomes
[17]	14/14	Cake	Corn bran (0–30%)	Cake formulation containing 0% corn bran	30	Physical parameters: • Crumb colour: L*, a* and b* • Crust colour: • L*, a* and b* • Texture (Kg F) Sensory attributes: flavour, crust colour, crumb colour, taste and overall acceptability.	Trained taste panellists (T)	The inclusion of 5–10% modified corn bran in the cake formulation gave rise to the production of a functional cake. Above 10% fibre addition resulted in increased firmness and gumminess, but reduced springiness and cohesiveness. The colour of cakes significantly ($p < 0.5$) increased as corn bran incorporation levels rise above 5%. Best acceptability of corn bran was at 5% followed by 10% supplementation levels. 30% incorporation of corn bran obtained the lowest acceptability.

Reference	Downs and Black quality assessment	Food vehicle	Fibre type and % incorporation	Control	n	Analysis parameters	Panela	Outcomes
[49]	13/14	Banana cakes	Sugarcane bagasse (0–6%)	Oat flour	50	Physical parameters: • Crumb colour: L*, a* and b* • Crust colour: • L*, a* and b* • Texture (Kg F) Sensory attributes: crust colour, taste and overall acceptability.	Untrained panel members: mixture of staff and students (UT)	The addition of sugarcane bagasse resulted in no distinct alteration in the chemical composition of the cakes. Increased dietary fibre incorporation levels in cake resulted in increased firmness of the cake. There was no interference in the sensorial acceptance of the product when dietary fibre was added between 3 and 6%. Cakes with 6% sugarcane bagasse incorporation were considered acceptable and resulted in the preparation of desirable high fibre cakes.

Reference	Downs and Black quality assessment	Food vehicle	Fibre type and % incorporation	Control	n	Analysis parameters	Panel[a]	Outcomes
[23]	12/14	Cakes	Corn bran (0–30%)	Cake flour	25	Physical parameters: • Crumb colour: L*, a* and b* • Crust colour: • L*, a* and b* • Texture (Kg F) Sensory attributes: flavour, crust colour, crumb colour, taste and overall acceptability.	Untrained panellists (UT)	The increasing levels of corn bran replacement in cake batter showed no influence on the hardness and springiness of cakes. Substitution of flour with 20% corn bran resulted in cakes with favourable sensory scores in the attributes of texture, taste and overall acceptability of the cakes.

[a] *Description is given by author. Trained (T), semi-trained (ST) and untrained (UT).*

In another study, protein contents in corn fibre enriched biscuits showed a reduction, as reported in most other studies ash, moisture and total dietary fibre contents were higher than in control biscuits [36]. Jauharah, Wan Ishak, and Robert [46] indicated that supplementation of biscuits with corn fibre up to 30% resulted in enhanced energy value. Contrastingly, more recently Sandhu, Bains, Singla, and Sangwan [36] revealed that biscuits supplemented with corn fibre up to 40% produced enriched biscuits with reduced energy value in comparison to control (a reduction from 499 Kcal to 486 Kcal). Moreover, there was evidence of reduced carbohydrates contents in corn fibre incorporated biscuits [36]. A recent study of sugarcane bagasse incorporation of cookies up to 10% Vijerathna et al. [38] indicated the presence of enhanced phenolic content and significantly higher ash and moisture contents, marginally similar fat composition in control and enriched cookies, but protein contents in enriched cookies were significantly lower than in the control sample. In 2011, Sangeetha, Mahadevamma, Begum, and Sudha indicated that sugarcane bagasse supplemented biscuits up to 15% biscuits possessed reduced total fat and protein contents, meanwhile, there were minimal variances in moisture content, ash and acid-insoluble contents in comparison to control.

There were two (2) studies, namely [17, 23] relating to corn bran supplemented cakes using similar incorporation percentages, namely 0–30% and that did not conduct any nutritional analysis, hence no data nutritional data were extracted from these studies. Rice bran supplemented bread up to 15% resulted in a significant (0.05) increase in moisture, protein, crude fibre, crude fat, ash and several minerals and vitamins in comparison to control bread [31, 47]. Importantly, it was reported in a study of enriched bread with rice bran that the sodium content was significantly (0.05) reduced, the composition of carbohydrates was reduced in enriched bread [47]. Further, a significant increase ($p < 0.05$) was reported in ash, moisture, proteins, lipids and minerals for example magnesium, potassium, zinc, manganese and iron; carbohydrates and energy value (Kcal/100 g) was significantly reduced ($p < 0.05$) and interestingly phytic acid contents of enriched bread using maize bran at 30% supplementation [48]. More recently, Gil-López et al. [44] stated that supplementation of chapatti using sugarcane bagasse resulted in enhanced total fibre 7.4 ± 0.5–11.7 ± 0.6 g/100 g, promote inhibition activity of the 2–2 diphenyl-1 picrylhydrazyl (DPPH) and also reduced the levels of crude fat, protein and ash content in enriched bread in comparison to control samples.

DISCUSSION

Roles of dietary fibre in foods

Fortification of foods

Dietary arabinoxylan-based sources namely, rice bran, corn, and sugarcane bagasse can be used to improve the nutritional contents of baked foods and making them become functional products. In recent studies, Haghighi-Manesh & Azizi [17] reported that the inclusion of 5–10% modified corn bran in the cake formulation gave rise to the production of a functional cake with reduced cohesiveness and springiness higher gumminess, darkness, and favourable sensory properties. This suggests that corn bran can be added to baked food products to enhance nutritive value, textural qualities and sensory qualities. Moreover, incorporation of food with sugarcane bagasse at 8% supplementation and other dietary fibre arabinoxylan sources into food systems such as bread tend to enhance its nutritional value and influence changes in the physical, rheological, and sensorial characteristics of foods [24, 44]. Also, previously, Amna, Bhatti, Anwaar, & Randhawa [45] indicated that enhance moisture, protein, fat and minerals (calcium, manganese, and magnesium) composition in cookies was reported as the level of rice bran incorporation increase in biscuits. The evidence suggests that 20% fibre incorporated rice bran cookies may possess a significant increase in moisture, protein and minerals (zinc and Iron) contents. Meanwhile, in bread, Pauline et al., [48] indicated a significant increase ($p < 0.05$) in water, ash, lipids, proteins, fibres and phytic acid contents between the control sample and enriched maize bran bread. The incorporation of 8.2 g of AXE per 100 g of available carbohydrates into bread facilitated the criteria for the health claim of the reduction of post-prandial glycaemic response [7, 50]. This suggests that enrichment using arabinoxylans-based sources can improve the nutritive value of ordinary food products.

Water retention capacity

Arabinoxylans dietary fibre sources tend to affect the moisture content of foods positively. Several studies reported the increase of moisture content in the enrichment of cookies/biscuits, bread and cakes, in comparison to the control samples, in some cases moisture was increased significantly ($p < 0.005$) and other cases marginally [31, 36, 38, 48, 49]. Javaria [20] indicated that moisture content in cookies/ biscuits can consider a fundamental quality since it tends to influence both end quality and shelf life of these food products. Arabinoxylan based sources during addition to dough tend to enhance viscosity and water absorption; improved dough development, reduce starch retrogradation and reduce the firmness of foods due to the presence of bound water facilitating less stiff gluten and starch network [22, 51]. This suggests that arabinoxylan dietary fibres supplementation into foods can enhance its moisture content and affect other properties in foods namely its storage properties and shelf life. However, in bread, it was proven that arabinoxylan has a beneficial effect on water activity and thus favourably influenced bread freshness [22]. Moreover, Jauharah, Wan Ishak, & Robert [46] stated that moisture in the range of 1–5% is considered a benchmark for cookies/biscuits and that fresh corn fibre may lead to too high moisture levels in foods and consequently result in food spoilage. It can be suggested that excess moisture may result in micro-bacterial activities which can impact shelf life unfavourably. Similarly, in the enrichment of bread with 20% corn bran high, bread quality was significantly affected by the water content in the composite formula [52]. Taken together, water retention capacity is fundamental in influencing baked foods' end quality.

Dietary fibre enriched food qualities and functional food ingredients

Arabinoxylans sources, namely rice bran, corn fibre and sugarcane bagasse tend to influence various qualities of enriched foods. Arabinoxylans (AXE) contribute essentially in determining the physical and chemical properties of the final quality of food products and made up a high percentage of the cell walls of cereal grains and also present in sugarcane bagasse [16, 22, 53]. In earlier studies, Foschia, Peressini, Sensidoni, and Brennan [13] indicated also that the quality and nutritional aspects of cereal products are influenced by the addition of dietary fibres into foods.

Importantly, rheological properties among arabinoxylan fibres such as corn fibres displayed differences that may influence the quality of food products after their incorporation during food processing [54]. It can be hypothesised that rheological factors can contribute to some of the undesirable features in dietary fibre enrichment of baked foods. Numerous studies have conducted dietary fibre supplementation of cookies/biscuits, bread and cakes using rice bran, corn fibre and sugarcane bagasse at different rates of incorporation and with varying results, see **Table 12** [17, 23, 38, 39, 47]. This suggests that fibre incorporation using rice bran, corn fibre and sugarcane bagasse tend to produce both desirable and undesirable physical and sensorial features in baked food products.

Moreover, because of the favourable nutritive value impact of dietary fibres on enriched foods, and the health-related benefits that be obtained upon consumption in a prescribed manner, it can be suggested that these ingredients can potentially be considered as functional ingredients in baked foods. In support of the suggestion, Zidan and Eldemery [55] found that the incorporation of 5–10% defatted black rice bran was found to possess acceptable sensory features and resulting in the nutritive value of bread being enhanced with minerals such as phosphorus, potassium, iron, copper, zinc and calcium. Contrastingly, Zhang et al. [40] found that arabinoxylan fortification of bread with 10% arabinoxylan fibre illustrated undesirable physical qualities in bread such as decreased specific volume, harder crumb, darker crust colour, and a coarser crumb structure. Overall dietary fibres facilitate the production of enriched foods, which may possess both desirable and undesirable sensorial and physical qualities, which can be attributed to the rheological behaviour of dietary fibre arabinoxylan when used in food supplementation.

Effects of the incorporation of dietary fibre derived from selected cereals (rice bran and corn fibre) and sugarcane bagasse on the physical and sensory qualities of baked food products: cookies/ biscuits, bread and cakes

The incorporation of sugarcane bagasse at both 5% and 10% in cookies/biscuits resulted in significantly undesirable ($p < 0.05$) overall acceptance in sensory qualities, namely appearance, taste and texture, (see **Table 2**). It can be suggested that the increased firmness of the cookies/biscuits increased as the rate of supplementation of sugarcane bagasse increased affected its overall acceptance negatively, (see **Table 3**). Javaria [20] indicated that the firmness in cookies tends to influence their overall acceptance by the panellists. Sugarcane bagasse incorporation in cookies / biscuits showed no significant improvement in the physical qualities of the enriched product in comparison to the control. This can be attributed to the rheological properties of dietary arabinoxylans within the food matrix of the enriched biscuits. Kale, Pai, Hamaker, and Campanella [54] found that the rheological properties of extensional viscosity and solution viscosity of arabinoxylans fibres in the food system tend to influence its quality. This suggests that as the competition may have increased between the dietary fibre and the protein network, it resulted in a rearrangement of the food matrix network. The incremental addition of sugarcane bagasse resulted in at least 50% rise in the firmness of 10% enriched cookies/ biscuits. This can be attributed to the stiffness of the gluten and starch matrix may have been destroyed due to the competition of unextracted arabinoxylan dietary fibre and proteins for bound water. In a previous study of biscuits, Sozer, Cicerelli, Heiniö, and Poutanen [56] found that the hardness of biscuits was increased due to the particle size reduction of bran, while the starch hydrolysis index of biscuits decreased due to particle size reduction of bran.

In rice bran enriched cookies/biscuits incorporation levels of 5–15% showed a significant difference in comparison to control cookies/biscuits. Interestingly, a low fibre incorporation of rice bran of 5% displayed significantly ($p < 0.05$) undesirable acceptance, similar to 10% and 15% rice bran enriched cookies/biscuits, (see **Table 4**). These results agree with previous studies that reported a reduction in overall acceptance scores in enriched foods as the fibre incorporation increased [20, 45]. Contrastingly, the addition of 5% dietary fibre was acceptable based on sensory evaluation results in wheat bran enriched biscuits. Evidence suggests that rice bran significantly ($p < 0.05$) enhanced the thickness of cookies/biscuits, however, other physical qualities such as width and spread factor (SF%) showed no significant differences ($p > 0.05$), (see **Table 5**). Similar, findings were reported in some other fibre supplementation studies conducted. This can be as a result of the enhanced moisture absorption capacity of the dietary fibre used in the enriched cookies. In a recent study, Vijerathna et al. [38] reported that the enhanced size of the hydrophilic starch granules increased the thickness of fibre enriched cookies.

Corn fibre supplemented cookies displayed desirable overall acceptance at 20% incorporation rate and was significantly different ($p < 0.05$) in comparison to control cookies/biscuits, as a result, enriched corn fibre cookies/biscuit obtained the highest sensory score, (see **Table 6**). This suggests that corn fibre can be incorporated at a higher rate and produced desirable results possibly due to its high water retention capacity and its particle. In previous studies, Mishra & Chandra [57] reported a similar trend in results using a higher incorporation rate in cookies/ biscuits. Similarly, supplementation of cookies/biscuits at 10% and 20% were significantly different ($p < 0.05$) to control, 20% corn fibre cookies/biscuits followed by 10% supplemented corn fibre cookies/biscuits obtained the best scores during panel sensory evaluation. This desirable overall acceptance at 10%-20% incorporation may have been influenced by cookies/biscuits increase moistures that facilitated texture 1.8 ± 0.5 (Kg Force) and 1.9 ± 0.4 (Kg Force) at 10% and 20% respectively; and colour of the enriched product. Cookies/biscuits incorporated with corn fibre at 30% obtained the lowest overall acceptance score. This suggests that increased firmness and other sensory attributes such as flavour, colour, appearance and texture acceptance which were significantly undesirable ($p < 0.05$) based on sensory panel evaluation contributed to a low overall acceptance of the food product.

In terms of physical parameters, results suggest that the firmness of corn fibre enriched cookies/biscuits increased with increasing incorporation of corn fibre up to 30%. Jia et al. [58] indicated that water-insoluble proteins in flours that possessed minimal gluten, namely glutenin and gliadin combined to influence elasticity and structural strength of dough (glutenin) and viscosity and fluidity of dough (gliadin) tend to intertwine and formed a strong protein network structure in the food matrix. This suggests that the degree of firmness affected the protein network in the food matrix negatively possibly due to low moisture levels which contributed greatly to the high texture of the cookies/biscuits [52]. Cookies/biscuits (51.9 ± 5.0 mm) supplemented at 30% were thicker than the control (49.1 ± 24.5), however, there was no significant differences ($p > 0.05$), (see **Table 7**). This can be attributed to the high moisture absorption capacity of corn fibre, a satisfactory source of arabinoxylan dietary fibre which possibly clings to the bound water within the food matrix network, as the viscosity of the solution rises in the food system, leaving the protein components to form strong aggregates. The polysaccharide, arabinoxylan tends to possess various physicochemical properties, such as the high capacity to retain water and display an inclination to form high viscosity solutions when incorporated into the food matrix [22].

Incorporation of rice bran into bread at 15% showed a significant undesirable overall acceptance ($p < 0.05$) in comparison to control samples for all sensory attributes, except crust colour, (see **Table 8**). This can be attributed to a possible darker crumb in comparison to the control. Ortiz de Erive, Wang, He, and Chen [52] indicated that due to the enrichment of bread with dietary fibre their colour is likely to be darker in comparison to the non-enriched sample, as a result of the lower baking temperature of the crumb, thus affecting caramelization or Maillard's reaction. Arabinoxylans significantly affected protein network formation influencing the food matrix features and aroma of incorporated food products [59]. The aroma of enriched bread up to 10% was significantly different ($p < 0.05$) from the control. Enriched rice bran at 15% obtained the lowest overall acceptance score by the sensory evaluation panellists. In other studies, rice bran enriched bread was found to be acceptable at 5–10% [55]. This may be due to the sensory panel members' dietary familiarity with wheat flour bread. There was no significant difference ($p > 0.05$) in the specific volume (ml/g) between enriched rice bran bread and the control, however, specific volume reduced as rice bran incorporation percentage increase in the enriched bread sample, (see **Table 9**). This can be attributed to the water absorption capacity of the rice bran, which absorbs potentially most of the moisture, its ability to form viscous solutions, thus leaving the gluten component in the bread matrix inadequately hydrated. Pauline et al. [48] reported that rice bran enriched showed a lower specific volume in comparison to the enriched bread.

Moreover, as rice bran is being incorporated into the enriched bread, due to the high-water capacity of rice bran, the enriched bread texture is likely to increase and become denser and less porous, thus resulting in the continual reduction of the specific volume from 10.7 ± 9.1 at 0% rice bran incorporation to 9.8 ± 6.6 at 10% rice bran incorporation [52]. It can be suggested that textural qualities of bread tend tends to be affected by its moisture content.

Corn bran enriched cakes up to 20% fibre incorporation displayed significantly ($p < 0.05$) more desirable sensory qualities than control cakes, (see **Table 10**). This can be attributed to the increased moisture content levels caused by the addition of corn bran during the enrichment process, which allowed the corn bran supplemented cakes to be adequately hydrated. There was a significant difference in crust colour of corn bran enriched cakes and control at the 20% fibre incorporation level. This can be attributed to the possible pale colour of the enriched cakes at the 20% level of fibre incorporation. Generally, dietary fibre enriched cakes take the inherent colour of the fibre being utilised in the process of enrichment, until Maillard's reaction occurs [52]. The increasing levels of corn bran replacement in cake batter showed no influence on the hardness and springiness of cakes [23]. In terms of physical qualities, the firmness of cakes increased incrementally with the increase in corn bran supplementation, (see **Table 11**). Moreover, there was no significant difference ($p < 0.05$) in physical qualities, except in crust L* at 25% and 30% enrichment. This can be attributed to the possible darker colour of the corn bran enriched cakes crust in comparison to the control. The high baking temperature will tend to affect the fibre enriched cakes' crust due to caramelization or Maillard's reaction and the inherent colour of the fibre [52]. In summary, the incorporation of dietary fibres, rice bran and sugarcane bagasse affected the sensory and physical qualities of baked food products unfavourably. However, corn fibre and corn bran produced desirable sensory and physical effects at 20% fibre supplementation in both cookies/biscuits and cakes.

Beneficial implications of dietary fibres in foods

Nutritive value of enriched foods

The fundamental evidence of this review suggests that as the level of fibre incorporation increased the nutritive value of the food products cookies/biscuits, bread and cakes were significantly ($p < 0.05$) enhanced, resulting in increased moisture, ash, protein, minerals, vitamins, dietary fibres and crude fats, essentially in cereal-based incorporated food products. Corn bran enriched bread was suggested to contain significant ($p < 0.05$) contents of phytic acid contents. More recently, Ekpa, PalaciosRojas, Kruseman, Fogliano, & Linnemann [4] indicated that food enrichment is an essential processing technique to enhance the nutrient content of staple foods including baked food products, thus, this suggests that fibre enriched foods can potentially provide convenience to consumers and ultimately facilitates improved food nutrition guidelines for the society and foster food security locally. Similarly, rice bran possessed vital antioxidants, which comprised of the well-recognised immune system enhancing compound, namely phytosterols; polysaccharides; minerals and trace minerals including magnesium, selenium, zinc, vitamin E, omega-3 fatty acids and many other phytonutrients [28]. This suggests that the strategy of fibre enrichment of popular food products can potentially be utilised more widely by stakeholders in the food industry in collaboration with health care agencies of various age groups, stakeholders of feeding programmes in schools and policymakers in governmental public health agencies. In another recent study, corn fibre was found to be an effective antioxidant than wheat bran and this was attributed to its elevated ferulic acid content and polyamine-conjugates. Interestingly, hydroxycinnamates may not be necessary for the antioxidant effect [30]. Contrastingly, in sugarcane bagasse supplemented food products, there were instances of reduced total fat, protein and marginally increased in moisture content of food products. Reduction in protein and fats contents was also reported in other fibre supplementation research [60]. Overall, the nutritive value of food products can be enhanced by fibre enrichment.

Health benefits

It has been widely established that numerous health benefits are associated with the consumption of enriched food products in recommended administered quantities. Moreover, dietary fibre potentially can contribute to several health benefits such as enhanced bowel function, reduced levels of cholesterol in the body, better weight maintenance and assisted in controlling blood sugar levels in the human body [24]. Chen et al. [5] indicated that there are several beneficial impacts on various physiological processes which can be attributed to arabinoxylans and consequently health functions and prebiotic effects are being influenced by arabinoxylans structures, which in turn depends on its method of extraction and source. This suggests that rice bran, corn fibre and sugarcane bagasse, which are well-established sources of dietary fibre arabinoxylans, can potentially be utilised to produce a wider range of enriched foods which can confer various types of health benefits to the consumers. However, the consumption of dietary fibre should adhere to the established guideline of at least 25 g/day for adults [6].

Dietary fibre supplemented foods with rice bran, corn fibre and sugarcane bagasse, namely cookies/biscuits, bread and cakes enriched foods was significantly ($p < 0.05$) higher in nutritive value, and contrastingly tend to contain significantly ($p < 0.05$) reduced energy value (Kcal/100 g) in comparison to control samples. Evidence suggests that fibre enriched cookies/biscuits, bread and cakes were significantly enhanced ($p < 0.05$) by supplementation, thus they can be considered potentially as functional foods. Moreover, further evidence indicated that supplementation of chapatti using sugarcane bagasse resulted in enhanced total fibre 7.4 0.5–11.7 ± 0.6 g/100 g and contributed to elevated inhibition activity of the 2–2 diphenyl-1 picrylhydrazyl (DPPH) free radical values. Diets with enhanced fibre contents are associated with desirable effects on the health of consumers [3]. In general, fibre enriched food products have shown a reduction in energy levels in comparison to the control samples. This suggests that enriched foods can be used for individuals with varying health conditions and risk factors such as coeliac patients and other different target groups.

Food policy

Dietary fibre enriched foods can potentially be utilised in various governmentalbased food policies. Recently, Lockyer and Spiro [61] reported that the average fibre intake in the United Kingdom can be considered fairly below the level recommended level. This suggests that a fibre enrichment strategy involving the relevant stakeholders can potentially facilitate an improvement in dietary fibre intake within the population. Previously, a comprehensive review study was conducted for modern dietary and policy priorities for cardiovascular diseases, obesity, and diabetes and found that there are complex influences of different foods on long-term weight regulation and recommended implementing an evidencebased strategy, including policy approaches, for lifestyle changes [62]. Importantly, fibre enriched foods contain rice bran, corn fibre or sugarcane bagasse all can be used in producing functional foods suitable for the particular health setting, thus this indicates enriched foods can be utilised in different food policies with different outcomes.

Consumer acceptance

Despite, the promising nutritional enhancement of enriched food products, rice bran and sugarcane bagasse at increasing incorporation levels tend to be associated with undesirable sensory and physical qualities of baked foods [55]. Evidence of this review suggests that lower incorporation rates can potentially result in improved overall acceptance of fibre enriched food products, further results suggest that the level of overall acceptance of baked products cookies/biscuits, bread and cakes were considered unacceptable above 20% incorporation for rice bran and sugarcane bagasse any type particular baked product. In support of evidence, [39] stated that incorporation of wheat flour using 10% rice bran protein concentrate (RBPC) resulted in the production of protein-enriched biscuits with favourable overall acceptability. It can be suggested that consumers tend to have a potentially better awareness of the health benefits of dietary fibre enriched foods, and also some of the most popular products are being enriched with dietary fibre thus creating a potential health trend among consumers. Similarly, Gul, Yousuf, Singh, Singh, & Wani [32] stated that consumer's attitude towards healthy foods is improving and thus presents potential opportunities for further development of functional foods on the world markets. This suggests dietary fibre food ingredients can potentially find wide applications in the other fields, thus increasing consumer awareness of dietary fibre enriched foods. In summation, using popular foods such as enriched bread and other potentially widely consumed food products as the benchmark for the enrichment of other products can facilitate further acceptance of consumers.

Limitations of study

There were three (3) main limitations that were experienced in the conduct of this review, namely the framework of the search strategy which influenced the availability of an adequate number and relevant primary research articles relating to the incorporation of sugarcane bagasse in baked food products such as cookies/biscuits, bread and cakes; the appropriateness of primary research articles to meet the selection criteria for inclusion in the review for analysis and the high percentage of primary research articles relating to dietary fibre supplementation in baked foods originating from predominantly less developed and developing countries based in Asia such as India, Pakistan, Bangladesh, Malaysia, Iran and Sri Lanka. The framework of the search strategy returned a relatively high percentage of primary research relating to wheat bran incorporation in foods, other cereal fibres such as barley, rye and oats, fibres from fruits and vegetable parts among others. Approximately less than 5% of the primary research articles were related to the incorporation of sugarcane bagasse in baked foods.

The appropriateness of primary research articles to meet the selection criteria for inclusion in the review for analysis resulted in some potentially interesting primary research studies not being include in the review, namely [7, 35, 59, 63]. Moreover, some of these studies was unable to be placed into sub-groups to facilitate statistical analysis such as [44, 48, 49] among others and therefore there were excluded from the list of included studies of the review. Moreover, primary research studies which utilised arabinoxylan extract as the source of incorporation in baked foods were extremely minimal, namely [35, 63], however both failed to achieved the selection criteria and thus were not selected. Therefore, the analysis of studies with arabinoxylan extract supplemented at a lower percentage was not possible.

The majority of primary research articles originated from developing and less developing countries located in Asia such as [20, 37–39, 45] among others. Thus, studies were able to be analysed from a wider cross-section of laboratory settings.

CONCLUSION

This systematic review demonstrates the utilisation of a comprehensive research methodology in the selection and examination of fourteen (14) dietary fibre food supplementation primary research studies to provide relevant and impartial new insights of the effects of the incorporation of dietary fibre derived from selected cereals (rice bran and corn fibre) and sugarcane bagasse on the physical and sensory qualities of baked food products: cookies/biscuits, bread and cakes. Arabinoxylansbased dietary fibre sources' roles in food supplementation involves enhancing the nutritive value of ordinary food products; influencing the end quality of baked foods and potentially improving the sensory and physical qualities of baked foods.

The supplementation of sugarcane bagasse at both 5% and 10% and rice bran up to 15% into cookies/biscuits resulted in significantly undesirable acceptance ($p < 0.05$). Corn fibre was supplemented into cookies/biscuits up to 20% and had a favourably significant ($p < 0.05$) impact on its sensory qualities. It was suggested that enhance moisture content from corn fibre incorporation in combination with its particle size contributed to this desirable outcome.

Sugarcane bagasse incorporation negatively affected the physical qualities of cookies/ biscuits. The incremental addition of sugarcane bagasse resulted in a 50% rise in the firmness of 10% enriched cookies/biscuits, from 5.7 ± 5.4 (Kg Force) to 13.0 ± 3.9 (Kg Force). Rice bran significantly increase ($p < 0.05$) the thickness of cookies/biscuits, from 8.6 ± 1.0 (mm) to 9.9 ± 0.3 (mm), the width and spread factor were similar to control. Corn fibre cookies supplementation did not significantly affect its physical qualities. Rice bran incorporation into bread at 15% showed a significant ($p < 0.05$) undesirable effect on its sensory qualities. However, the was no significant adverse effect on its physical quality. Corn bran enriched cakes up to 20% fibre incorporation displayed a significant ($p < 0.05$) favourable effect on the sensory properties of cakes; contrastingly, it resulted in a significant undesirable physical effect on the crust colour of corn bran enriched cakes.

There were four (4) main beneficial implications of dietary fibre food fortification using rice bran, corn fibre and sugarcane bagasse, namely the evidence suggests the enhancement of the nutritive value of foods; allows for the potential production of fibre enrichment foods to cater for particular target groups; dietary fibre enriched foods can be utilised in various food policies with particular outcomes, such as increasing the fibre intake within the population and using enriched breads and other potentially widely consumed food products as the benchmark for enrichment of other products can facilitate further acceptance consumers. Future research should assess the effects of derived arabinoxylans from selected cereal fibres (rye, sorghum, rice and corn) and energy crop sugarcane fibres on the rheological, sensory and physical effects in muffins production. Updated and accurate rheological properties food supplementation data is important for the food industry.

ABSTRACT

The supplementation of baked foods, namely cookie/biscuits, bread and cakes with agricultural by-products from cereal based fibres (rice bran and corn fibre) and sugarcane bagasse at rates of 0% - 15%; 0% - 30% and 0% - 10% respectively can significantly improve its nutritive value and enhanced its physical and sensorial qualities. This chapter aims to review the role of dietary fibres derived from selected cereals (rice bran and corn fibre) and sugarcane bagasse in baked foods, namely cookies/biscuits, bread and cakes; evaluate their effects on the physical and sensory qualities of these baked food products and to critically assess their beneficial impacts in baked foods. These enriched food products can potentially be utilised in shaping health policies, contribute to the dietary fibre needs of consumers and facilitate the development of functional foods. Fibre enriched foods potentially can assist in improving various physiological functions of the human body. A Keyword- based search strategy was utilised to conduct a comprehensive search for articles catalogued in ScienceDirect, Web of Science, PubMed, Medline, CINAHL and Google Scholar that were published between January 1, 2010 and August 1, 2020. Applicable aspects of the Preferred Reporting Items for Systematic Reviews and Meta-Analyses (PRISMA) guidelines provided the framework of this review. Four- teen (14) studies met the inclusion/extraction criteria and was placed into sub- groups by food types and fibre used in supplementation. Only eleven (11) studies were suitable for statistical data analysis. The supplementation of sugarcane bagasse at both 5% and 10% and rice bran up to 15% into cookies/biscuits significantly undesirable acceptance ($p < 0.05$). Corn fibre enriched cookies/biscuits up to 20% showed a significantly ($p < 0.05$) favourable impact on the sensory qualities of the food product. The physical qualities of sugarcane bagasse supplemented cookies/ biscuits were negatively affected. The incremental addition of sugarcane bagasse resulted in at 50% rise in the firmness of 10% enriched cookies/biscuits, from 5.7 ± 5.4 (Kg Force) to 13.0 ± 3.9 (Kg Force). Corn fibre cookies supplementation did not significantly affect its physical qualities. Rice bran incorporation of 15% in bread showed a significant ($p < 0.05$) undesirable effect on its sensory qualities. How- ever, the was no significant adverse effect on its physical quality. Corn bran enriched cakes up to 20% fibre incorporation displayed a significant ($p < 0.05$) favourable effect on the sensory properties of cakes.

REFERENCES

1. Alan, P. A., Ofelia, R. S., Patricia, T., & Maribel, R. S. R. (2012). Cereal bran and wholegrain as a source of dietary fibre: technological and health aspects. *International Journal of Food Sciences and Nutrition*, *63*(7), 882-892. https://doi. org /10.3109/09637486.2012.676030

2. Aydogdu, A., Sumnu, G., & Sahin, S. (2018). Effects of addition of different fibers on rheological characteristics of cake batter and quality of cakes. *Journal of food science and technology*, *55*(2), 667-677. https://doi.org/10.1007/ s13197-017-2976-y

3. Dhingra, D., Michael, M., Rajput, H., & Patil, R. T. (2012). Dietary fibre in foods: a review. *Journal of Food Science and Technology, 49*(3), 255-266. https:// doi.org/10.1007/s13197-011-0365-5

4. Ekpa, O., Palacios-Rojas, N., Kruseman, G., Fogliano, V., & Linnemann, A. R. (2019). Sub-Saharan African Maize-Based Foods-Processing Practices, Challenges and Opportunities. *Food Reviews International, 35*(7), 609-639. https:// doi.org/10.1080/ 87559129.2019.1588290

5. Chen, Z., Li, S., Fu, Y., Li, C., Chen, D., & Chen, H. (2019). Arabinoxylan structural characteristics, interaction with gut microbiota and potential health functions. *Journal of Functional Foods, 54*, 536-551. https://doi.org/10.1016/j.jff. 2019.02.007

6. European Food Safety Authority Panel on Dietetic Products, N., & Allergies. (2010). Scientific Opinion on Dietary Reference Values for carbohydrates and dietary fibre. EFSA Journal, 8(3), 1462. https://doi.org/ 10.2903/j.efsa.2010.1462

7. Li, W., Hu, H., Wang, Q., & Brennan, C. S. (2013). Molecular Features of Wheat Endosperm Arabinoxylan Inclusion in Functional Bread. *Foods (Basel, Switzerland), 2*(2), 225-237. https://doi.org/10.3390/food s2020225

8. Scientific Advisory Committee on Nutrition (2015). *SACN carbohydrates and health report.* Retrieved from https://www.gov.uk/government/ publications/ sacn-carbohydrates-andhealth-report

9. Buttriss, J. L. (2015). Fibre-Need to increase intake according to new recommendations. *Nutrition Bulletin, 40* (4), 291-295. https://doi.org/10.1111/ nbu.12174

10. European Commission. (2020). *Dietary fibres.* European Commission. Retrieved September 1, 2020 from https://ec.europa.eu/jrc/en/healthknowledge-gateway/promotionprevention/nutrition/fibre

11. World Health Organisation. (2020). The Ottawa Charter for Health Promotion. World Health Organisation. Retrieved September 1, 2020 from https://www. who.int/health promotion/ conferences/previous/ottawa/en/ index1.html

12. Shahbandeh, M. (2019). Size of the global dietary fibers market from 2018 to 2025. Statista. Retrieved August 18, 2020 from https://www.statista.com/ statistics/1053431/dietary-fibersmarket-value-global

13. Foschia, M., Peressini, D., Sensidoni, A., & Brennan, C. S. (2013). The effects of dietary fibre addition on the quality of common cereal products. *Journal of Cereal Science, 58*(2), 216-227. https://doi.org/10.1016/j.jcs. 2013.05.010

14. Casas, G. A., Lærke, H. N., Bach Knudsen, K. E., & Stein, H. H. (2019). Arabinoxylan is the main polysaccharide in fiber from rice coproducts, and

increased concentration of fiber decreases in vitro digestibility of dry matter. *Animal Feed Science and Technology, 247*, 255-261. https://doi.org/ 10.1016/j.anifeedsci.2018.11.017

15. Fadel, A., Mahmoud, A. M., Ashworth, J. J., Li, W., Ng, Y. L., & Plunkett, A. (2018). Health-related effects and improving extractability of cereal arabinoxylans. *International journal of biological macromolecules, 109*, 819-831. https://doi.org/10.1016/j.ijb iomac.2017.11.055

16. Dimopoulou, M., & Kontogiorgos, V. (2020). Soluble dietary fibres from sugarcane bagasse. *International Journal of Food Science & Technology, 55*(5), 1943-1949. https://doi.org/10.1111/ijf s.14445

17. Haghighi-Manesh, S., & Azizi, M. H. (2018). Integrated extrusionenzymatic treatment of corn bran for production of functional cake. *Food science & nutrition, 6*(7), 1870-1878. https://doi.org/10.1002/fsn3.738

18. Yadav, M. P., Kale, M. S., Hicks, K. B., & Hanah, K. (2017). Isolation, characterization and the functional properties of cellulosic arabinoxylan fiber isolated from agricultural processing by-products, agricultural residues and energy crops. *Food Hydrocolloids, 63*, 545-551. https://doi. org/10.1016/j.foodhyd.2016.09.022

19. Bultum, L. E., Emire, S. A., & Wolde, Y. T. (2020). Influence of full fat rice bran from Ethiopian rice milling industries on nutritional qualities, physicochemical and sensory properties of bread and biscuits. *Journal of Food Measurement and Characterization, 14* (4), 2253-2261. https://doi.org/10.1007/ s11694-020-00472-7

20. Javaria, S. (2017). Formulation and quality evaluation of aglutenics biscuits supplimented with rice bran for coeliac patients. *Pure and Applied Biology, 6*. https://doi.org/10.19045/bspab.2017. 600137

21. Mena, B., Fang, Z., Ashman, H., Hutchings, S., Ha, M., Shand, P. J., & Warner, R. D. (2020). Influence of cooking method, fat content and food additives on physicochemical and nutritional properties of beef meatballs fortified with sugarcane fibre. *International Journal of Food Science & Technology, 55*(6), 2381-2390. https:// doi.org/10.1111/ijfs.14482

22. Rosicka-Kaczmarek, J., Komisarczyk, A., Nebesny, E., & Makowski, B. (2016). The influence of arabinoxylans on the quality of grain industry products. *European Food Research and Technology, 242*(3), 295-303 https://doi.org/10.1007/ s00217-015-2549-0

23. Singh, M., Liu, S. X., & Vaughn, S. F. (2012). Effect of corn bran as dietary fiber addition on baking and sensory quality. *Biocatalysis and Agricultural Biotechnology, 1*(4), 348-352. https://doi. org/10.1016/j.bcab.2012.02.005

24. Jarosław Wyrwisz, M. K. (2015). The Application of Dietary Fiber in Bread Products. Journal of Food Processing & Technology, 06(05). https://doi.org/ 10.4172/2157-7110.1000447

25. Ozturk, O. K., & Mert, B. (2018). The effects of microfluidization on rheological and textural properties of gluten-free corn breads. *Food Research International, 105*, 782-792. https://doi. org/10.1016/j.foodres.2017.12.008

26. Sairam, S., Krishna, A. G. G., & Urooj, A. (2011). Physico-chemical characteristics of defatted rice bran and its utilization in a bakery product. *Journal of Food Science and TechnologyMysore, 48*(4), 478-483. https://doi.org/ 10.1007/s13197-011-0262-y

27. Mendis, M., Leclerc, E., & Simsek, S. (2016). Arabinoxylans, gut microbiota and immunity. *Carbohydrate polymers, 139*, 159-166. https://doi.org/ 10.1016/j. carbpol.2015.11.068

28. Park, H.-Y., Lee, K.-W., & Choi, H.-D. (2017). Rice bran constituents: immunomodulatory and therapeutic activities. *Food & Function, 8*(3), 935-943. https://doi.org/10.1039/c6f o01763k

29. Phimolsiripol, Y., Mukprasirt, A., & Schoenlechner, R. (2012). Quality improvement of rice-based gluten-free bread using different dietary fibre fractions of rice bran. *Journal of Cereal Science, 56*(2), 389-395. https://doi.org/ 10.1016/j.jcs.2012.06.001

30. Bauer, J. L., Harbaum-Piayda, B., Stöckmann, H., & Schwarz, K. (2013). Antioxidant activities of corn fiber and wheat bran and derived extracts. LWT-Food Science and Technology, 50(1), 132-138. https://doi.org/10.1016/j. lwt.2012.06.012

31. Begum, R., Ahmed, S., Hakim, M., & Sen, J. (2019). Comparative Study among Composite Breads Incorporated with Full Fatted and Defatted Rice Bran. *Journal of Environmental Science and Natural Resources, 11*(1-2), 43-52. https://doi. org/10.3329/jesnr. v11i1-2.43364

32. Gul, K., Yousuf, B., Singh, A. K., Singh, P., & Wani, A. A. (2015). Rice bran: Nutritional values and its emerging potential for development of functional food—A review. *Bioactive Carbohydrates and Dietary Fibre, 6*(1), 24-30. https://doi. org/10.1016/j.bcdf. 2015.06.002

33. Hussein, A., & Ibrahim, G. (2019). Effects of various brans on quality and volatile compounds of bread. *Foods and Raw Materials, 7*, 35-41. https://doi.org/ 10.21603/2308-4057-2019-1-42-50

34. Rose, D. J., Inglett, G. E., & Liu, S. X. (2010). Utilisation of corn (*Zea mays*) bran and corn fiber in the production of food components. *Journal of the Science of Food and Agriculture, 90*(6), 915-924. https://doi.org/10.1002/jsfa.3915

35. Koegelenberg, D., & Chimphango, A. F. A. (2017). Effects of wheat-bran arabinoxylan as partial flour replacer on bread properties. *Food Chemistry, 221*, 1606-1613. https://doi.org/10.1016/j. foodchem.2016.10.130

36. Sandhu, P., Bains, K., Singla, G., & Sangwan, R. (2018). Utilization of Corn Fibre and Pectin Gel for the Development of Low Calorie and High Fibre Biscuits. *Current Journal of Applied Science and Technology, 25*(5), 1-9. https://doi. org/10.9734/cjast/2017/ 38966

37. Sangeetha, A. V., Mahadevamma, S., Begum, K., & Sudha, M. L. (2011). Influence of processed sugarcane bagasse on the microbial, nutritional, rheological and quality characteristics of biscuits. *International Journal of Food Sciences and Nutrition, 62*(5), 457-464. https://doi.org/10.3109/ 09637486.2010.549819

38. Vijerathna, M. P. G., Wijesekara, I., Perera, R., Maralanda, S. M. T. A., Jayasinghe, M., & Wickramasinghe, I. (2019). Physico-chemical Characterization of Cookies Supplemented with Sugarcane Bagasse Fibres. *Vidyodaya Journal of Science, 22* (1), 29. https://doi.org/10.4038/vjs. v22i1.6062

39. Yadav, R. B., Yadav, B. S., & Chaudhary, D. (2011). Extraction, characterization and utilization of rice bran protein concentrate for biscuit making. *British Food Journal, 113*(9), 1173-1182. https://doi.org/10.1108/ 00070701111174596

40. Zhang, L., Van Boven, A., Mulder, J., Grandia, J., Chen, X. D., Boom, R. M., & Schutyser, M. A. I. (2019). Arabinoxylans-enriched fractions: From dry fractionation of wheat bran to the investigation on bread baking performance. *Journal of Cereal Science, 87*, 1-8. https://doi.org/10.1016/j.jc s.2019.02.005

41. Moher D, Liberati A, Tetzlaff J, Altman DG, The PRISMA Group (2009). Preferred reporting items for systematic reviews and MetaAnalyses: The PRISMA statement. *PLoS Med 6*(7): e1000097. https://doi.org/10.1371/ journal. pmed1000097

42. [42] Downs, S. H., & Black, N. (1998). The feasibility of creating a checklist for the assessment of the methodological quality both of randomised and nonrandomised studies of health care interventions. *Journal of epidemiology and community health, 52*(6), 377–384. https://doi.org/10.1136/jech.52.6.377

43. Hooper, P., Jutai, J. W., Strong, G., & Russell-Minda, E. (2008). Agerelated macular degeneration and lowvision rehabilitation: a systematic review. Canadian Journal of Ophthalmology, 43(2), 180-187. https:// doi.org/10.3129/i08-001

44. Gil-López, D. I. L., Lois-Correa, J. A., Sánchez-Pardo, M. E., DomínguezCrespo, M. A., Torres-Huerta, A. M., Rodríguez-Salazar, A. E., & OrtaGuzmán, V. N. (2019). Production of dietary fibers from sugarcane bagasse and sugarcane tops using

microwaveassisted alkaline treatments. *Industrial Crops and Products, 135*, 159-169. https:// doi.org/10.1016/j.indcrop.2019.04.042

45. Amna, Y., Bhatti, M. S., Anwaar, A., & Randhawa, M. A. (2011). Effect of rice bran supplementation on cookie baking quality. *Pakistan Journal of Agricultural Sciences, 48*(2), 133-138.

46. Jauharah, M. Z., Wan Ishak, W. R., & Robert, S. (2014). Physicochemical and Sensorial Evaluation of Biscuit and Muffin Incorporated with Young Corn Powder. Sains Malaysiana, 43, 45-52.

47. Ameh, M. O., Gernah, D. I., & Igbabul, B. D. (2013). Physico-Chemical and Sensory Evaluation of Wheat Bread Supplemented with Stabilized Undefatted Rice Bran. Food and Nutrition Sciences, 04(09), 43-48. https://doi.org/10.4236/fns.2013.49a2007

48. Pauline, M., Roger, P., Sophie Natacha Nina, N. E., Arielle, T., Eugene, E. E., & Robert, N. (2020). Physicochemical and nutritional characterization of cereals brans enriched breads. *Scientific African, 7*, e00251. https://doi.org/10.1016/j.sciaf. 2019.e00251

49. Silva, A. S., Correa, L. G., Kanai, R. S. S., & Shirai, M. A. (2020). Effect of sugarcane bagasse addition on physical, chemical, and sensory properties of oat flour and banana cake. *Journal of Texture Studies*. https://doi.org/10.1111/jtxs.12542

50. European Food Safety Authority Panel on Dietetic Products, N., & Allergies. (2011). Scientific Opinion on the substantiation of health claims related to arabinoxylan produced from wheat endosperm and reduction of postprandial glycaemic responses (ID 830) pursuant to Article 13(1) of Regulation (EC) No 1924/2006. EFSA Journal, 9(6), 2205. https://doi.org/10.2903/j.efsa. 2011.2205

51. Saeed, F., Pasha, I., Anjum, F. M., & Sultan, M. T. (2011). Arabinoxylans and Arabinogalactans: A Comprehensive Treatise. Critical Reviews in Food Science and Nutrition, 51(5), 467-476. https://doi.org/10.1080/ 10408391003681418

52. Ortiz de Erive, M., Wang, T., He, F., & Chen, G. (2020). Development of high-fiber wheat bread using microfluidized corn bran. Food Chemistry, 310, 125921. https://doi.org/ 10.1016/j.foodchem.2019.125921

53. Leang, Y. H., & Saw, H. Y. (2011). Proximate and functional properties of sugarcane bagasse. *Agro Food Industry Hi-Tech, 22*(2), 5-8.

54. Kale, M. S., Pai, D. A., Hamaker, B. R., & Campanella, O. H. (2010). Structure-function relationships for corn bran arabinoxylans. *Journal of Cereal Science, 52*(3), 368-372. https:// doi.org/10.1016/j.jcs.2010.06.010

55. Zidan, N., & Eldemery, M. (2016). Utilization of defatted black rice bran in wheat bread preparation for enhancing nuritional and functional properties. *Journal of Food and Dairy Sciences, 7*(2), 107-117. https://doi.org/10.21608/ jfds.2016.42819

56. Sozer, N., Cicerelli, L., Heiniö, R.L., & Poutanen, K. (2014). Effect of wheat bran addition on in vitro starch digestibility, physico-mechanical and sensory properties of biscuits. *Journal of Cereal Science, 60*(1), 105-113. https:// doi. org/10.1016/j.jcs.2014.01.022

57. Mishra, N., & Chandra, R. (2012). Development of functional biscuit from soy flour & rice bran. *International Journal of Agricultural and Food Science, 2*, 14-20.

58. Jia, M., Yu, Q., Chen, J., He, Z., Chen, Y., Xie, J., . . . Xie, M. (2020). Physical quality and in vitro starch digestibility of biscuits as affected by addition of soluble dietary fiber from defatted rice bran. *Food Hydrocolloids, 99*, 105349. https://doi. org/10.1016/j. foodhyd.2019.105349

59. Döring, C., Nuber, C., Stukenborg, F., Jekle, M., & Becker, T. (2015). Impact of arabinoxylan addition on protein microstructure formation in wheat and rye dough. *Journal of Food Engineering, 154*, 10-16. 10.1016/j. jfoodeng.2014.12.019

60. El-Gammal, R., & El Kewawy, H. (2014). Effect of addition of stabilized rice bran on physical, rheological and chemical characteristics of pan bread. *Journal of Food and Dairy Sciences, 5*(11), 813-825. https://doi.org/10.21608/jfds. 2014.53231

61. Lockyer, S., & Spiro, A. (2020). The role of bread in the UK diet: An update. *Nutrition Bulletin, 45*(2), 133-164. https://doi.org/10.1111/nbu.12435

62. Mozaffarian, D. (2016). Dietary and Policy Priorities for Cardiovascular Disease, Diabetes, and Obesity. *Circulation, 133*(2), 187-225. https://doi. org/10.1161/ circulationaha.115.018585

63. Ayala-Soto, F. E., Serna-Saldívar, S. O., & Welti-Chanes, J. (2017). Effect of arabinoxylans and laccase on batter rheology and quality of yeast-leavened gluten-free breads. *Journal of Cereal Science, 73*, 10-17. https://doi.org/ 10.1016/j. jcs.2016.11.003

Chapter 3

HONEY PRODUCTION PROCESS

INTRODUCTION

Honey consumption has a very old history for humans. It has been used as a sweetener and flavoring in countless food and beverages. Since ancient times, honey has been known for its nutritious and therapeutic aspects. The most important components of honey are carbohydrates, which are found in the form of fructose, glucose, disaccharides, and oligosaccharides, and components such as maltose, isomaltose, maltulose, sucrose provide the sweet taste to honey. It also contains enzymes such as amylase, oxidase peroxide, catalase and acid phosphorylase, including anderosis and panoz. Also, honey is rich in amino acids, minerals, antioxidants and various phytochemicals [1]. Many of the reported biological properties of honey, such as antioxidant, antibacterial, antifungicidal, anti-inflammatory, hypotensive, antiproliferative, hepato-protective properties of these components are associated with presence of these properties. However, the composition of honey largely depends on a number of factors, such as flower source, geographical region, climatic conditions, harvesting season, processing and storage conditions.

There are studies that report that honey, administered alone or in combination with traditional therapy, may be useful in the treatment of chronic diseases that are commonly associated with oxidative stress and the state of inflammation [2]. Honey is classified according to various criteria. In this classification, honey is classified as secretion honey (such as pine honey, oak honey, fir honey, leaf honey) and flower honey (linden honey, cotton honey, trirose honey, thyme honey, mashed honey, acacia honey, heather honey, etc.). According to the form of marketing, framed comb honey, natural comb honey, partial comb honey, cut-comb honey, strained honey, crystallized honey, creamed honey, pressed honey, chunk honey (strained with comb or comb with strained), filtered honey and baker's honey. According to the moisture content, honey is classified as grade 1 honeys (humidity below 17.8%), grade 2 honeys (humidity up to 18.6%) and grade 3 honeys (humidity up to 20%). According to their color, honey is classified as white, golden, amber and dark. The color of honey can vary from light water white to black warehouse [3]. The physical and chemical properties, antimicrobial effects, which are of great importance for public health, and GMP and HACCP systems applied in the production process and microbiological dangers will be addressed in this section.

Physical and chemical properties of honey

Honey contains about 200 substances and is a nutrient consisting of substances such as carbohydrates, water, enzymes, free amino acids, essential minerals, vitamins, phenolic compounds, volatile compounds (monoterpenes, benzene derivatives) and some other solids. Carbohydrates in honey are mainly monosaccharides, glucose and fructose. This is followed by disaccharides and trisaccharides. They contribute mainly to the energy value. Proteins include enzymes such as invertase, diastase, glucose oxidase, catalase, peroxidase and acid phosphatase, and their content varies from 0.1% to 3.3% depending on the type of honey. It contains essential and non-essential amino acids, but the most common amino acid in honey is proline, which accounts for 1% of honey components [2]. Honey contains tocopherol (E), anti-hemorrhagic vitamin (K), ascorbic acid (C), thiamine (B_1), riboflavin (B_2), niacin (B_3), pantothenic acid (B_5) and a small amount of vitamin pyridoxine (B_6). Vitamins of the B complex and vitamin C are mainly derived from pollen and can be affected by commercial and industrial processes such as filtration or oxidation reactions [4].

Honey has a slight acid reaction due to about 0.57% organic acids. Acids contribute to the aroma and antimicrobial activity of honey. The predominant acid in honey is gluconic acid, it is followed by aspartic citric, acetic, formic, fumaric, galacturonic, malonic, formic, acetoglutaric, glutamic, butyric, glutaric, propionic, pyruvic, glioxia, 2-hydroxybutyric, a-hydroxiglutaric, isocyric, lactic, malic, methylmalonic, kynic, succinic, tartaric, oxalic acid [2]. The mineral content in honey ranges from 0.04% in light honey and 0.2% in dark honey. Potassium is the most abundant element. But the main bioactive molecules contained in honey are represented by polyphenols. Polyphenols are a heterogeneous chemical compound that can be divided into flavonoids (flavonols, flavones, flavanols, flavanones, anthocyanin, calcones and isoflavones) and non-flavonoid (phenolic acids). The profile of polyphenolic compounds in honey is thoroughly studied and includes vanillin, caffeic, syringing, p-gamic, ferulic, ellagic, 3-hydroxybenzoic, chlorogenic, genistic, gallic and benzoic acids and contains different phenolic acids, such as different flavonoids, mainly quercetin, kaempferol, myricetin, chrysin, galangin, hesperetin. The amount and type of polyphenols largely depends on the flower source or the variety of honey. In addition, it is known that there is a strong relationship between antioxidant activity and total phenolic content [5].

The importance of honey in terms of health and its antimicrobial effect

Honey is a food that has been used in therapeutic treatments for thousands of years. Among other useful properties to health, this product has been reported as a promising agent for wound healing, including leg ulcers and eyes, skin disorders by in vitro and clinical studies. In studies on New Zealand manuka honey, unique to the New Zealand, positive effects were observed on the viability of potentially useful *Lactobacillus reuter* and *Bifidobacterium longum* found in the human intestine. Moreover, it was found that *Salmonella enterica* Typhimurium, an enteric pathogenic bacterial type, showed a 65% reduction in their proliferation. In this sense, it has been established that manuka honey has a beneficial effect on the intestine by producing acid metabolites that reduce the intestinal pH and prevent pathogenic colonization and hence support the growth of bifidobacteria and lactic acid bacteria. Honey has been reportedly able to modulate oxidative stress and also has anti-proliferative, pro-apoptotic, anti-inflammatory and anti-metastatic properties. The anticancer effect of honey is connected to the presence of natural bioactive compounds, mainly such as pinobanksin, pinocembrin, luteolin, chrysin, salicylic acid and 3.4 dihydroxybenzoic acid [6, 7]. Some of the vitamins contained in honey are ascorbic acid, pantothenic acid, niacin and riboflavin. Moreover, it is a food that also contains minerals such as calcium, copper, iron, magnesium, manganese, phosphorus, potassium and zinc. Its rich variety of vitamins and minerals also plays a role in increasing the antioxidant characteristics of honey. The presence of free radicals and reactive oxygen types is responsible for pathogenesis of aging, as well as cellular dysfunction, metabolic and cardiovascular diseases. Consumption of foods rich in antioxidants can protect against these pathological changes, preventing the pathogenesis of chronic ailments [8].

Various parameters such as low water activity, high sugar content, acidity and hydrogen peroxide (H_2O_2) content, phytochemicals, peptides, non-peroxidase glycoeptides and proteins make up the antibacterial potential of honey. Water activity of honey varies from 0.56–0.62. These values might be considered low enough to prevent the development of bacteria or other microorganisms [9].

Although previously it was believed that the only responsible agent for the antibacterial effect of diluted honey was H_2O_2 and that this antibacterial effect can be completely eliminated through catalysis, it has been found out that bacteria can also be affected via the existence of pythochemical elements present in honey [10]. As it suppresses the activities of bacteria causing infections in urinary systems, such as *E. coli* and *Proteus* species and *Streptococcus faecalis*, diluted honey is used to treat urinary system infections and it inhibits toxin production [11]. Undiluted honey hinders the reproduction and development of bacteria due to the content of sugar, which exerts osmotic pressure on bacterial cells and causes water to flow out of bacterial cells through osmosis. Thus, the cells shrink due to dehydration, and they cannot remain alive in hypertonic sugar solution. The optimal pH necessary for the development of most microorganisms ranges from 6.5–7.5. The pH value of honey is between 3.2–4.5, and this value is a very distinctive feature of its antibacterial activity. This acidity is caused by the presence of certain important organic acids, especially gluconic acid-in 0.5% (a/h) concentration. Glycogenic acid is produced from glucose oxidation by an endogenous enzyme of glucose oxidase and is an extremely powerful antibacterial agent. In undiluted pure honey, low pH can contribute to antibacterial action, but when the product is diluted pH alone is not enough to prevent the development of bacteria [9]. The formation of H_2O_2 is a dominant mechanism in which honey exerts bacteriostatic and bactericidal activity. It provides antibacterial activity of honey and is produced enzymatically. The enzyme glucose oxidase is inherently inactive in honey due to low pH conditions, and glucose oxidase is activated when honey is diluted. However, it is known that concentrations of H_2O_2 are adversely affected by various minor components, such as nectar, pollen, and yeast. It has also been reported as having high sensitivity to light and light sources [12]. Honey contains relatively small amounts of proteins, whose molecular weights range from 20 to 80 kDa, ranging from approximately 0.1% to 0.5%. These proteins contain many enzymes involved in sugar metabolism, such as alpha and beta glucosidase, glucose oxidase and amylase. Numerous studies have shown that important royal jelly proteins have antimicrobial and anticancer activity and anti-inflammatory potential [13].

Honey shows antibacterial activity against a large number of bacteria in different environments. Natural components of honey have antifungal, antiviral, antibacterial activities. It has been reported that the antibacterial activity of honey is also likely to depend on the pasture, climatic conditions, and also on the natural composition of flower nectar. Honey has excellent antibacterial activity against methicillin-resistant *Staphylococcus aureus* (MRSA), often associated with wound and burn infections, and *Pseudomonas* spp. Many studies have shown that honey is also effective against hemolytic streptococci and vancomycin resistant enterococci. Twenty-one kinds of honey tested for antibacterial activity against *Staphylococcus aureus (S. aureus)* and *Pseudomonas aeruginosa (P. aeruginosa)*, and it has been established that they have a positive effect due to H_2O_2 and polyphenolic content levels. The effectiveness of free radical cleansing is observed in all kinds of honey. In addition, honey tested by freezing, drying and powdering has been reported to show antioxidant activity in each form [9, 14]. Flavonoids contained in the natural composition of honey is also known to be effective against microorganisms that are present in the tissue of chronic wounds, in particular *S. aureus*, *P. aeruginosa*,

as well as *Escherichia coli (E. coli*). Flavonoids are often recommended as a natural source to control chronic inflammatory diseases, the incidence of which increases significantly. Despite the fact that the topical application of honey for medicinal purposes is old, there are a small number of studies that address its anti-inflammatory activity at the cellular level. Although flavonoids are small components of honey, their anti-inflammatory effect is extraordinary compared to other natural compounds [15]. Honey was found to have a preventive effect on about 60 bacteria such as, *Bacillus anthracis, Corynebacterium diptheriae, Haemophilus influenzae, Klebsiella pneumoniae, Shigella, Mycobacterium tuberculosis*, and many aerob and anaerob bacterial types. In vitro studies of *Helicobacter pylori* in the human digestive system have shown that when using honey, its activity decreases by 20%. It has been reported that honey can be used in combination with antibiotics to produce a synergistic effect of bactericidal activity against *Helicobacter pylori*. The main difference of honey with antibiotics is that it does not develop antibiotic-resistant bacteria, so it can be used continuously without such risk [16].

OTHER BEEKEEPING PRODUCTS

Pollen

Pollen is the only source of protein found in nature for bees. The amino acids contained in its composition are isolosin, arginine, lysine, histidine, leucine, methionine, treonine, phenylalanine, tryptophan and valine. It is essential for adequate development of their muscles, tissues, secretory glands and other organs in the upbringing of honeybees and its young stages. It is a nutritional source rich in vitamins, proteins, sterols, minerals and lipids. It has been reported that pollen collected by honeybees may have differences in their general chemical composition as a result of supplying from different plants [17].

Nectar

Bees have two stomach and they use one of them to perform normal body functions whereas the other to store the nectar they collect. In order to collect nectar found in flowers, bees use rod-like, tubular long tongues. It has been reported that bees can contain about 70 mg of nectar in the stomach they store nectar, and that they should visit 100 to 1500 different flowers to fully fill their honey stomach [17].

Propolis

Propolis is recognized as a therapeutic agent due to several reported functional effectiveness. It is known that honey contains phenolic compounds. Propolis contains a higher content of phenolic compounds than honey and shows significantly higher antimicrobial and antioxidant activities. Today it is used in industry as a component of confectionery, biopharmaceuticals and cosmetics. It is gaining popularity as a natural preservative and helps to improve shelf life and consumer health as a source of bioactive compounds for food and drinks. However, propolis has a strong and bitter taste, which changes the sensory properties of food due to the high concentration of phenolic compounds. Therefore, the acceptance of foods containing propolis by consumers must be determined by its propolis concentration, which has to be carefully researched so as not to adversely change the sensory properties of such foods [18].

Bee milk

The importance of bee milk, one of bee products, was noticed in the 1600s and was given the name "*Royal Jelly*", which means excellent food in English. Bee milk is secreted from the upper jaw (mandibular) and throat glands (hypopharyngeal) of young worker bees of 5–15 days of age. All larvae only in their first three-day period, and the larvae that will become the queen bee are fed with royal jelly during the entire larval and adult periods. Bee milk can be described as food with a peculiar smell and a bitter taste and a mush-like form with bone-like color. It is collected from the cells of larvae of 3–4 days of age of the future queen, or from the cells of queen bees where larvae of 1–2 days are laid after 48–72 hours. It is quite flowing and has yogurt-like consistency but is a homogeneous substance. It has a light beige and yellowish whitish color, a sharp phenolic smell and a distinctive sour taste. Its density is approximately 1.1 g/cm^3 and is soluble in water [19].

MICROBIOLOGICAL RISKS IN HONEY AND HONEY PRODUCTS

Although honey is considered a low-risk food due to its antimicrobial and bacteriostatic effects, studies disprove this view. In addition to primary contamination, staff, tools and equipment used in beekeeping and honey production are also a potential source of secondary contamination. In addition, honey, which has the potential to contain many microorganisms as a result of cross-contamination, is among the important nutrients and can indirectly threaten public health. Despite the fact that some types of honey contain H_2O_2 and benzoic acid and phenolic compounds such as some flavonoids, it can constitute risks for consumer health due to minimal hygiene rules. It is reported that pathogens can be found as causative agents in honey produced without food safety systems. Food-borne pathogens are recognized as an important risk factor for public health in developed and developing countries due to their spread around the world. Viruses, bacteria, fungi, parasites and mites are the most common disease factors in beekeeping. Fecal-oral route is an important way of transmission of these diseases. Agents that pollute bees through water and food can be transmitted to larvae by infected bees. Another contamination that may occur in honey is secondary contamination caused by secondary contamination sources such as personnel, tools and equipment [20].

The presence of strains *Bettsya alvei*, *Acosphaera apis* and *Acosphaera* major in honey production can be indicative of improper beehive management practices. Different types of microorganisms such as *Acinetobacter* spp., *Bacillus* spp., *Clostridium* spp., *Corynebacterium* spp., *Pseudomonas* spp. are bacteria that are widely found in the soil. *Brochothrix* spp., *Citrobacter* spp., *Enterobacter* spp., *Erwinia* spp., *Flavobacterium* spp., *Lactobacillus* spp., *Lactococcus* spp., *Leuconostoc* spp., *Listeria* spp. and *Pediococcus* spp. are other bacteria that are likely to be found in plants and plant products. On the other hand, among yeast strains *Saccharomyces*, *Schizosaccharomyces* and *Torula* species predominate in high humidity sugars.

Bacterial spores, especially *Bacillus* and *Clostridium*, can be seen in honey. *Clostridium* is an indicator organism that provides evidence of contamination or pollution in honey. *Clostridium botulinum* (*C. botulinum*) spores are usually found at low levels in honey. The presence of *clostridium* spores can be dangerous, especially for children under one year of age. It is known that infant botulism is mainly caused by the consumption of honey contaminated with *C. botulinum* [21, 22]. *C. botulinum* forms 4 different types of neuroparalytic diseases in humans. In addition to infant botulism, they are classified as food-borne botulism, wound botulism, and yet unclassified latent botulism. The most important of them is infant botulism, which occurs in newborn babies of 3–20 weeks. Infant botulism is diagnosed with isolation of *C. botulinum* and toxin in feces. Decreases in sucking and swallowing reflexes of infants can be observed, which is very rarely fatal [23].

One of the animal products that have been the focus of food warnings due to the presence of chemical hazards such as antibiotics or pesticides in recent years are honey and honey products. The source of these residues in honey is mainly due to bee parasites, such as European offspring rot (*Streptococcus pluton*) or American offspring rot (*Bacillus larvae*) and are veterinary drugs that are necessary to treat bacterial diseases. It is known that chemical residues caused by these drugs used to eliminate microbiological risks, lead to such adverse conditions on human health as allergic reactions, bacterial resistance, along with changes of reproductive toxicity [24, 25].

HACCP IN HONEY PRODUCTION

Food safety can be ensured by systematic implementation of all activities in line with a plan. The Hazard Analysis Critical Control Points (HACCP) system, as a preventive system for ensuring food safety, controls production at various points throughout the food production, thereby ensuring that the final product complies with legislation. Preliminary Requirement Program must be created first in order to establish the HACCP system in any food business. In this context, the deficiencies of the infrastructure and processes such as water, energy, warehouse, cleaning and sanitation, personnel, environment and equipment hygiene, personnel training and pest control should be addressed. However, it is necessary to plan the process management by writing down the procedures. The processes that need to be addressed afterwards can be sorted as follows; identification of the HACCP team and a clear definition of the task descriptions by making the managerial organization chart, determination of food safety policy by business management, making an understandable description of the products to be produced, determination of the intended usage method, preparation of a flow diagram and placement plan by HACCP team and verification of this plan at site, analyzing hazards and risks, identification of critical control points, making up of critical limits and monitoring procedures, determination of corrective activities for cases where it is necessary, and the proving or verification of the effectiveness of the system [26].

Codex Alimentarius Standard and the European Commission allows nomenclature for honeys produced from certain botanical sources if the product comes from the specified origin and has anticipated physicochemical, organoleptic and microscopic properties. The fact that there are different varieties of honey and each has its own production steps, leads to an increase in the limits that need to be controlled. For the import of food products of animal origin, such as honey, EU legislation requires a number of health and national residue monitoring procedures such as HACCP during the production and processing of honey. These requirements are known to be independent of whether honey is organic or traditional. Thus, imported products are intended to meet the standards required for production and trade within EU member countries. Costs, lack of qualified personnel, misinterpretation of EU legislation, lack of laboratory in international standards and improper infrastructure are the main obstacles to being accredited by the EU [27, 28].

Although honey is a product that is part of the low-risk group due to its high sugar content, it should be carefully examined for physical, chemical and biological hazards. In general, the hygiene of the processing area, tool and equipment and personnel in contact with food should be observed as it should be in all food enterprises. Physical hazards such as soil, plant materials, glass materials, tools and equipment are defined as potential hazards to honey. Traces of pesticides and herbicide, beekeeping drugs and antibiotics are chemical hazards. Soil originated *C. botulinum*, the most important biological danger in honey production, is eliminated by the provision of hygienic conditions in the production of honey [29].

General hygiene rules should be applied effectively to prevent physical, chemical, and microbiological hazards. In hives, legally approved preservatives should be used. Insects and mice should be kept away from hives. During transportation, the vehicles should be cleaned well, in case of the presence of dirt left from the previous use. It is necessary to effectively clean the equipment and work area before and after use. Persons involved in the process should wear a separate clothing to protect the product from contamination caused by clothing or individuals. Especially before and after use, cleaning control of filters must be carried out effectively. In the HACCP plan, the purpose of conducting hazard analysis must be effectively controlled. All potential hazards in each step of the workflow process must be identified and the risk and severity of each identified hazard should be assessed.

At this point, it is also necessary to determine the sources of dangers. In a study, corrective activities to prevent and/or eliminate hazards were determined and two critical controls points, "filtration/unloading" and "packaging" were pointed out. Examples of forms used in each HACCP plan and all procedures of the HACCP plan must be provided and monitored [26, 30].

CONCLUSIONS

HACCP system, which is successfully implemented in the food industry, is the most effective quality system in terms of the supply of safe products. The purpose of the use of HACCP system is to provide reliable food to the consumer with the desired characteristics and quality. Honey production, which is suitable to be affected by climatic conditions, should be made systematic and controllable by removing traditional methods that are difficult to trace. In this context, the creation of honey workflow process, determining potential hazards, and analyzing hazards, taking necessary precautions, recording the system, providing internationally reliable product guarantee is of great importance for public health as well as for

the economies of countries. Effective implementation of the HACCP system in enterprises is inevitable so that retrospective monitoring and recall models can be used in the event of any negativity. In the production processes of foods with high nutritional value as honey, all necessary food safety requirements must be met to protect and improve public health.

ABSTRACT

Honey has been considered as a very important and superior nutrient in human nutrition since ancient times due to its ability to be consumed by humans without processing, easy digestibility, nutritional properties and biological benefits.

Although honey contains many desired bioactive and antibacterial substances, which may be sufficient for antimicrobial activity, it cannot be produced in sufficient quantities due to low water activity under normal conditions. This causes various food and bee-borne spores/non-spores pathogens going viral. Hence, it may cause the risk of parasitological and fungal agents to be found. In honey production, "Hazard Analysis Critical Control Point (HACCP)" must be applied meticulously and completely. Current technologies in honey production will be explained in this section.

REFERENCES

1. Meo SA, Al-Asiri SA, Mahesar AL, Ansari MJ. Role of honey in modern medicine. Saudi Journal of Biological Sciences. 2017;24:975-978. DOI: http:// dx.doi.org/10.1016/j.sjbs.2016.12.010
2. Terzo S, Mulè F, Amato A. Honey and obesity-related dysfunctions: a summary on health benefits. Journal of Nutritional Biochemistry. 2020;82:108401. DOI: https://doi. org/10.1016/j.jnutbio.2020.108401
3. Doğan H. Determination of chemical, physical and antimicrobial properties of nectar honeys [thesis]. Atatürk University; 2014.
4. Cianciosi D, Forbes-Hernández TY, Afrin S, Gasparrini M, ReboredoRodriguez P, Manna PP, Zhang J, Lamas LB, Florez SM, Toyos PA, Quiles JL, Giampieri F, Battino M. Phenolic compounds in honey and their associated health benefits: A review. Molecules. 2018; 23:2322. DOI:10.3390/ molecules23092322
5. Miguel MG, Antunes MD, Falerio ML. Honey as a Complementary Medicine. Integrative Medicine Insights. 2017; 12:1-15. DOI: https://doi. org/10.1177/117863371770286
6. Seraglio SKT, Schulz M, Gonzaga VL, Fett R, Costa ACO. Current status of the gastrointestinal digestion effects on honey: A comprehensive review. Food Chemistry. 2021; 357: 129807. DOI: https://doi.org/10.1016/j. foodchem.2021.129807
7. Afrin S, Haneefa SM, Fernandez-Cabezudo MJ, Giampieri F, Al-Ramadi BK, Battino M. Therapeutic and preventive properties of honey and its bioactive

compounds in cancer: an evidence-based review. Nutrition Research Reviews. 2020;33(1):50-76. DOI: https://doi.org/10.1017/ S0954422419000192

8. Ajibola A, Chamunorwa JP, Erlwanger KH. Nutraceutical values of natural honey and its contribution to human health and wealth. Nutrition & Metabolism. 2012;9(1):1-12.

9. Almasaudi S. The antibacterial activities of honey. Saudi Journal of Biological Sciences. 2021;28:2188-2196. DOI: https://doi.org/10.1016/j. sjbs.2020.10.017

10. Albaridi NA. Antibacterial Potency of Honey. International Journal of Microbiology. 2019. DOI: https://doi. org/10.1155/2019/2464507

11. Eteraf-Oskouei T, Najafi M. Traditional and Modern Uses of Natural Honey in Human Diseases: A Review. Iranian Journal of Basic Medical Sciences. 2013; 16(6): 731-742.

12. Brudzynski K. A current perspective on hydrogen peroxide production in honey. A review. Food Chemistry. 2020;332:127229. DOI: https://doi. org/10.1016/j. foodchem.2020.127229

13. Tonks AJ, Cooper RA, Jones KP, Blair S, Parton J, Tonks A. Honey stimulates inflammatory cytokine production from monocytes. Cytokine 2003;21:242-247. DOI: doi:10.1016/ S1043-4666(03)00092-9

14. Stagos D, Solitsiotis N, Tsadila C, Papaeconomou S, Arvanitis C, Ntontos A, Karkanta F, Adamou-Androulaki S, Petrotos K, Spandidos DA, Kouretas D, Mossialos D. Antibacterial and antioxidant activity of different types of honey derived from Mount Olympus in Greece. International Journal Of Molecular Medicine. 2018 42:726-734. DOI: 10.3892/ijmm.2018.3656

15. Silva B, Biluca FC, Gonzaga LV, Fett R, Dalmarco EM, Caon T, Costa ACO. In vitro anti-inflammatory properties of honey flavonoids: A review. Food Research International. 2021;14:110086. DOI: https://doi. org/10.1016/j. foodres.2020.110086

16. Ilia G, Simulescu V, Merghes P, Varan N. The health benefits of honey as an energysource with antioxidant, antibacterial andantiseptic effects. Science & Sports. DOI: https://doi. org/10.1016/j.scispo.2020.10.005

17. Sahin G. Research on the chemical composition of some honey [thesis]. Ankara University; 2019.

18. Osés SM, Pascual-Maté A, Fernández-Muiño MA, López-Díaz TM, Sancho MT. Bioactive properties of honey with propolis. Food Chemistry. 2016;196:1215-1223. http://dx.doi. org/10.1016/j.foodchem.2015.10.050

19. Akyol E, Baran Y. Arı sütünün yapısı, insanlar ve arılar için önemi. Uludag Bee Journal. 2015; 15(1):16-21.
20. Bayrakal GM, Ekici G, Akkaya H, Sezgin FH, Dümen E. Detection and Molecular Examination of Pathogens in Honey and Bees in the Northern Marmara Region, Turkey. Journal of Kafkas University of Veterinary Faculty. 2020;26(3):313-319. DOI: 10.9775/ kvfd.2019.22845
21. Finola MS, Lasagno MC, Marioli JM. Microbiological and chemical characterization of honeys from central Argentina. Food Chemistry. 2007;100:1649-1653. DOI:10.1016/j. foodchem.2005.12.046
22. Vázquez-Quinones CR, MorenoTerrazas R, Natividad-Bonifacio I, Quinones-Ramírez EI, Vázquez-Salinas C. Microbiological assessment of honey in México. Revista Argentina De Microbiologia. 2018;50(1):75-80. DOI: https://doi.org/10.1016/j. ram.2017.04.005
23. Nevas M, Lindström M, Hörman A, Keto-Timonen R, Korkeala H. Contamination routes of *Clostridium botulinum* in the honey production environment. Environmental Microbiology. 2006;8(6):1085-1094. DOI:10.1111/j.1462-2920.2006.01000.x
24. Juan-Borras M, Periche A, Domenech E, Escriche I. Routine quality control in honey packaging companies as a key to guarantee consumer safety. The case of the presence of sulfonamides analyzed with LC-MS-MS. Food Control. 2015;50:243-249. DOI: http://dx.doi.org/10.1016/j. foodcont.2014.08.021
25. El-Nahhal Y. Pesticide residues in honey and their potential reproductive toxicity. Science of the Total Environment. 2020;741:139953. DOI: https://doi.org/10.1016/j. scitotenv.2020.139953
26. Suna S, Özcan Sinir G, Anlar D. Bal Üretim Prosesinde HACCP Uygulaması. Arıcılık Araştırma Dergisi. Arıcılık Araştırma Dergisi. 2013;5(10):17-21.
27. Bogdanov S, Fluri P, Imdorf A, Charrierre JD, Kilchenmann V. Self control sytstem for the production of high quality honey: The Swiss example. Apimondia Journal. 2005;40:28-33.
28. Oddo LP, Bogdanov S. Determination of honey botanical origin: problems and issues. Apidologie. 2004;35:2-3. DOI: 10.1051/ apido:2004044
29. Al-Waili N, Salom K, Al-Ghamdi A, Ansari MJ. Antibiotic, Pesticide, and Microbial Contaminants of Honey: Human Health Hazards. The ScientificWorld Journal. 2012;9: 930849. DOI:10.1100/2012/930849.
30. Mutinelli F, Costa C, Lodesani M, Baggio A, Medrzycki P, Formato G, Porrini C. Honey bee colony losses in Italy. Journal of Apicultural Research 2010;49(1):119-120. DOI: 10.3896/ IBRA.1.49.1.24

Chapter 4

CHOCOLATE: HEALTH, PROCESSING, AND FOOD SAFETY

INTRODUCTION

The cocoa bean is the seed of the cacao tree, it is a tropical plant indigenous to the equatorial regions of the Americas, but currently, it is cultivated in many countries with a warm, tropical climate. Cocoa is the main raw material to produce chocolate, which can value more than $100 billion [1]. Chocolate history started with the Maya, who were the first people who cultivated the cocoa plant in South America. During that period, chocolate was a cocoa drink prepared with hot water, and it can be flavored usually with cinnamon and pepper [2].

The Republic of Côte d'Ivoire, a country on the southern coast of West Africa, is the biggest producer of cocoa beans with more than 2 million tons annually. The national economy of Ivoirians depends on their export revenue from cocoa. Its population is more than 26 million, almost 6 million work in cocoa beans production, and almost 2,034,000 tons were produced in 2021 [1]. Many countries such as Ghana (883,652 tons), Indonesia (659,776 tons), Nigeria (328,263 tons), Cameron (295,028 tons), Brazil (235,809 tons), Ecuador (205,955 tons), Peru (121,825 tons), Dominican Republic (86,599 tons), and Colombia (56,808 tons) [3] that are considered top producers of cocoa come after Cote d'Ivoire.

This chapter is focused on the health benefits of chocolate, processing, type of chocolate, and finally, the safety aspect of chocolate production. Furthermore, the negative impact of chocolate on human health has been discussed.

HEALTH BENEFITS

There are numerous health benefits associated with the consumption of chocolate, which are due to its nutritional contents [4, 5]. A bar of chocolate with a higher number of nutrients has high health benefits, for example, dark chocolate [5], which contains soluble fiber and is rich in minerals. In more detail, a 100-gram bar of dark chocolate with 70–85% cocoa consists of 11 grams of fiber, 67% of the daily value (DV) for iron, 58% of the DV for magnesium, 89% of the DV for copper, 98% of the DV for manganese. So, dark chocolate is the best type with regard to the number of nutrients [6].

Chocolate possesses antioxidant activity, which is very beneficial for human health [5] and can be attributed to the presence of polyphenols, flavanols, and catechins, among others [6]. According to Nora et al. [7], antioxidants play an important role in the enhancement of immunity and for preventing cardiovascular diseases and metabolic diseases including obesity, diabetes mellitus, and some types of cancer. According to Harvard T.H. Chan School of Public Health [8], numerous studies showed a link between consuming chocolate and health benefits. Studies showed that consumption of 6 grams of chocolate daily leads to reducing the risk of heart disease and mortality, which could be due to reduced blood pressure and inflammation.

It is considered that chocolate is one of the 30 best anti-inflammatory foods. It has been observed that inflammation could lead to many health problems, and chocolate works as an anti-inflammatory effect in the human system, subsequently protecting humans against diseases. A study carried out by Creveling [9] found that cocoa polyphenols help to regulate the bacterial composition of the intestine. The author observed that cocoa polyphenols increase the number of good gut bacteria and that trigger this anti-inflammatory response [9].

Consuming chocolate could help in maintaining cardiovascular health. According to a study, carried out on 49 adults (32 women, 17 men) to assess the effectiveness of consuming dark chocolate on cholesterol levels, the results indicated that consuming chocolate that contains plant sterols and cocoa flavanols as a low-fat diet could contribute to cardiovascular health by reducing cholesterol and improving blood pressure [10].

Another study proved that the flavanols in dark chocolate can boost the endothelium, the lining of arteries, which lead to producing nitric oxide that works to send signals to the arteries to relax, which lowers the resistance to blood flow, and as a result of this, it reduces blood pressure [6]. Moreover, a study from Walden University's School of Nursing has been done on participants who eat dark chocolate, and it has been found that consumption of dark chocolate reduces blood pressure [11].

Consumption of dark chocolate improves the function of the brain. A study that was carried out on healthy volunteers confirmed that consuming high flavanol cocoa for 5 days contributes to improved blood flow to the brain. Furthermore, it also helps in improving cognitive function in older adults with mild cognitive impairment [6]. Nordqvist [5] referred that researchers from Harvard Medical School indicated that drinking two cups of hot chocolate a day for older people could help in keeping the brain healthy and reduce memory decline. They have found that hot chocolate helped in improving blood flow to parts of the brain where it was needed. A similar study was published in 2014 showed that a cocoa extract that is called lavado could help in reducing or preventing nerve pathways found in patients with Alzheimer's disease and could contribute to slow symptoms such as cognitive decline [5].

Consumption of chocolate also helps to protect human skin against skin disease because of cocoa flavanols [6, 11, 12]. Also, chocolate contains theobromine, which helps pregnant women against complications of pregnancy, which is known as preeclampsia. According to a study that was done on pregnant women, as they were given a higher amount of chocolate, they had a 40% less chance of getting preeclampsia [13]. Studies also showed that consuming chocolate can help in protection against cancer as Brook [14] indicated in a study, which was published in the Journal of the American Society of Hypertension, that people who consume many flavonoids or antioxidant-rich chocolate have lower chances of developing cancer, compared with those who do not.

SIDE EFFECTS ASSOCIATED WITH CHOCOLATE CONSUMPTION

As discussed in the last section, there are numerous health benefits associated with the consumption of chocolate; however, there are also some negative effects on human health such as childhood hyperactivity, migraine, and headaches. According to a study conducted in the Bogalusa Heart, to identify the presence of caffeine in children's snacks, the results indicated the presence of a large amount of caffeine in children, which could be due to eating foods such as chocolate [15]. Bruso [16] referred to that according to the University of Maryland Medical Center, approximately 8% of children in the US younger than 18 have hyperactivity at some point. These cases are often treated by behavioral therapy and medication, but some parents resort to changing children's diets by avoiding chocolate in their children's meals. The Medical Center reported that this change is an effective method but only in a small percentage of children [16].

According to Nowaczewska et al. [17], there are 25 studies have indicated the prevalence of chocolate as a migraine trigger. Two studies of them indicated that chocolate is not a migraine trigger because no participant has reported that chocolate is a trigger factor. Also, other studies considered chocolate a migraine trigger in a small percentage of participants (ranging from 1.3 to 33) Nowaczewska et al. [17]. According to Kelishadi [15], there are some authors who say migraine and tension-type headache occur due to the consumption of some type of foods: one of them is chocolate. Also, another study indicated that chocolate could provoke a migraine attack in some patients who think themselves sensitive to it. [15]. An increase in migraines in some people upon eating chocolate could be due to cocoa's tyramine, histamine, and phenylalanine content [5]. It has been also reported that consuming more chocolate could lead to bone health issues. Nordqvist [5] mentioned in a study, which was published in The American Journal of Clinical Nutrition, low bone density in older women who consume chocolate every day. Also, it has been reported that some chocolate includes a high amount of heavy metals such as cadmium and lead, which can affect the kidneys and bones [5]. A study found that almost all samples (43 chocolate products) contain more than 0.3 mcg of cadmium per serving, which means more than the recommended level identified by the World Health Organization [5].

PRODUCTION OF CHOCOLATE

The cacao tree is considered a sensitive plant, and it requires protection from wind and needs shade in its early years [18]. The cocoa tree needs a semi-shaded environment (50% light and 50% shade). In the cocoa plantations, there are many tall trees beside cocoa trees to provide the required shade and shall not prevent the light [19]. The cocoa plant also needs high rainfall and temperatures to grow. Therefore, it can only grow in a narrow band of countries between 20 degrees north and south of the equator [20]. The suitable temperature for growing is between 25 and 27°C (between 77 and 81°F). The tree cannot resist the high dry or wet. The perfect rainfall should be between 1250 and 2500 mm per year. The cocoa tree requires fertile, slightly acidic, and well-drained soil [21].

The wild cocoa tree can grow up to 12–15 m but to make harvesting easier, most farmers grow it to less than 5 m. Regarding blooms, it blooms and bears fruit all year round, which means that the cocoa tree has both flowers and fruits at the same time [21]. Moreover, the most producing period is from October to February and from May to August [22].

The products could be different in many properties such as shape, texture, and size. The length can be started from 15 cm to over 35 cm. One ripe fruit contains 20–75 cocoa beans, the length of each one can be from 1 cm to 3 cm, embedded in a white pulp. The nutritious cocoa bean is considered very high beneficial as it contains fat (50%) and carbohydrates (25%), as well as proteins, theobromine, niacin, vitamins (including A, B1, B2, and B6), and minerals (calcium, iron, potassium, magnesium, sodium, and phosphorus) [21].

Varieties of cocoa

There are three types of chocolate, which are Criollo, Forastero, and Trinitario. These types are dominant varieties [23]. It was named officially in 1753 by the Swedish scientist Carl von Linné [24]. According to the chocolate society [25], the most productive and common type is Forastero beans, which are because of their high yielding in Brazil and W Africa. On the other hand, Criollo and Trinitario beans are better in quality, but they are lower yielding, which made them very expensive.

Trinitario

This type of cocoa exists on the Island of Trinidad after a hurricane nearly destroyed the local Criollo crops in 1727. As a result, farms are replanted with the Forastero type, which was brought from Venezuela, and cross-fertilized with the native Criollo type [25]. Therefore, a new type appears, which is known now as Trinitario. This type merged the best two types, which are Forastero (which is featured by the hardiness and is very high in yielding) and the Criollo type (which is featured by great taste) [21]. As a result, the Trinitario variety is classified as fine flavor cocoa, but it is less intense than the Criollo type [26]. The properties of this type are usually varied because of the parents as they have different characters, and it can be grown where Criollo cocoa was once grown including in Mexico, the Caribbean islands, Colombia, Venezuela, and parts of Southeast Asia [21]. The percentage of the world's production is uneven regarding the Walter Matter [26] approximately only 5% of the world's production, and according to Bar & Cocoa, proved that the production is less than 10% globally. However, according to the chocolate society [25], this type represents about 12% of the world's cocoa production, and in 2020, the Ministry of Foreign Affairs confirmed that the production represents approximately 10–15%.

Criollo

This type was cultivated by the Maya more than 2500 years ago. It is considered a high-quality variety compared with others. Also, it is very complex with regard to cultivation and handling; consequently, it represents less than 5% globally, and it is the rarest type [27]. It is cultivated in Central America, northern South America, the Caribbean, and Sri Lanka. It has a delicate and sweet flavor, so it is often mixed with different types because of its rarity and high cost [28]. According to its characteristics, the pods are red or green before ripeness with less than ½ in length, and it has a very accentuated tip at the lower end, marked with 10 deep furrows. Its surface is smooth and round-shaped. Moreover, it is plump with fresh white cotyledons, and it is very easy to ferment [27].

Forastero

This is the most common type of cocoa, its native most likely to the Amazon basin. These days, it is grown mostly in Africa, Ecuador, and Brazil. It is represented almost 80% of the world's cocoa production, because of its properties, which include hardness, and resistance to diseases. There are many subspecies of Forastero such as Cundeamor and Calabacillo [21]. This type was discovered by the Spanish in the heart of the Amazon [27]. It grows in West Africa and Brazil [26].

Diseases of cocoa fruit

Black pod

It is mostly cultivated cocoa in small labor-intensive farms for less than 2 hectares (5 acres) to protect the product against disease and pests. Even though, still, there are many losses of production globally, which approximately range from 30 to 100%. The most popular diseases of the cacao tree are pod rots. It is called a black pod as a result of fungus (Phytophthora), which spreads very quickly on the pods under rainy and humid climate conditions, with a lack of sunshine, and temperature under 21°C (70 °F). Therefore, control is required by using copper-containing fungicides, and the infected pods should be removed to reduce spread.

Witches' broom

Witches' broom, which is caused by *Moniliophthora perniciosa*, has a very dangerous effect on products [29]. It is one of the most risky diseases throughout South America. High temperature and high humidity are good environments for this disease. Therefore, it must be following good sanitation and remove any infected materials, which are difficult because it could not show any visible symptoms. Also, fungicides must be used to protect the cocoa [30].

Frosty pod

This disease is caused by *Moniliophthora roreri* fungus, and it is considered a highly infectious disease that impacts the production of cocoa. It is spread rapidly by water movement, wind, or the movement of pods [30].

Swollen shoot

This one is only found in West Africa, which is a huge problem in Togo, Ghana, Cote d'Ivoire, and Nigeria. It is transmitted by mealybugs. This natively results from trees that grew in the rain forests of West Africa, so it is not native to the cocoa plants [30].

Table 1 includes the main estimated losses of cocoa production annually because of diseases [31].

Pest

There are over 1500 different insects that feed on cocoa; however, there is only 2% of economic importance. Mirid bugs such as Helopeltis are the most common and important insects in attacking cocoa, and the main pest in Malaysia and Indonesia is the cocoa pod borer. The most popular insect pests are including broad mite, flower-eating caterpillars, Helopeltis, and yellow peach moth [31].

Table 1. *Estimated losses of cocoa production annually because of diseases.*

Disease	Region	Estimated world production loss (tonnes)
Black pod (phytophthora fungus)	Africa, Brazil, Asia	450,000
Witches broom (fungus)	Latin America	250,000
Frosty pod rot (fungus)	Latin America	30,000
Swollen shoot virus	Africa	50,000
Vascular streak dieback (virus)	Africa	30,000

Source: https://www.daf.qld.gov.au/business-priorities/agriculture/plants/fruit-vegetable/fruit-vegetable-crops/ cocoa/pests-and-diseases-of-cocoa.

PROCESSING OF COCOA

Harvesting cocoa

The coco plant life on average is 25 years, and the product needs range from 150 and 165 days for complete maturation. There is no specific season for ripening as it can be harvested all year round [19]. Beans grow in pods that sprout out of the trunk and branches of cocoa trees. The size of the pods is almost the same as the football size. The pods start green and then convert to orange, which means they are ripe and ready for harvesting [32].

When the pods are ripe, harvesting should be done manually, as using machines will lead to damage to the tree, the clusters of flowers and pods that grow from the trunk [32], as the tree is weak, which makes the picking a hard job for workers [18]. For picking, workers use short, hooked blades mounted on long poles to reach the highest products [32]. After picking cocoa from trees, it is collected in baskets and transported to the braking operation [32]. In this step, it is split, and the cocoa beans are removed. Pods can contain upward of 50 cocoa beans each [18]. The pods are opened, and the beans are removed within a week to 10 days after harvesting to avoid germinating [33].

Fermentation of cocoa

Fermentation of cocoa after harvesting is a critical process [34]. This process leads to getting the aromas to the cocoa and knowing the difference between many types of cocoa. However, for some traditional dishes, unfermented beans are used in parts of Mexico and Central America. The fermentation method could take 2–8 days; it depends on the type of cocoa [18]. It is fermenting the pulp of Forastero type for 5–7 days, and the pulp of Criollo type for 1–3 days [29].

During fermentation, the cocoa beans are placed in large, shallow, heated trays or covered with large banana leaves. If the weather is good, it is heated by the sun with the importance of noticing that it is necessary to move them up frequently to allow the beans to come out for fermentation equally [32]. In this method, the juicy sweating of the pulp is drained and the germ in the seed is killed by the heat, and the flavor will be improved. As well as the color will be changed to brown. This step is implemented in sun-dried or kiln-dried to reduce the moisture content to 6–7 % [29]. Throughout the fermentation of cocoa beans, the top layer of the beans is covered with banana leaves, because the bottom part of the banana leaf consists of natural yeast and microorganisms, which help to strengthen the natural fermentation [18].

Drying

After fermentation, a drying process must be implemented, which is an important step to enhance the cacao flavor [35]. The drying is done by the sun, and it should turn the beans as well in this step to be equally dried. It can dry on wooden floors, which can be covered by a sliding roof during rainy weather. Electric dryers are used on large farms [18]. The drying stage period could take between 2 and 10 days in the nature processing [36]. During drying, the color of cocoa beans is changed from reddish-brown to dark brown [18]. As a result of this process, the humidity of the cocoa is reduced from 60% to 7% [35].

In this method, the polyphenol content of cocoa beans is reduced, many numbers have been reported in a review study, which was done by [37], and depending on many studies include a reduction from 77 to 44%, a 72% reduction, a 30% reduction, and a 26% reduction. In addition, sun drying reduces the phenolic content and antioxidant activity of cocoa beans.

However, according to a study that was done by [36] to compare the methods of drying cocoa beans to evaluate antioxidant activity, to determine phenolic compounds and methylxanthine content, and to determine the presence of ochratoxin A, four methods are used:

- Dryer with stainless steel platform and plastic roof with UV protection (DP): Artificial dryer, using the wooden platform with an artificial heat source, through forced and heated air circulation with electric resistance and temperature controlled by a thermostat at 60°C (AD);
- Traditional dryer in the barge with wooden platform and drying by direct sunlight (TD);
- Mixed dryer with stainless steel platform and mobile plastic roof with UV protection for drying coverage and exposure to sunlight (MD).

It has been found that the best method is the traditional drying method, in which drying is by the sun directly to maintain the phenolic compound content in the cocoa beans, as the phenolic compound content is reduced in different methods, particularly in the artificially oven-dried method. Also, the traditional method is the best in retention of cocoa antioxidant activity and the methylxanthine content in dry seeds, which confirm that these chemical compounds are impacted by changing temperature. The reason for the traditional method being the best could be because it uses less than milder temperatures.

Manufacturing process of chocolate

Cleaning

The first process that must be done in the industry is cleaning the cocoa beans probably by using sieves and brushes [19]. There are many cleaning steps to remove any contamination, for example, twigs, stones, and dust [29]. In general, the manufacturing operations are different slightly because of the various types of cocoa trees; however, most industries use similar machines to convert cocoa beans to cocoa butter and chocolate [32].

Roasting

The roaster step is very important in processing chocolate, and that is done by drying beans at a temperature of 100°C, then roasting beans at a temperature range between 100°C and 160°C. It could be different depending on the type. The balance in roasting is very necessary as increasing it will lead to the bitterness of the beans and less than required will not improve the aromas [19]. A study [38] has proved that roasting at high temperature influences cocoa beans and has no influence on roasting time. Moreover, roasting at 160°C leads to undesirable burnt odor and flavor and a low acceptability score by consumers. However, roasting at temperature range from 90 to 110°C was acceptable by consumers in appearance, aroma, flavor, texture, and overall quality attributes.

This process develops the aromas and contributes to reducing acidity and astringency, reducing moisture content, deepening color, and facilitating shell removal [29]. Also, roasting is killing any organisms that exist on the bean, which appear in the fermentation process [18]. In this process, it has been found to decrease the flavanols and phenolic content of cocoa beans [37].

Grinding

The roasting method makes the shells of the cocoa brittle, so it is winnowed to remove the shells of the bean and only cocoa nibs pass through a series of sieves, and this method is called winnowing [29]. In the grinding process, the nibs are ground in a granite stone mill, which crushes the grain and releases the fat or cocoa butter [29]. Cocoa butter is the main ingredient of chocolate [19]. Chocolate consists of both cocoa solids and cocoa butter in almost the same ratio [18].

Coaching

This process is completed by conche machines. Moreover, this process is developing flavor, aerating, and emulsifying. The required time is 4–72 hours (depending on the desired results, and the machine type plays a major as well). The temperature is between 55 and 88°C (130 and 190 °F) and is controlled regarding the desired flavor and uniformity [29]. In this process, chocolate could lose almost 80% of volatile substances [19].

Types of chocolate

There are many types of chocolate that are different in ingredients and characteristics.

1. Dark chocolate, which contains the highest percentage of cocoa bean solids (up to 80% of the total weight) and cocoa butter compared with other varieties. This type is featured by the strong aroma of cocoa. Its quality depends on the percentage of cocoa. Most of the human benefits of health have been linked to consuming this type.
2. Another type is called Gianduja chocolate, which contains hazelnuts, cocoa, and sugar. Its color is brown.
3. Milk chocolate consists of cocoa butter, sugar, milk powder, lecithin, and cocoa of 20–25%. Its properties are including a bright appearance, an intense, persistent aroma, and a sweet taste with a slightly bitter accent of cocoa [2]. According to FDA Standards of chocolate identity, milk chocolate must include at least 10% chocolate liquor and 12% milk solids. Also, cocoa butter and milk are the only fat allowed in this type [39].
4. The last type is white chocolate, which contains cocoa butter, milk, and sugar with no cocoa solids. It is a sweet, pleasant taste [2]. To meet the FDA standards, this type must have at least 20% cocoa butter and 14% milk solids [39].

FOOD SAFETY OF CHOCOLATE

Food contamination is referring to food that is contaminated microbiologically, chemically, or physically. That can be done at any stage of food processing such as during storage or transportation [40]. Also, food allergies can be considered a contamination factor [41].

Food safety issues are considered a huge challenge in the health sector compared to those of malaria or tuberculosis [42]. As the person who consumes contaminated food gets food poisoning and starts the poisoning symptoms in hours, and it could need to visit a hospital, particularly the high-risk group, which includes older adults, pregnant women, infants, and young children, and people with chronic disease [43]. Food poisoning includes many common symptoms, which are upset stomach, stomach cramps, nausea, vomiting, diarrhea, and fever [44]. There are many foods that are linked to food poisoning, which are poultry, raw fruits and vegetables, fish and shellfish, rice, deli meats, unpasteurized dairy, and eggs [45]. Many cases of outbreaks of food poisoning in the United States have been linked to fruits and vegetables in 2018 [46].

Microbial contamination of food

Microbial contamination is the most common type of food contamination. It can be done if the food is contaminated by microorganisms such as bacteria, viruses, mold, fungi, and toxins. Many reasons lead to microbial contamination that including undercooking chicken, storing and preparing raw foods near to ready-to-eat food, and this leads to cross-contamination [41]. Also, the vectors of biological contamination include food handlers and that can be caused by aerosol droplets from coughing near the production line. Moreover, vectors include packaging materials, equipment, and tools used; an example of this is using the same cutting board and knife in raw food and ready-to-eat food. Contaminated water is also an important factor to cause microbial contamination. Furthermore, pests such as insects and rodents are factors of contamination [40]. Therefore, food hygiene practices must be implemented to prevent contamination that including personal hygiene, separating raw and ready-to-eat food at all stages of processing, washing raw fruits and vegetables, and pest control on the premise [41].

For a long period, it was referred to low-moisture foods, such as chocolate, as a safe product against microbial contamination because of their properties, as the water activity (aw) is below 0.6, which is not an active area for microbial growth. However, the early 1970s were linked to the first outbreaks of Salmonella to low-moisture products such as chocolate, oat cereals, peanut butter, and infant formula. Salmonella was the main pathogen of concern for all of those products [47]. Also, another study that was to identify the emerging hazards related to chocolate products between 2013 and 2018 has concluded that microbiological hazard is 16,49%, and it is the second hazard following chemicals in cocoa from Africa and South America as well [48]. According to the most recent data, it has been shown that a Salmonella outbreak in a brand of chocolate wafers from Poland between December 2020 and early April 2021affected 32 people as announced [49].

Moreover, many reasons cause biological contamination linked to chocolate products, for example, pathogenic microorganisms can be found because of using contaminated ingredients such as milk and sugar. Also, pathogens can be presented because of damage or soiled packaging material because of mishandling at the supplier level. Furthermore, it can be found in returned chocolate products because of mishandling in any area such as retail or during transportation [50].

Chemical contamination of food

Chemical contamination is any food contaminated by a chemical agent. For example, food is contaminated by cleaning agents and that can happen when using chemical cleaning during processing or when using chemical cleaning on tools and not cleaning well. Furthermore, there is natural chemical contamination such as toxins in some fish, such as methylmercury [51]. Also, contamination can happen in agriculture production when they spray fertilizers and pesticides close to food when it is growing. Therefore, it must be ensured that chemical agents are away from the food area. Always follow the manufacturers' instructions before using chemicals. Food must be covered when cleaning and dealt only with approved chemical suppliers [41].

There are many potential chemical contaminations related to chocolate that can happen at any stage of the food chain, which includes the existence of environmental contaminants such as pesticides above the allowance level, as well as presence of food additives in dark chocolate or heavy metals or addition of food additives, which are not permitted for use by government regulations [50].

This type of contamination, either industrial contaminants or pesticides, is most commonly related to chocolate products in Africa, South America, and Asia [48]. Another study proved that some cocoa products exceed the European Union and Chinese Maximum Contaminant Level regarding arsenic, cadmium, lead, and mercury, which could affect human health [52].

Physical contamination of food

Physical contamination of food is any food that contains a foreign object at any stage of the production process. That foreign substance causes health issues for consumers, such as broken teeth or choking [41]. There are many food products recalled every year because of physical contamination, which causes consumers health issues and costs less. There are many real examples from the Food and Drug Administration (FDA) 2016 product recall list such as "small metal shavings in apple coffee cakes, metal fragments in sugar used in Asian sauce, plastic mesh screen fragments in flour, and pieces of rubber in baby food" [53].

An example of physical contamination in almond processing is that, when the crop is ready for harvesting, the tree is shacked by a tractor machine, which leads to falling of almonds on the ground, which are left for many days to dry. Then, another machine sweeps them into rows so a harvester can pick them up with a series of belts. The almonds are cleaned by the machine and dumped into a bucket in the back of the truck. Then the truck is transferred to the industry to remove the shells and sticks and clean the product. The industry machines can be damaged and worn out, which could lead sometimes to the transfer of small pieces of that machinery into a package. This could include bolts and washers. Therefore, using technology in processing plants is an important role such as using X-rays and sensors and food metal detectors to identify any foreign substance to protect humans' health [53].

Allergenic contamination of food

Food allergies are an individual adverse reaction to some food. Food allergens are proteins that can be found in food in huge amounts [54]. Even a small amount of food is enough to impact a person who has a food allergy [41]. According to the Food Safety Authority of Ireland [55], there are 14 named allergens (**Table 2**).

Allergic food contamination can happen when food from the list of food allergies enters or contacts another food. An example of this is using the same container to store pasta that is already selected to store peanuts. Therefore, it must be ensured to deal only with approved suppliers who take critical control of allergenic contamination. Always separate tools, equipment, and clothes that are used for allergenic foods from other foods. Separate allergen in storage from other foods. Cleaning must be implemented properly in the area after the use of one of the 14 allergen products.

Table 2. *Food allergies products.*

Food allergen products
1. Cereals containing gluten such as wheat
2. Crustaceans and products thereof
3. Eggs and products thereof
4. Fish and products thereof

Food allergen products	
5.	Peanuts and products thereof
6.	Soybeans and products thereof
7.	Milk and products thereof
8.	Nuts, which are almonds, hazelnuts, walnuts, cashews, pecan nuts, Brazil nuts, pistachio nuts, macadamia or Queensland nuts, and products thereof
9.	Celery and products thereof
10.	Mustard and products thereof
11.	Sesame seeds and products thereof
12.	Sulfur dioxide and sulfites at concentrations of more than 10 mg/kg
13.	Lupin and products thereof
14.	Molluscs and products thereof

Many food safety concerns can be avoided when implementing an effective prerequisite program that includes cleaning and sanitation, staff training, maintenance, chemical control, waste management, and storage and transportation [56]. Hygiene must be implemented in the chocolate industry to provide safe products to protect consumers [57].

It should provide maintenance of food equipment and infrastructure; this can be done by regularly auditing all equipment used in the chocolate factory and fixing any problem, which could post hygiene issues and food safety. Also, hygiene must be maintained during maintenance and the equipment cleaned after that and sanitized before using it again, and the product carefully protected from any foreign body. For the infrastructure part, it should follow a regular check to protect the production area from any issues such as soil and dust to keep the environment safe and hygiene. Any repair must be done outside of production periods, but if that is not possible, a separation wall must be set to protect the production area [57].

Also, it must follow cleaning and disinfecting daily in the industry and removing any undesirable substances such as residues and foreign bodies, which could affect the hygiene of products or equipment. Also, it must separate chemical agents apart from food products to prevent cross-contamination, and all storage should be cleaned regularly [57].

In general, food industries, particularly the chocolate industry attracts insects and rodents through the odors as this place is seen as a great environment for insects because of the presence of food and water and shelter. Therefore, pest control must be implemented effectively, and auditing inspection should be followed regularly and correct any failure related to that. Moreover, it must protect the chocolate production from any contamination, so it should set rules for staff working and visitors such as regularly washing their hands and wearing gloves, mouth masks, hairnets, and clean disinfected clothing [57]. Regarding peanuts and other nuts, they should be segregated as much as possible. It should clean any machine correctly during changing from nut products to other products and must refer to the label that this industry is processing nuts. A further step that should be implemented is using a metal detector to ensure that the product is free from any foreign body such as plastic or other [58].

Furthermore, applying a Hazard Analytical Critical Control Point is required to prevent any type of contamination as this system is to reduce or prevent any risks early before it happens [56]. A Critical Control Point (CCP) is referred to as a step that can be applied, and it is a necessity to prevent or reduce food safety hazards. Most steps include a heating, cooking, or cooling stage. Examples of critical control points in chocolate production include roasting cocoa at temperatures between 105°C and 120°C and at specific times to eliminate pathogens. The further step is metal detection.

Also, there are many control points in chocolate production, for example:

- Receiving, which is required visual inspection, and use of chemical-free materials
- Splitting; use sanitized equipment and personal hygiene.
- The fermentation that is required visual inspection, and regular maintenance of fermentation tanks.
- Drying, which needs visual inspection, observation of time, and temperature.
- Roasting is very important to follow the correct time and temperature to kill pathogens.
- Grinding required also visual inspection and sanitization of equipment.
- Coaching that includes visual inspection and proper maintenance of equipment.
- Tempering—visual inspection, sanitized equipment
- Molding—ensure molds are clean.
- Packing—metal detection, food-grade inks, appropriate packaging materials, correct labeling [59].

CONCLUSIONS

In conclusion, cocoa is an important product globally as it provides many health benefits for people such as reducing the risk of heart disease, reducing blood pressure, and working as an anti-inflammatory substance. However, consuming too much could lead to many negative impacts on human health such as diabetes or overactivation for children because of the presence of caffeine and weak bone for elderly people, subsequently should avoid eating a lot of chocolate. Also, more research is needed to establish more health benefits. Chocolate is considered a lowrisk product, but there are many contaminants linked to it, which could cause many health issues; therefore, good hygiene practices and the HACCP system should be followed to provide safe products.

ABSTRACT

Chocolate is a popular food product internationally, and it is consumed daily.

Consuming chocolate has been linked to many human health benefits such as lower cholesterol levels, but there are some negative impacts such as weight gain because of its sugar content. Moreover, food safety issues related to chocolate have existed, and it can be contaminated by any biological, chemical, or physical hazards, which lead to many health issues. Regarding that, this chapter will discuss the benefits and negative impacts of consuming chocolate and provide the process of manufacturing the product.

REFERENCES

1. Shvili J. The Top Cocoa-Producing Countries. 2020. Available from: https:// www.worldatlas.com/articles/top-10cocoa-producing-countries.html [Accessed: September 20, 2021]

2. Montagna et al. Chocolate, "Food of the Gods": History, Science, and Human Health. 2019. Available from: https:// www.mdpi.com/1660-4601/16/24/4960 [Accessed: September 20, 2021]

3. World Population Review. Cocoa Producing Countries. 2021. Available from: https://worldpopulationreview. com/country-rankings/cocoa-producingcountries [Accessed: September 20, 2021]

4. Cleveland Clinic. Heart Healthy Benefits of Chocolate. 2021. Available from: https://my.clevelandclinic.org/ health/articles/16774-heart-healthybenefits-of-chocolate [Accessed: August 5, 2021]

5. Nordqvist J. Health benefits and risks of chocolate. 2018. Available from: https:// www.medicalnewstoday.com/ articles/270272 [Accessed: August 4, 2021]

6. Gunnars K. 7 Proven Health Benefits of Dark Chocolate. 2021. Available from: https://www.healthline.com/ nutrition/7-health-benefits-darkchocolate [Accessed: August 4, 2021]

7. Nora et al. Fruits and Vegetables in the Management of Underlying Conditions for COVID-19 High-Risk Groups. 2021. Available from: https:// www.mdpi.com/2304-8158/10/2/389 [Accessed: August 20, 2021]

8. Harvard T.H. Chan School of Public Health Dark Chocolate. 2021. Available from: https://www.hsph.harvard.edu/ nutritionsource/food-features/darkchocolate/ [Accessed: August 7, 2021]

9. Creveling M. 14 Health Benefits of Eating Chocolate. 2021. Available from: https://www.eatthis.com/chocolatehealth-benefits/ [Accessed: August 7, 2021]

10. Allen et al. Daily Consumption of a Dark Chocolate Containing Flavanols and Added Sterol Esters Affects Cardiovascular Risk Factors in a Normotensive Population with Elevated Cholesterol. 2008. Available from: https://academic.oup.com/jn/ article/138/4/725/4665024 [Accessed: August 5, 2021]

11. Mertes A. Is Chocolate Good for You? The Pros & Cons of Eating Chocolate. 2021. Available from: https:// www.qualitylogoproducts.com/blog/ health-benefits-of-chocolate/ [Accessed: August 4, 2021]

12. Realbuzz Team. 7 Health Benefits Of Chocolate. 2021. Available from: https:// www.realbuzz.com/articlesinterests/nutrition/article/7-healthbenefits-of-chocolate/ [Accessed: August 15, 2021]

13. Locke R. 20 Health Benefits Of Chocolate. 2021. Available from: https:// www.lifehack.org/articles/lifestyle/20health-benefits-chocolate.html [Accessed: August 4, 2021]

14. Brooks A. 8 Healthy Reasons to Eat Dark Chocolate. 2019. Available from: https:// www.everydayhealth. com/diet-and-nutrition-pictures/ delicious-reasons-to-eat-darkchocolate.aspx [Accessed: August 8, 2021]

15. Kelishadi R. Cacao to cocoa to chocolate: Healthy food?. 2005. Available from: https://sacredchocolate. com/docs/sacredpdf/cacao-cocoachocolate-arya-journal.pdf [Accessed: August 8, 2021]

16. Bruso J. Chocolate's Influence on Hyperactivity in Children. 2011. Available from: https://healthfully. com/159485-chocolates-influence-onthe-hyperactivity-of-a-child.html [Accessed: August 10, 2021]

17. Nowaczewska et al. To Eat or Not to eat: A Review of the Relationship between Chocolate and Migraines. 2020. Available from: https://www.ncbi.nlm. nih.gov/pmc/articles/PMC7146545/ [Accessed: August 10, 2021]

18. Science of Cooking. How is Chocolate Made?. 2021. Available from: https:// www.scienceofcooking.com/ chocolate/how-is-chocolate-made.htm [Accessed: July 7, 2021]

19. Cioccolat Italiani. Production of chocolate, from the bean to packaging. 2021. Available from: https://www. cioccolatitaliani.it/en/blog/productionof-chocolate-from-the-bean-topackaging.html [Accessed: July 7, 2021]
20. Cocoa Life. COCOA GROWING. 2021. Available from: https://www. cocoalife.org/in-the-cocoa-origins [Accessed: July 7, 2021]
21. Barry Callebaut. Theobroma cacao, the food of the gods. 2021. Available from: https://www.barry-callebaut.com/ en/group/media/press-kit/theobromacacao-food-gods [Accessed: July 5, 2021]
22. Singh R. Cacao. 2021. Available from: https://www.britannica.com/topic/cocoa-food [Accessed: July 10, 2021]
23. Espinet R. The story of Trinitarios. 2019. Available from: https://newsday.co.tt/2019/06/26/the-story-oftrinitarios/ [Accessed: July 10, 2021]
24. Willie's Cacao Ltd. The different cocoa varieties. 2021. Available from: https://www.williescacao.com/worldcacao/different-cacao-varieties/ [Accessed: July 13, 2021]
25. The Chocolate Society. The different varieties of cocoa beans: Criollo, Trinitario, & Forastero. 2021. Available from: https://www.chocolate.co.uk/ blogs/news/the-different-varieties-ofcocoa-beans-criollo-trinitario-andforastero [Accessed: July 10, 2021]
26. Walter Matter (2021) Varieties of Cocoa. Available from: https://www. wama.ch/en/varieties-of-cocoa.html [Accessed: August 15, 2021]
27. Chococlic. Criollo Cocoa. 2021. Available from: https://en.chococlic.com/?s=Criollo [Accessed: July 13, 2021]
28. Ministry of Foreign Affairs. The European market potential for speciality cocoa. 2020. Available from: https:// www.cbi.eu/market-information/ cocoa-cocoa-products/speciality-cocoa/ market-potential [Accessed: July 13, 2021]
29. Cook L. Cacao. 2021. Available from: https://www.britannica.com/ plant/cacao [Accessed: July 10, 2021]
30. Plant Village. Cocoa. 2021. Available from: https://plantvillage.psu.edu/ topics/cocoa-cacao/infos [Accessed: July 14, 2021]
31. Queensland Government. Pests and diseases of cocoa. 2013. Available from: https://www.daf.qld.gov.au/businesspriorities/agriculture/plants/fruitvegetable/fruit-vegetable-crops/cocoa/ pests-and-diseases-of-cocoa [Accessed: July 14, 2021]

32. Tyler et al. The World Atlas of Chocolate. 2003. Available from: https:// www.sfu.ca/geog351fall03/groupswebpages/gp8/intro/intro.html [Accessed: July 3, 2021]

33. Cocoa and Heart. How are cocoa beans harvested. 2021. Available from: https:// www.cocoaandheart.co.uk/ how-are-cocoa-beans-harvested/ [Accessed: July 18, 2021]

34. Vuyst L, Weck S. The cocoa bean fermentation process: from ecosystem analysis to starter culture development. 2016. Available from: https:// sfamjournals.onlinelibrary.wiley.com/ doi/pdf/10.1111/jam.13045 [Accessed: July 18, 2021]

35. Guevara J. Cacao Harvesting & Processing: A Step By Step Explanation. 2018. Available from: https:// perfectdailygrind.com/2018/02/stepstep-explanation-cacao-harvestingprocessing/ [Accessed: July 18, 2021]

36. Deus et al. Influence of drying methods on cocoa (Theobroma cacao L.): Antioxidant activity and presence of ochratoxin A. Food Science and Technology. 2018;**38**:278-285

37. Di Mattia et al. From cocoa to chocolate: The impact of processing on in vitro antioxidant activity and the effects of chocolate on antioxidant markers in vivo. Frontiers in Immunology. 2017;**8**:1207

38. Rocha et al. Effect of the roasting temperature and time of cocoa beans on the sensory characteristics and acceptability of chocolate. 2017. Available from: https://www.scielo.br/j/ cta/a/6NvBzynyYXyHKXpxsfxxqbG/?l ang=en [Accessed: July 19, 2021]

39. Vishnefske J. FDA Standard Definition of Chocolate. 2017. Available from: https://www. santabarbarachocolate.com/blog/fda-standard-definition-of-chocolate/ [Accessed: July 22, 2021]

40. Abdolshahi A, Yancheshmeh B. Food Contamination. 2020. Available from: https://www.intechopen.com/ books/mycotoxins-and-food-safety/ food-contamination [Accessed: May 30, 2021]

41. Collier E. What are the Four Types of Food Contamination?. 2019. Available from: https://www.highspeedtraining. co.uk/hub/four-types-contamination/ [Accessed: May 29, 2021]

42. Hoffmann V. Addressing food safety questions along the value chain. 2019. Available from: https://www.ifpri.org/ blog/addressing-food-safety-questionsalong-value-chain [Accessed: May 30, 2021]

43. Mayo Clinic Staff. Food poisoning. 2020. Available from: https://www. mayoclinic.org/diseases-conditions/food-poisoning/symptoms-causes/syc-20356230 [Accessed: May 29, 2021]

44. CDC. Food Poisoning Symptoms. 2021. Available from: https://www.cdc. gov/ foodsafety/symptoms.html [Accessed: May 29, 2021]

45. Coyle D. Top 9 Foods Most Likely to Cause Food Poisoning. 2018. Available from: https://www.healthline.com/ nutrition/foods-that-cause-foodpoisoning [Accessed: July 1, 2021]

46. Moreb et al. Fruits and Vegetables in the Management of Underlying Conditions for COVID-19 High-Risk Groups. 2021. Available from: https:// www.mdpi. com/2304-8158/10/2/389 [Accessed: June 25, 2021]

47. Winkler. Microbiological safety of chocolate confectionery products. 2009. Available from: https://www. newfoodmagazine.com/article/189/ microbiological-safety-of-chocolateconfectionery-products/ [Accessed: July 6, 2021]

48. Foodakai. Case study: Chocolate Product. 2019. Available from: https:// agroknow.com/wp-content/uploads/2020/07/Foodakai_Case_Study_ Chocolate_Industry_Report.pdf [Accessed: July 8, 2021]

49. Whitworth J. Salmonella outbreak in Sweden linked to chocolate wafers. 2021. Available from: https://www. foodsafetynews.com/2021/04/ salmonella-outbreak-in-sweden-linkedto-chocolate-wafers/ [Accessed: July 28, 2021]

50. Food Ready. Food Safety / HACCP Plan—Chocolate. 2021. Available from: https://foodready.ai/food-safety-haccpguidance/plan-guidance-chocolate/ [Accessed: July 25, 2021]

51. Australian Institute of Food Safety. Food Safety and the Different Types of Food Contamination. 2021. Available from: https://www.foodsafety.com.au/ blog/ food-safety-and-the-differenttypes-of-food-contamination [Accessed: July 20, 2021]

52. Anyimah-Ackah et al. Exposures and risks of arsenic, cadmium, lead, and mercury in cocoa beans and cocoa-based foods: a systematic review. 2019. Available from: https://academic.oup. com/fqs/article/3/1/1/5365839?searchre sult=1#133623096 [Accessed: July 25, 2021]

53. Schultz M. What are physical contaminants in food manufacturing?. 2018. Available from: https://www. thermofisher.com/blog/food/what-arephysical-contaminants-in-foodmanufacturing/ [Accessed: May 29, 2021]

54. Popping B, Bourdichon F. Allergen management in the food industry: it's all about cross-contact and contamination prevention. 2017. Available from: https://www.newfoodmagazine.com/ article/73985/allergen-management/ [Accessed: May 29, 2021]

55. Food Safety Authority of Ireland. List of 14 Allergens. 2014. Available from: https://www.fsai.ie/legislation/ food_legislation/food_information/14_allergens.html [Accessed: May 31, 2021]

56. Albandary and Moran. A Study of the Effectiveness of a Software HACCP System FoodCheck for the Management of Food Safety. 2019. Available from: https://www.actascientific.com/ASNH/ ASNH-04-0593.php [Accessed: July 20, 2021]

57. Crystal Chocolatier. 4 Steps to Maintain Good Hygiene in a Chocolate Factory. 2013. Available from: https:// www.crystalchocolatier.com/4-stepsmaintain-good-hygiene-chocolatefactory [Accessed: July 27, 2021]

58. Johnston I. Food Safety Protocols for Chocolate Companies. 2021. Available from: https://smallbusiness.chron.com/ food-safety-protocols-chocolatecompanies-74340.html [Accessed: July 25, 2021]

59. Aliw. Chocolate production hazards – HACCP Easter everyone!. 2018. Available from: https://www. vwa.co.uk/blog/news/5970/ [Accessed: July 25, 2021]

Section 3

PACKAGING

Chapter 5

PECTIN AND ITS APPLICABILITY IN FOOD PACKAGING

INTRODUCTION

Packaging plays a fundamental role in all our daily lives purposes, especially in food industries for food products (fruit and vegetables) during post-harvest and processing due to their different uses, such as; 1) Separate the food from the surrounding environment, 2) Preserving food quality, and safety, 3) Protecting from the spoilage factors, 4) Maintaining the nutritional value of products, 5) Extending shelf life, and 6) Provides information about the products for the consumers [1–3]. The packaging process was defined as the art, science, and technology to deliver the different goods to the consumers at economical prices, securely, and high quality [4].

Fossil fuels (include coal, petroleum, natural gas, oil shales, bitumens, tar sands, and heavy oils) are considered the main source for commercial polymers used in food packaging and producing plastic packaging [2, 5]. This is due to several reasons, such as ease of processing, low cost, lightweight, high energy effectiveness, and flexibility. The use of plastic packaging increases worldwide, owing to increasing population growth, which leads to increasing demand for processed food [6–8]. The plastic packaging production reached 380 million metric tons worldwide, approximately 60% of all plastic packaging used for foods and beverages, and the remainder covers nonfood applications [9].

Despite the different advantages of plastic polymers, it considered highly hazardous due to containing a large variety of chemical additives which using during manufacturing; thus, it has a severe threat to consumer's health and environmental pollution [10–15] such as 1) Produce a large of wastes which can be toxic to biological life [16–18]. 2) Migration might result in accumulation of undesired chemical compounds from packaging materials to foods, such as Bisphenol A (BPA) [12, 19]. 3) Lead to toxic and harmful effects on wild and human life because it is nonbiodegradable and not reusable [6, 20, 21].

In 2015, global plastic waste volume reached around 6.3 billion metric tons and is expected to increase to 12 billion metric tons by 2050 [22].

Therefore the food industry has classified the conventional plastic packaging material as a source of pollution and social concerns due to its nonbiodegradability and poor waste management, due to the accumulation of huge amounts of plastic waste in the environment, and also rapid depletion of fossil reserves and increases in the cost of petroleum, the food packaging industry toward the development and application of eco-friendly materials and biodegradable materials [23].

Various technologies have been used in food packaging to preserve the fruit and vegetable's quality and safety and prolong their shelf life, such as ultraviolet irradiation, ozonation, changed atmosphere, and biodegradable films [24]. Edible packaging is considered a promising alternative and received much attention to tackle the plastics packaging problems [25–27].

Production of environmental-friendly packaging such as biodegradable and renewable films represents an interesting alternative to conventional polymers. Polysaccharides and proteins address those requirements because of their desirable film-forming properties, and, as most of them are also edible, they can be used as edible films and coatings. Edible films and coatings can extend the shelf life of food products by improving the mechanical properties and reducing the transfer of moisture, lipids, flavors, or gases between the food and the surrounding environment [17]. Edible coatings modify the atmosphere around fruit and/or vegetables, altering oxygen levels inside the fruit, retarding the production of ethylene and thus, limiting the physiological decay of fruits; This also reduces the ripening-induced quality degradation in terms of texture or loss of bioactive compounds during storage [28].

Edible coating materials efficiency and performance depend on i) Properties of coating materials (type, concentration, viscosity). ii) Methods of coating (dipping, spraying, and dripping) [29–31]. A coating with excellent barrier and poor mechanical properties, bad flavor, or high cost will not be of commercial interest for using, and selecting a proper coating for the fruits also is not easy. So, it is necessary to study the film's physicochemical and structural characteristics [32, 33].

Pectin as edible films/coating used to increase food quality and extend the shelf life of food products which is considered an alternative to packaging materials from synthetic polymers, thus preventing environmental pollution from nonbiodegradable plastic materials [34].

Younis and Zhao [35] reported that pectin polymers are considered promising polysaccharide biomaterials in developing eco-friendly films due to their filmforming, biodegradable, and nontoxic characteristics.

PECTIN

Pectin (PEC) is a natural polymer and complex anionic heteropolysaccharides [36–39], consisting mainly of α-(1-4) linked D-galacturonic acid residues and neutral sugars, which are partially esterified with methyl alcohol or acetic acid at the carboxylic acid [37, 40].

Pectin is a family of plant polysaccharides accounting for up to 35% of primary cell walls in certain species. It has been considered the most structurally complex polysaccharide in nature. The term "pectin", in fact, describes a group of oligosaccharides and polysaccharides that share common features but are highly diverse in their fine structure, except that they all comprise at least 65% galacturonic acid (GalA), which is linked at the O-1 and O-4 position [41].

Zhou et al. [36] noted that the pectin polymer consists of three polysaccharide domains: Homogalacturonan (HGA), Rhamnogalacturonan-I (RG-I), and Rhamnogalacturonan-II (RG-II), where Homogalacturonan (HG) considered the most abundant cell wall pectic polysaccharide, about 50–90% of total pectin, meanwhile rhamnogalacturonan I (RG-I) is the second most abundant type comprising between 20% and 35% of total pectin [42, 43].

Pectin is a white, amorphous, and colloidal carbohydrate of high molecular weight occurring in ripe fruits, especially in apples, currants, etc., and used in fruit jellies, pharmaceuticals, and cosmetics for its thickening and emulsifying properties and ability to solidify to a gel. These properties and applications have put pectin in the biopolymers market with great potential for future developments [44].

Pectin sources

Several studies have documented that pectin is one of the most abundant polysaccharides in the primary cell wall and middle lamella of all plant tissues [39, 45]; the pectin production comes from two ways i) Commercial pectin comes from citrus fruit peels [46, 47], apple pomace [47, 48] pomegranate peels [49, 50], mango peel [3, 51], lemon peel [52], sugar beet pulp and potato pulp [3, 51]; ii) Noncommercial pectin obtained from cocoa husks [53], mulberry branch bark [54], peach pomace

[42], sisal waste [42], pumpkin [51], banana peel [55, 56] watermelon rind, and soy hull [38].

Pectin extraction

There are different techniques for pectin extraction which is considered highly efficient and eco-friendly such as conventional solvent extraction based on stirring and heating; microwave-assisted extraction (MAE); ultrasound-assisted extraction [57, 58]; subcritical water extraction, and enzyme-assisted extraction [59], each of them has advantages and disadvantages.

Wicker et al. [60] noted that the pectin polymer could easily be extracted using different acids by conventional heating extraction. It consumes more time, the hot solution leads to pectin degradation, produces a large volume of effluent, and causes environmental pollution concerns. The MAE method is considered an excellent alternative to conventional extraction methods and offers significant advantages 1) Requires shorter processing times and less solvent, 2) Higher extraction rates, 3) Low in cost [52, 59].

The microwave-assisted extraction method (MAE) has been reported as the preferred extraction method of pectins from natural sources such as dragon fruit peels, bagasse, and pomace obtained from Mexican lime fruit, pomelo, mango, and papaya peels under different operation conditions. MAE conditions are dependent on different factors, such as microwave power, pH, time, and S:L ratio. MAE methods show significant advantages over conventional extraction techniques, such as reducing the amount of the extraction solvents, low energy consumption, high recoveries, good reproducibility, short extraction times (minutes rather than hours), and minimal sample manipulation [44].

The percentage of galacturonic acid group of pectin is esterified using methyl, or acetyl groups are termed the degree of esterification [61]. According to the degree of esterification (DE), PEC is commonly categorized into: High methoxyl pectin (HMP) (DE > 50%), and low methoxyl pectin (LMP) (DE < 50%) [57, 62, 63]. The degree of methyl-esterification can be defined as the percentage of carboxyl groups esterified with methanol of the pectin or the number of moles of methanol per 100 moles of galacturonic acid [64].

Pectins are usually extracted by hot dilute mineral acids at pH 1.5–2.5, taking 2–4 h with further precipitation with ethanol or isopropyl alcohol, separation to remove impurities, drying, grinding, and blending with other additives. The

extracted pectins can be categorized into two major types, depending on their DM degree, and different factors like pH, temperature, time, and solvent: liquid (S:L) ratio are usually studied to optimize the extraction conditions [44].

Valdés et al. [44] showed that the different natural pectins extracted by solvent extraction methods; such as tomato, banana peel, and sugar beet, the tomato peel pectin can be extracted through two steps by ammonium oxalate and oxalic acid at 90°C (24 and 12 h) and pectin yield extraction (PYE) of about 32.0%, while Sugar beet pulp pectin produced by citric acid, pH 1, 166 min at 99°C, 1:20 g·mL^{-1} 23.95%. Banana peel using a citric acid solution with pH 2.0 at 160 min, 87°C, 1:20 g·mL^{-1} give a lower PYE 13.89%, the PYE can be increased to 16.54% using citric acid and HCl, pH 1.5, 4 h, 90°C, but the pectin from mango peel can extract using sulfuric acid in water with pH 1.5, and 2.5 h at 90°C which give a higher extraction yields about >70%.

The HMP and LMP pectin have different physicochemical characteristics and thus different applications for each of them [53]; therefore, the degree of esterification plays an important role and is a good parameter for the biochemical, physical properties of pectin applications [16, 65]. Cho et al. [66] reported that the HMP pectin could be converted into LMP using two ways: 1) Chemical de-esterification by alkali; 2) Enzymatic treatment by pectin methylesterase. The LM-pectin is mainly produced by the chemical de-esterification of HM-pectin with acid, alkali, and alcoholic/ aqueous ammonia [67].

Depending on the raw material and the extraction method, pectins have various molecular weights, structures, and functional groups. These characteristics influence and determine the techno-functional properties of pectin molecules, such as gelling properties [43].

However, the industrial production of LM-pectin by enzymatic de-esterification of HM-pectin remains a challenge. For successful industrial application, it is necessary to enhance pectin methylesterase (PME) productivity by genetic modification and optimizing culture conditions [67].

Properties and advantages of pectin

Several studies have reported that the pectin polymers have several properties and advantages such as renewability, biodegradability and biocompatibility [20, 68, 69], amphiphilic properties (hydrophobic/hydrophilic nature) [48, 49], pectin-based films showing high tensile strength, and water resistance [70], but the applications of pectin is limited due to their poor chemo-physical properties and has poor water barrier properties if it used alone [3, 71]. Younis et al. [34] noted that the pure pectin films encountered several defects when used alone without any other polymers, such as soluble in water, high water permeability, poor waterproof, and unsatisfactory mechanical properties. These defects heavily limited the application of pectin films, especially in an atmosphere with high humidity or to package high-moisture foods.

Pectin applications

Pectin is considered a good source for human health, where its consumption by large volume in our life due to its ability to decrease blood cholesterol levels [37, 42, 51], and has been widely used in the food and nonfood industries; 1) Food and beverage industries, as edible films [3], gelling and thickener, texturizer [39, 49, 72], emulsifier agent [38, 49], colloidal stabilizer in food products such as jams, yogurt drinks, dairy products, and ice cream [51, 72, 73], also carrier polymer for the encapsulation of food ingredients [74]. 2) Pharmaceutical and biomedical applications [51, 75], including drug delivery, gene delivery [69], cosmetics [39], wound healing, and tissue engineering hydrogels as carriers for tissue regeneration [68, 76].

PECTIN IN FOOD INDUSTRY AND FOOD PACKAGING

For the food industries, several studies showed that the pectin has an excellent potential to be an edible film for food packaging and biocomposite materials [3, 20]; the presence of methoxyl groups in the pectin structure leads to an increase in the hydrophobic nature of pectin molecules that help the pectin to use in different food applications such as emulsifier agent and stabilizer [48]. The gels from the HM pectin are formed under sugar or acidic conditions, while from LM pectin formed through ionotropic gelation with low-valence ions, such as calcium ions, due to its containing a large number of ionizable carboxylic groups; therefore, they have a strong affinity for calcium ions [68], where the carboxyl group linked with the D-galacturonic acid residue exists in two essential forms: 1) carboxylate salt; 2) neutral methoxylated or ester [77]. Natural hydrocolloid pectin is widely utilized for thickening, stabilization, and encapsulation in the food & beverage, cosmetic, and pharmaceutical industries [67].

Pectins are hydrocolloids typically used as thickening and gelling agents, e.g., sauces and jams [43]. The use of pectin as a thickening and gelling agent, for example, sauces and jams, also gaining interest as a potential food emulsifier of natural origin; it can be used as a multifunctional ingredient in different food applications [78]. Einhorn-Stoll et al. [79] found that citrus pectin can be used as a thickening and gelling agent in a wide range of foods. At the same time, Liu et al. [80] showed that the pectin produced from sugar beet pulp (SBP) is the most prominent for stabilizing oil-in-water (O/W) emulsions and has received increasing attention in the food industry as a polysaccharide-based emulsifier.

The most recent trends in the field of pectins coating applications include; the shelf-life extension of fresh-cut highly perishable food, the application of pectin coatings as a pre-frying treatment to reduce the oil consumption in deep-fat fried products, and the use of pre-dried treatments to improve the retention of nutrients and quality characteristics of dehydrated and lyophilized food [44].

Due to its biodegradability, biocompatibility, edibility, chemical and physical properties, pectin is considered an applicable polymeric matrix for elaborating edible films intended as active food packaging [40]. Pectin-based biofilms could extend the shelf life of avocado fruits over a month compared to uncoated fruits by decreasing the oxygen absorption and thus delaying texture and color change [81].

Sucheta et al. [82] noticed that the performance of the commercial pectin films was affected by the incorporation of cornflour and beetroot; significant differences were found in tensile strength (TS), water solubility (WS), and the thickness film, where the TS increased when the pectin added to cornflour and beetroot from 1.36 to 7.47 MPa and 3.79 MPa. The WS decreased from 97.8% to 70.7%, also decreased film thickness from 0.24 to 0.06 mm and 0.09 mm, respectively. The effect of pectin with cornflour and beetroot as a coating has been studied of tomatoes fruits by Sucheta et al. [28] and find that the best coating for tomatoes was pectin/cornflour, it showed a significantly reduced the weight loss %, also delayed the respiration rate with retention of internal quality during the storage period (30 days) improved the shelf life of tomatoes and showed minimum shrinkage during the end of storage.

Meerasri and Sothornvit [83] find that the pectin polymer could provide a combination with a plasticizer (glycerol) and bioactive compounds (Gamma-aminobutyric acid [GABA]) and affect its performance; GABA incorporation showed an excellent enhancement in the barrier water than the glycerol addition where the WVP for the pectin decreased from 3.59×10^{-10} to 2.54×10 g^{-10} $g{\cdot}m^{-1}{\cdot}s^{-1}{\cdot}Pa^{-1}$ (pectin/GABA). In contrast, the glycerol incorporation led to increasing the WVP to 3.73×10 g^{-10} $g{\cdot}m^{-1}{\cdot}s^{-1}{\cdot}Pa^{-1}$, the mechanical properties of pectin films were significantly affected by glycerol and GABA incorporation, the TS for the pure pectin film 6.41% decreased to 2.14 and 2.17% for glycerol and GABA, respectively, while the Elongation at break (EB) increased from 8.78 (pectin film) to 15.11 and 27.99 MPa, respectively.

Sucheta et al. [28] studied the effect of pectin-corn flour-based coating and observed that it was the best treatment that reduced weight loss and decay, delayed respiration with retention of biochemical quality, and improved the shelf life of tomatoes.

Chitosan/poly (vinyl alcohol)/pectin ternary film was prepared by solution casting method; the films showed the antimicrobial activity against different pathogenic bacteria, such as *Escherichia coli*, *Staphylococcus aureus*, *Bacillus subtilis*, *Pseudomonas spp* [84].

Lorevice et al. [85] studied the effect both of HMP (High methoxyl pectin) and LMP (Low methoxyl pectin) with chitosan nanoparticles (DD = 94%) as edible films using water as solvent and find that the addition of chitosan nanoparticles (CSNPs) to pectin films did not cause remarkable visual changes but improved the mechanical and physical properties when compared with pure pectin films as a result of stronger interactions between the polymer matrices; The pure LMP films recorded a lower values than HMP films for the thickness and the mechanical properties, where the thickness, Elongation at break (EB) and Tensial strength (TS) of pectin films was remarkably increased after the addition of CSNPs; For the TS increased from 30.81 MPa to 46.95 MPa for HMP pectin films; whereas LMP films increased from 26.07 MPa to 58.51 MPa, the EB recorded a higher value for the LMP/CSNP 2.91% compared with pure LMP pectin (0.94%), the enhanced in the EB values of LMP films incorporated with CSNPs may be explained by the higher number of interactions between this pectin and CSNP surface, where it has more carboxylic groups that interact with CSNP amine groups.

The thermal properties performed by differential scanning calorimetry (DSC) analysis and noted that the heat absorbed (ΔH, change in enthalpy) by pectin films was larger than by pectin powder and PEC/CSNPs films where it increases from 401.6 → 444.7 → 547.9 $J \cdot g^{-1}$ for HMP powder, HMP films and HMP/CSNP film, respectively. Regarding the endothermic peak temperatures (EPT) property, it decreased when CSNPs were added from 121.8 → 105.1 → 105.9°C for HMP powder, HMP films, and HMP/CSNP film, respectively. Also, for the LMP, the EPT decreased from 108.9 → 97.8°C, for LDM pectin powder, LMP films, respectively, while by the CSNP addition, the temperature increased to 102.1°C. this can be attributable to reduced natural hydration degrees of pectin films due to CSNP presence, the lower hydration degrees may result in lower chain mobility, the lower EPTs, and higher ΔH values suggest that CSNPs were occupying water sites within pectin matrices and led to absorb more thermal energy than pure pectin films.

For the barrier properties, the addition of CSNPs into the LMP films made a significant difference and decreased the water vapor transmission rate and permeance property from 118.67 to 96.56 $g \cdot h^{-1} \cdot m^{-2}$ and 81.04 to 54.06 $g \cdot kPa^{-1} \cdot h^{-1} \cdot m^{-2}$, respectively, the effect of reducing permeance caused by CSNP presence was more remarkable than in HMP since LMP has a higher content of free hydroxyl groups than HMP [85].

Ngo et al. [86] evaluated the films from pectin (2% w/v) and nano chitosan (2% w/v) with different ratios (100:0, 75:25, 50:50, 25:75, and 0:100 w/w) and find that the blending ratio 50:50 increased the tensile strength and decreased the water solubility, water vapor permeability, and oxygen permeability; Also, these films showed antimicrobial activity against *C. gloeosporioides*, *S. cerevisiae*, *A. niger*, and *E. coli*.

Ranganathan et al. [87] noted that pectin polymer has the great potential to form a polymer composite and makes it flexible film due to its anionic property, whereas; Tsai et al. [88] found that the addition of pectin into chitosan with a high degree of deacetylation increased the viscosity of the matrix solution, and making it more difficult to prepare homogeneous solutions; also mentioned if the densities of positive charges on the polycations and negative charges on polyanions are not equitable, the solution will be water-soluble and form homogeneous.

Chen et al. [89] showed that when chitosan was mixed with pectin, the mechanical properties and hydrophilicity increased for the chitosan/pectin films, where they prepared the chitosan/pectin membranes successfully via using a freezegelation method and proved that the combination of pectin into chitosan could significantly enhance the properties of the film including and increased the tensile strength from 8 $N\cdot g^{-1}$ (CH films) to 32 $N\cdot g^{-1}$ (CH: PEC films) also decreased the water contact angle from 85° (CH films) to 45° (CH: PEC films); therefore it can be used as a carrier for food or drug.

Norcino et al. [16] used the low molecular weight chitosan with the degree of deacetylation of 80% and pectin from the peel of citrus fruits to prepare different CH/PEC films by solution blending and observed the following; The contact angle values were clearly visible for pure CH film, this can be related with the sorption of water on the surface, and film swelling, followed by water absorption, the water contact angle for CH/ PEC blends 50/50 and 25/75 recorded a higher value (87°) while for the pure CH, PEC, and CH/PEC 75/25 films were ~ 64°, 75°, and 74° respectively and remained practically constant even after 60 s. The ionic cross-linking effect restricted the CS/PEC blend film swelling and reduced their water sorption.

For the mechanical properties, the tensile strength increased for the CH/PEC blend films. It is evident that the tensile strength of CH/PEC blends presented a large positive deviation from addition law; While pure CH and PEC films showed lower values for the tensile strength 41 and 48 MPa, respectively, while it increased for CH/PEC blend 75/25 was 70 MPa and reached 81 MPa for CH/PEC blends 50/50 and 25/75.

Ionic crosslinking between the chitosan ammonium groups (NH^{+}) and pectin carboxylate groups (COO-) and thus playing an important role on the physical properties of CH/PEC polyelectrolyte based-films where the electrostatic interactions were responsible for increasing water resistance and mechanical properties of chitosan/pectin films, which allows modulating their thermo-mechanical and other physical properties for specific applications like medicine, agriculture, and food packaging [16].

Younis and Zhao [35] prepared the different films from chitosan (deacetylation degree 67.9%) and high methoxyl apple pectin by *casting* method. They noted that the tensile strength increased from 1.22 MPa (CH) to 5.06 MPa (CH/ PEC), while the elongation at break for the chitosan did not affect by the pectin incorporation; For the water vapor permeability WVP values of the PEC, CH and blend films recorded a significant difference where the WVP value of PEC film ($5.53 \times 10\ g^{-12}\ g\ cm^{-1}\cdot s^{-1}\cdot Pa^{-1}$)

was much higher than that of PEC/CH film (3.91×10 g^{-12} g $cm^{-1} \cdot s^{-1} \cdot Pa^{-1}$). Still, the blending does not influence the barrier properties. The more porous structure of PEC film contributes to its higher WVP; the significant reduction in WVP makes the films more suitable for food packaging. Meanwhile, the surface morphology for the films was observed by scanning electron microscopy (SEM), and numerous small pores were evenly distributed in PEC film while big caves were unevenly scattered in CH film; by incorporating the pectin into chitosan, these cavities disappeared, as shown in **Figure 1.**

Chetouani et al. [90] prepared edible films by *casting* method from apple pectin (DE, 70–75%) and chitosan (viscosity of 800,000 cps) and compared it with chitosan/ oxidized pectin by sodium meta periodate ($NaIO_4$); the results showed the following; The X-ray diffraction, as shown in **Figure 2** showed that the chitosan films exhibit broad diffraction peaks observed at $(2\theta) = 9.8°$ and 19.9°, which are typical fingerprints of semi-crystalline chitosan. The pure pectin films showed sharp peaks centered at 9.2°, 12.8°, 18.5°, 28.1°, and 40.1° and were considered a semi-crystalline material. After the pectin's oxidation process, the X-ray diffraction analysis (XRD) is clearly different from that of pure pectin. The peaks at 16.5°, 24.1°, 27.5°, and 33.5° indicate less crystalline the oxidized pectin (OPEC).

From the above results, the films confirmed that CH's crystallinity is drastically decreased with the addition of PEC or oxidized PEC. Also, the addition of PEC to CH makes the material amorphous; this could be due to the interacts of chitosan with pectin through inter-molecular hydrogen bonding.

Chetouani et al. [90] noted that the pectin, CH, CH/pectin, and CH/oxidation pectin films do not affect the Gram-negative bacteria (*P. aeruginosa, E. coli*). In contrast, Gram-positive bacteria (*B. subtilis and S. aureus*) are more sensitive to the films. Also, there is a strong interaction between chitosan and oxidized pectin films, which influences the thermal decomposition; using these types of films can improve the antibacterial activity of chitosan, which demonstrates them to be promising materials for food packaging and biomedical applications.

Baron et al. [91] reported that the chitosan CH (blue crab waste) and PEC (orange peel) can produce edible films by *casting* method; they noted that the addition of high concentrations of chitosan with pectin polymer in the films formulations (75:25 ratio) led to producing films with lower solubility and lower moisture content 10.2% and 15.8% respectively, compared with the values of pure pectin; also reduced the water vapor permeability from 1.06×10 g^{-15} $g \cdot m^{-1} \cdot s^{-1} \cdot Pa^{-1}$ (pectin films) to 0.99×10 g^{-15} $g \cdot m^{-1} \cdot s^{-1} \cdot Pa^{-1}$ (75 CH:25 PEC), when the quantity of protonated free amino groups (NH^{3+}) in chitosan and the free anionic groups present in the pectin becomes higher in the films, the degree of swelling increased from 14.3% (PEC films) to 15.5% (CH: PEC), while for the mechanical properties, increasing chitosan proportion in biopolymers blend increased the tensile strength (TS) and become more elastic and flexible from 17 MPa (PEC) to 23 MPa (75 CH:25 PEC) also reduced deformability (EB) 37.7–27%, respectively.

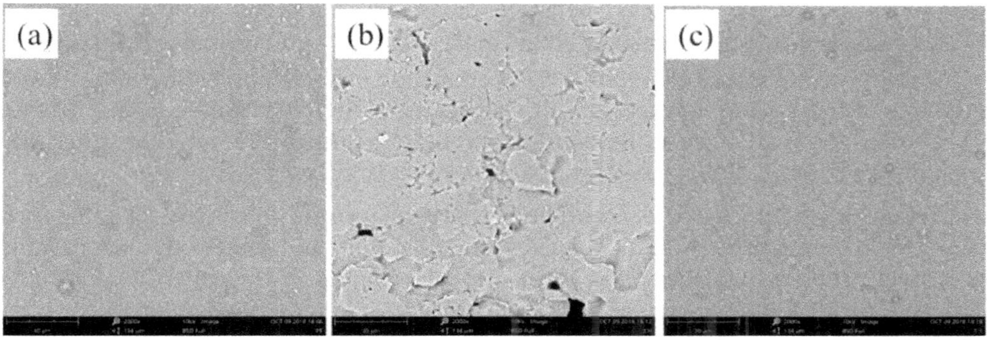

Figure 1. *SEM micrographs of the films surface a) pectin, b) chitosan, c) pectin/chitosan films [35].*

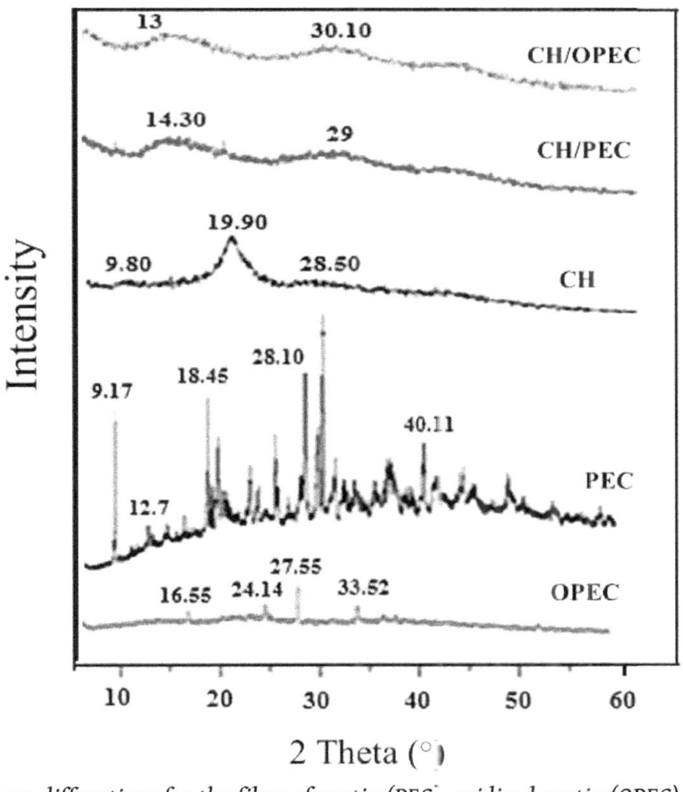

Figure 2. *X-ray diffractions for the films of pectin (PEC), oxidized pectin (OPEC), and chitosan (CH).*

CONCLUSION

Today, the food industry's important challenges are developing environmentally friendly, green, and intelligent food packaging materials to avoid environmental problems caused by synthetic polymer packaging materials. Edible coating of biodegradable materials is considered the most trusted technique for storing fruit and vegetables, an excellent alternative to chemically synthesized preservatives, more effective in prolonging various agricultural products' shelf life, and inhibiting disease growth on the fruit surface. Biodegradable materials have several characteristics, including biocompatibility, biodegradability, renewability, non-toxicity.

Therefore, in recent times the researchers have been focused on edible packaging such as pectin polymers. Pectin is a promising material that can avoid nonbiodegradable polymers problems when applied as an edible coating for different fruits due to its different properties.

ABSTRACT

Food packaging based on plastic films made from nonrenewable resources often causes environmental problems after disposal. Recently, researchers are increasingly focusing on alternative materials to reduce the use of nonbiodegradable and nonrenewable films. Generally, biomaterials are nontoxic, biocompatible, and renewable always presents reasonable film-forming ability. Thus, they are important for food safety, where undesired chemical compounds might migrate from chemicals migrate from the plastic packaging materials into foods. Pectin (PEC), as a natural carbohydrate polymer, belongs to the anionic heteropolysaccharide family and is often extracted from various residues from plant food processing, such as apple and citrus pomaces. The pectin molecules are highly branched with a backbone α-(1–4) linked D galacturonic acid. Among the naturally derived carbohydrate-based biopolymers, pectin was considered a promising substrate in fabricating edible films due to its diverse advantages, such as perfect film-forming ability, evidenced bioactivity, easy availability, and excellent quality biodegradability and biocompatibility, nontoxicity, and low cost. Pectin-based films have excellent oxygen barrier capacity and extend the shelf life for different fruits.

The properties of pure pectin films can enhance through combination with other polymers or nanoparticles/fibers.

REFERENCES

1. Ma D, Jiang Y, Ahmed S, Qin W, Liu Y. Physical and antimicrobial properties of edible films containing Lactococcus lactis. International Journal of Biological Macromolecules. 2019;**141**:378-386

2. Otoni CG, Moura MR, Aouada FA, Camilloto GP, Cruz RS, Lorevice MV, et al. Antimicrobial and physicalmechanical properties of pectin/papaya puree/

cinnamaldehyde nanoemulsion edible composite films. Food Hydrocolloids. 2014;**41**:188-194

3. Šešlija S, Nešić A, Ružić J, Kalagasidis Krušić M, Veličković S, Avolio R, et al. Edible blend films of pectin and poly(ethylene glycol): Preparation and physico-chemical evaluation. Food Hydrocolloids. 2018;**77**:494-501
4. Priyadarshi R, Rhim J-W. Chitosanbased biodegradable functional films for food packaging applications. Innovative Food Science & Emerging Technologies. 2020;**62**:102346
5. Pacheco N, Naal-Ek MG, Ayora-Talavera T, Shirai K, Román-Guerrero A, Fabela-Morón MF, et al. Effect of bio-chemical chitosan and gallic acid into rheology and physicochemical properties of ternary edible films. International Journal of Biological Macromolecules. 2019;**125**: 149-158
6. Ahmed J, Arfat Y, Bher A, Mulla M, Jacob H, Auras R. Active chicken meat packaging based on polylactide films and bimetallic Ag–Cu nanoparticles and essential oil. Journal of Food Science. 2018;**83**(5):1299-1310
7. Sartori T, Feltre G, do Amaral Sobral PJ, Lopes da Cunha R, Menegalli FC. Properties of films produced from blends of pectin and gluten. Food Packaging and Shelf Life. 2018;**18**:221-229
8. Gigli M, Negroni A, Zanaroli G, Lotti N, Fava F, Munari A. Environmentally friendly PBS-based copolyesters containing PEG-like subunit: Effect of block length on solid-state properties and enzymatic degradation. Reactive and Functional Polymers. 2013;**73**(5):764-771
9. Groh KJ, Backhaus T, Carney-Almroth B, Geueke B, Inostroza PA, Lennquist A, et al. Overview of known plastic packagingassociated chemicals and their hazards. Science of The Total Environment. 2019;**651**:3253-3268
10. Atarés L, Chiralt A. Essential oils as additives in biodegradable films and coatings for active food packaging. Trends in Food Science & Technology. 2016;**48**:51-62
11. Mahcene Z, Khelil A, Hasni S, Akman PK, Bozkurt F, Birech K, et al. Development and characterization of sodium alginate based active edible films incorporated with essential oils of some medicinal plants. International Journal of Biological Macromolecules. 2020;**145**:124-132
12. Rodríguez FJ, Galotto MJ, Guarda A, Bruna JE. Modification of cellulose acetate films using nanofillers based on organoclays. Journal of Food Engineering. 2012;**110**(2):262-268
13. Espitia PJP, Du W-X, Avena-Bustillos RJ, Soares NFF, McHugh TH. Edible films from pectin: Physical-mechanical and antimicrobial properties-A review. Food Hydrocolloids. 2014;**35**:287-296

14. Tahir HE, Xiaobo Z, Mahunu GK, Arslan M, Abdalhai M, Zhihua L. Recent developments in gum edible coating applications for fruits and vegetables preservation: A review. Carbohydrate Polymers. 2019;**224**:115141

15. Pellá MCG, Silva OA, Pellá MG, Beneton AG, Caetano J, Simões MR, et al. Effect of gelatin and casein additions on starch edible biodegradable films for fruit surface coating. Food Chemistry. 2020;**309**:125764

16. Norcino LB, de Oliveira JE, Moreira FKV, Marconcini JM, Mattoso LHC. Rheological and thermomechanical evaluation of bio-based chitosan/pectin blends with tunable ionic cross-linking. International Journal of Biological Macromolecules. 2018;**118**:1817-1823

17. Khodaei D, Oltrogge K, Hamidi-Esfahani Z. Preparation and characterization of blended edible films manufactured using gelatin, tragacanth gum and, Persian gum. LWT. 2020;**117**: 108617

18. Qin Y, Liu Y, Zhang X, Liu J. Development of active and intelligent packaging by incorporating betalains from red pitaya (Hylocereus polyrhizus) peel into starch/polyvinyl alcohol films. Food Hydrocolloids. 2020;**100**:105410

19. Homez Jara A, Daza L, Aguirre D, Muñoz A, Solanilla Duque JF, Váquiro H. Characterization of chitosan edible films obtained with various polymer concentrations and drying temperatures. International Journal of Biological Macromolecules. 2018;**113**:1233-1240

20. Gurram R, Souza Filho P, Taherzadeh M, Zamani A. A solventfree approach for production of films from pectin and fungal biomass. Journal of Polymers and the Environment. 2018;**26**:4282-4292

21. Nešić A, Gordić M, Davidović S, Radovanović Ž, Nedeljković J, Smirnova I, et al. Pectin-based nanocomposite aerogels for potential insulated food packaging application. Carbohydrate Polymers. 2018;**195**:128-135

22. Geyer R, Jambeck JR, Law KL. Production, use, and fate of all plastics ever made. Science Advances. 2017;**3**(7):e1700782

23. Haghighi H, Licciardello F, Fava P, Siesler H, Pulvirenti A. Recent advances on chitosan-based films for sustainable food packaging applications. Food Packaging and Shelf Life. 2020;**26**:100551

24. Tóth A, Halász K. Characterization of edible biocomposite films directly prepared from psyllium seed husk and husk flour. Food Packaging and Shelf Life. 2019;**20**:100299

25. Liu J, Liu S, Wu Q, Gu Y, Kan J, Jin C. Effect of protocatechuic acid incorporation on the physical, mechanical, structural and antioxidant properties of chitosan film. Food Hydrocolloids. 2017;**73**:90-100

26. Lima M, Carneiro LC, Bianchini D, Dias ARG, Zavareze ER, Prentice C, et al. Structural, thermal, physical, mechanical, and barrier properties of Chitosan films with the addition of Xanthan gum. Journal of Food Science. 2017;**82**(3):698-705

27. Khazaei N, Esmaiili M, Djomeh ZE, Ghasemlou M, Jouki M. Characterization of new biodegradable edible film made from basil seed *(Ocimum basilicum L.)* gum. Carbohydrate Polymers. 2014;**102**: 199-206

28. Sucheta, Chaturvedi K, Sharma N, Yadav SK. Composite edible coatings from commercial pectin, corn flour and beetroot powder minimize post-harvest decay, reduces ripening and improves sensory liking of tomatoes. International Journal of Biological Macromolecules. 2019;**133**:284-293

29. Jiang Y, Yu L, Hu Y, Zhu Z, Zhuang C, Zhao Y, et al. Electrostatic spraying of chitosan coating with different deacetylation degree for strawberry preservation. International Journal of Biological Macromolecules. 2019;**139**:1232-1238

30. Thakur R, Pristijono P, Scarlett CJ, Bowyer M, Singh SP, Vuong QV. Starchbased edible coating formulation: Optimization and its application to improve the postharvest quality of "Cripps pink" apple under different temperature regimes. Food Packaging and Shelf Life. 2019;**22**:100409

31. Mannozzi C, Tylewicz U, Chinnici F, Siroli L, Rocculi P, Dalla Rosa M, et al. Effects of chitosan based coatings enriched with procyanidin by-product on quality of fresh blueberries during storage. Food Chemistry. 2018;**251**:18-24

32. Olivas G, Dávila-Aviña J, Salas-Salazar NA, Molina FJ. Use of edible coatings to preserve the quality of fruits and vegetables during storage. Stewart Postharvest Review. 2008;**4**:1-10

33. Tapia-Blácido DR, Maniglia B, Martelli TM. Transport phenomena in edible films. Polymers for Food Applications. 2018:149-192

34. Younis HGR, Abdellatif HRS, Ye F, Zhao G. Tuning the physicochemical properties of apple pectin films by incorporating chitosan/pectin fiber. International Journal of Biological Macromolecules. 2020;**159**:213-221

35. Younis HGR, Zhao G. Physicochemical properties of the edible films from the blends of high methoxyl apple pectin and chitosan. International Journal of Biological Macromolecules. 2019;**131**:1057-1066

36. Zhou W, Liu W, Zou L, Liu W, Liu C, Liang R, et al. Storage stability and skin permeation of vitamin C liposomes improved by pectin coating. Colloids and Surfaces B: Biointerfaces. 2014;**117**:330-337

37. Xu Y, Zhang L, Bailina Y, Ge Z, Ding T, Ye X, et al. Effects of ultrasound and/or heating on the extraction of pectin from grapefruit peel. Journal of Food Engineering. 2014;**126**:72-81

38. Hosseini SS, Khodaiyan F, Ms Y. Optimization of microwave assisted extraction of pectin from sour orange peel and its physicochemical properties. Carbohydrate Polymers. 2016;**140**:59-65

39. Pasini Cabello SD, Takara EA, Marchese J, Ochoa NA. Influence of plasticizers in pectin films: Microstructural changes. Materials Chemistry and Physics. 2015;**162**:491-497

40. Yeddes W, Djebali K, Aidi Wannes W, Horchani-Naifer K, Hammami M, Younes I, et al. Gelatin-chitosan-pectin films incorporated with rosemary essential oil: Optimized formulation using mixture design and response surface methodology. International Journal of Biological Macromolecules. 2020;**154**:92-103

41. Wusigale, Liang L, Luo Y. Casein and pectin: Structures, interactions, and applications. Trends in Food Science & Technology. 2020;**97**:391-403

42. Santos JDG, Espeleta AF, Branco A, de Assis SA. Aqueous extraction of pectin from sisal waste. Carbohydrate Polymers. 2013;**92**(2):1997-2001

43. Saavedra Isusi GI, Bindereif B, Karbstein HP, van der Schaaf US. Polymer or microgel particle: Differences in emulsifying properties of pectin as microgel or as individual polymer chains. Colloids and Surfaces A: Physicochemical and Engineering Aspects. 2020;**598**:124793

44. Valdés A, Burgos N, Jiménez A, Garrigós CM. Natural pectin polysaccharides as edible coatings. Coatings. 2015;**5**(4):865-886

45. Guerreiro AC, Gago CML, Faleiro ML, Miguel MGC, Antunes MDC. The effect of edible coatings on the nutritional quality of 'Bravo de Esmolfe' fresh-cut apple through shelf-life. LWT. 2017;**75**:210-219

46. Seixas F, Turbiani F, Salomao P, Souza R, Gimenes M. Biofilms composed of alginate and pectin: Effect of concentration of crosslinker and plasticizer agents. Chemical Engineering Transactions. 2013;(32):1693-1698

47. Willats WGT, Knox JP, Mikkelsen JD. Pectin: New insights into an old polymer are starting to gel. Trends in Food Science & Technology. 2006;**17**(3):97-104

48. Mungure TE, Roohinejad S, Bekhit AE-D, Greiner R, Mallikarjunan K. Potential application of pectin for the stabilization of nanoemulsions. Current Opinion in Food Science. 2018;**19**:72-76

49. Yang X, Nisar T, Hou Y, Gou X, Sun L, Guo Y. Pomegranate peel pectin can be used as an effective emulsifier. Food Hydrocolloids. 2018;**85**:30-38

50. Pereira PHF, Oliveira TÍS, Rosa MF, Cavalcante FL, Moates GK, Wellner N, et al. Pectin extraction from pomegranate peels with citric acid. International Journal of Biological Macromolecules. 2016;**88**:373-379

51. Grassino AN, Barba FJ, Brnčić M, Lorenzo JM, Lucini L, Brnčić SR. Analytical tools used for the identification and quantification of pectin extracted from plant food matrices, wastes and by-products: A review. Food Chemistry. 2018;**266**:47-55

52. Guolin H, Shi J, Kai Z, Xiaolan H. Application of ionic liquids in the microwave-assisted extraction of pectin from lemon peels. Journal of Analytical Methods in Chemistry. 2012;**2012**:302059

53. Chan S-Y, Choo W-S. Effect of extraction conditions on the yield and chemical properties of pectin from cocoa husks. Food Chemistry. 2013;**141**(4):3752-3758

54. Liu L, Jiang T, Yao J. A two-step chemical process for the extraction of cellulose fiber and pectin from Mulberry branch Bark efficiently. Journal of Polymers and the Environment. 2011;**19**(3):568

55. Gopi D, Kanimozhi K, Bhuvaneshwari N, Indira J, Kavitha L. Novel banana peel pectin mediated green route for the synthesis of hydroxyapatite nanoparticles and their spectral characterization. Spectrochimica Acta Part A: Molecular and Biomolecular Spectroscopy. 2014;**118**:589-597

56. Oliveira TÍS, Rosa MF, Cavalcante FL, Pereira PHF, Moates GK, Wellner N, et al. Optimization of pectin extraction from banana peels with citric acid by using response surface methodology. Food Chemistry. 2016;**198**:113-118

57. Gharibzahedi SMT, Smith B, Guo Y. Pectin extraction from common fig skin by different methods: The physicochemical, rheological, functional, and structural evaluations. International Journal of Biological Macromolecules. 2019;**136**:275-283

58. Moorthy IG, Maran JP, Ilakya S, Anitha SL, Sabarima SP, Priya B. Ultrasound assisted extraction of pectin from waste Artocarpus heterophyllus fruit peel. Ultrasonics Sonochemistry. 2017;**34**:525-530

59. Su D-L, Li P-J, Quek SY, Huang Z-Q, Yuan Y-J, Li G-Y, et al. Efficient extraction and characterization of pectin from orange peel by a combined surfactant and microwave assisted process. Food Chemistry. 2019;**286**:1-7

60. Wicker L, Kim Y, Kim M-J, Thirkield B, Lin Z, Jung J. Pectin as a bioactive polysaccharide – Extracting tailored function from less. Food Hydrocolloids. 2014;**42**:251-259

61. Dubey V, Mohan P, Dangi JS, Kesavan K. Brinzolamide loaded chitosan-pectin mucoadhesive nanocapsules for management of glaucoma: Formulation, characterization and pharmacodynamic study. International Journal of Biological Macromolecules. 2020;**152**:1224-1232

62. Villanova JCO, Ayres E, Oréfice RL. Design, characterization and preliminary in vitro evaluation of a mucoadhesive polymer based on modified pectin and acrylic monomers with potential use as a pharmaceutical excipient. Carbohydrate Polymers. 2015;**121**:372-381

63. Hosseini S, Parastouei K, Khodaiyan F. Simultaneous extraction optimization and characterization of pectin and phenolics from sour cherry pomace. International Journal of Biological Macromolecules. 2020;**158**: 911-921

64. Seixas FL, Fukuda DL, Turbiani FRB, Garcia PS, Petkowicz CLO, Jagadevan S, et al. Extraction of pectin from passion fruit peel (Passiflora edulis f. flavicarpa) by microwave-induced heating. Food Hydrocolloids. 2014;**38**:186-192

65. Pérez CD, MD D'N, Rizzo SA, Gerschenson LN, Descalzo AM, Rojas AM. High methoxyl pectin–methyl cellulose films with antioxidant activity at a functional food interface. Journal of Food Engineering. 2013;**115**(1):162-169

66. Cho E-H, Jung H-T, Lee B-H, Kim H-S, Rhee J-K, Yoo S-H. Green process development for apple-peel pectin production by organic acid extraction. Carbohydrate Polymers. 2019;**204**:97-103

67. Talekar S, Vijayraghavan R, Arora A, Patti AF. Greener production of low methoxyl pectin via recyclable enzymatic de-esterification using pectin methylesterase cross-linked enzyme aggregates captured from citrus peels.

68. Food Hydrocolloids. 2020;**108**:105786 Zhao S, Yang F, Liu Y, Sun D, Xiu Z, Ma X, et al. Study of chemical characteristics, gelation properties and biological application of calcium pectate prepared using apple or citrus pectin. International Journal of Biological Macromolecules. 2018;**109**:180-187

69. Mallick SP, Singh BN, Rastogi A, Srivastava P. Design and evaluation of chitosan/poly(l-lactide)/pectin based composite scaffolds for cartilage tissue regeneration. International Journal of Biological Macromolecules. 2018;**112**: 909-920

70. Mellinas C, Ramos M, Jiménez A, Garrigós MC. Recent trends in the use of pectin from agro-waste residues as a natural-based biopolymer for food packaging applications. Materials. 2020;**13**(3):673

71. Gutierrez-Pacheco MM, OrtegaRamirez LA, Cruz-Valenzuela MR, Silva-Espinoza BA, Gonzalez-Aguilar GA, Ayala-Zavala JF. Chapter 50 Combinational Approaches

for Antimicrobial Packaging: Pectin and Cinnamon Leaf Oil. Antimicrobial Food Packaging. San Diego: Academic Press; 2016. pp. 609-617

72. Laurent M, Patrick B. Stabilization mechanism of acid dairy drinks (ADD) induced by pectin. Food Hydrocolloids. 2003;**17**:445-454

73. Ralet M-C, Lerouge P, Quéméner B. Mass spectrometry for pectin structure analysis. Carbohydrate Research. 2009;**344**(14):1798-1807

74. Perussello CA, Zhang Z, Marzocchella A, Tiwari BK. Valorization of Apple Pomace by extraction of valuable compounds. Comprehensive Reviews in Food Science and Food Safety. 2017;**16**(5):776-796

75. Meneguin A, Cury B, Evangelista R. Films from resistant starch-pectin dispersions intended for colonic drug delivery. Carbohydrate Polymers. 2014;**99**:140-149

76. Dranca F, Oroian M. Extraction, purification and characterization of pectin from alternative sources with potential technological applications. Food Research International. 2018;**113**:327-350

77. Veisi Z, Gallant ND, Alcantar NA, Toomey RG. Responsive coatings from naturally occurring pectin polysaccharides. Colloids and Surfaces B: Biointerfaces. 2019;**176**:387-393

78. Neckebroeck B, Verkempinck SHE, Vaes G, Wouters K, Magnée J, Hendrickx ME, et al. Advanced insight into the emulsifying and emulsion stabilizing capacity of carrot pectin subdomains. Food Hydrocolloids. 2020;**102**:105594

79. Einhorn-Stoll U, Kastner H, Fatouros A, Krähmer A, Kroh LW, Drusch S. Thermal degradation of citrus pectin in low-moisture environment – Investigation of backbone depolymerisation. Food Hydrocolloids. 2020;**107**:105937

80. Liu Z, Guo X, Meng H. Added ferulic acid enhances the emulsifying properties of pectins from different sources. Food Hydrocolloids. 2020;**100**:105439

81. Maftoonazad N, Ramaswamy HS. Effect of pectin-based coating on the kinetics of quality change associated with stored avocados. The Journal of Food Processing and Preservation. 2008;**32**(4):621-643

82. Sucheta, Rai SK, Chaturvedi K, Yadav SK. Evaluation of structural integrity and functionality of commercial pectin based edible films incorporated with corn flour, beetroot, orange peel, muesli and rice flour. Food Hydrocolloids. 2019;**91**:127-135

83. Meerasri J, Sothornvit R. Characterization of bioactive film from pectin incorporated with gammaaminobutyric acid. International Journal of Biological Macromolecules. 2020;(147):1285-1293

84. Tripathi S, Mehrotra GK, Dutta PK. Preparation and physicochemical evaluation of chitosan/poly(vinyl alcohol)/pectin ternary film for foodpackaging applications. Carbohydrate Polymers. 2010;**79**(3):711-716

85. Lorevice MV, Otoni CG, Moura MR, Mattoso LHC. Chitosan nanoparticles on the improvement of thermal, barrier, and mechanical properties of highand low-methyl pectin films. Food Hydrocolloids. 2016;**52**:732-740

86. Ngo TMP, Nguyen TH, Dang TMQ, Tran TX, Rachtanapun P. Characteristics and antimicrobial properties of active edible films based on pectin and nanochitosan. International Journal of Molecular Sciences. 2020;**21**(6):2224

87. Ranganathan P, Mutharani B, Chen S-M, Sireesha P. Biocompatible chitosan-pectin polyelectrolyte complex for simultaneous electrochemical determination of metronidazole and metribuzin. Carbohydrate Polymers. 2019;**214**:317-327

88. Tsai R-Y, Chen P-W, Kuo T-Y, Lin C-M, Wang D-M, Hsien T-Y, et al. Chitosan/pectin/gum Arabic polyelectrolyte complex: Process-dependent appearance, microstructure analysis and its application. Carbohydrate Polymers. 2014;**101**:752-759

89. Chen P-H, Kuo T-Y, Kuo J-Y, Tseng Y-P, Wang D-M, Lai J-Y, et al. Novel chitosan–pectin composite membranes with enhanced strength, hydrophilicity and controllable disintegration. Carbohydrate Polymers. 2010;**82**(4):1236-1242

90. Chetouani A, Follain N, Marais S, Rihouey C, Elkolli M, Bounekhel M, et al. Physicochemical properties and biological activities of novel blend films using oxidized pectin/chitosan. International Journal of Biological Macromolecules. 2017;**97**:348-356

91. Baron RD, Pérez LL, Salcedo JM, Córdoba LP, Sobral PJA. Production and characterization of films based on blends of chitosan from blue crab (Callinectes sapidus) waste and pectin from Orange (Citrus sinensis Osbeck) peel. International Journal of Biological Macromolecules. 2017;**98**:676-683

Chapter 6

UTILIZATION OF AGRO-INDUSTRIAL WASTES AS EDIBLE COATING AND FILMS FOR FOOD PACKAGING MATERIALS

INTRODUCTION

Majority of food packaging are made of synthetic materials derived from nonrenewable sources and based on petrochemicals, having advantages of being available in large quantities at a low cost and having excellent barrier and resistance properties. Nevertheless, these are causing serious environmental issues due to the production of large amounts of non-biodegradable solid wastes [1]. Apart from its primary function of containing food, delaying deterioration, and extending shelf life, packaging also plays a critical role in regulating food-environment interactions [2, 3]. Environmental concerns and consumer demands for high-quality eco-friendly products that are similar to those found in nature (natural products), has prompted the development of technologies for novel packaging materials, such as the production of biodegradable films from renewable polymers [1, 3, 4]. As a result, consumer demand for packaging materials has switched to safe and environment friendly biodegradable materials, particularly from renewable agriculture by-products and food processing industry wastes. Polysaccharides including starch, cellulose, sodium alginate, pectin, chitosan, and gums, as well as proteins like whey, soy, gluten, and gelatin, are among the most commonly used biopolymers in the manufacture of biodegradable films [5]. Because of their abundance in nature, biodegradability, and edibility, these natural biopolymers are widely employed. Casting, pressing, and extrusion, followed by blowing, are some of the procedures utilized in the production of these films [6]. Plant-derived bioactive substances, such as essential oils, vitamins, minerals, polyphenols, and carotenoids, are extensively distributed in nature in addition to biopolymers. Because of their biological nature, different parts of plants, such as leaves, flowers, seeds, and roots, can possibly be employed in the manufacture of environment friendly films with functional features [7]. Some bioactive substances have antioxidant and antibacterial properties [8–10]. Bioactive films with new and/or improved properties, such as antioxidant [8, 9] and antimicrobial [10] effects, innovative colors [11, 12], and customized barrier and mechanical properties, have been developed using biopolymers and natural bioactive compounds [8, 11, 13]. Some of the techniques used for their production include the use of inherently bioactive biopolymer-based materials [14, 15], as well as the direct or sprinkling incorporation of free or encapsulated bioactive compounds into the film-forming solutions [8, 12].

Some polymers generated from renewable agro-waste sources are edible and have played important roles in food throughout history as well as in the food, pharmaceutical, and other industries. Biopolymers can be used alone or in combination with other biopolymers to produce an edible coating or film material [16, 17]. In comparison to non-edible polymeric packaging, edible coatings and films offer a number of advantages. They can simplify food packaging and, even if they are not consumed with the packaged product, they can assist to environmental pollution reduction due to their biodegradable nature. Material fragmentation and subsequent mineralization are the mechanisms through which polymers degrade in a bioactive environment. Outer temperature and moisture, as well as the enzyme activity of microorganisms degrade polymer, resulting in fragmented polymer residues. These polymer fragments are only considered biodegradable if they are consumed as food and for energy by microorganisms and converted into carbon dioxide (CO_2), water (H_2O), and biomass under aerobic conditions and hydrocarbons, methane, and biomass under anaerobic conditions at the end of the degradation process [18].

AGRO-WASTE BASED RENEWABLE SOURCES USED IN DEVELOPMENT OF EDIBLE COATINGS AND FILMS

Bio-packaging films are made of materials derived from renewable resources that degrade completely. These can be made directly by biological systems (for example, plants, animals, algae, and microbes) or by polymerizing bio-based monomers (e.g., polylactic acid). Classification and life cycle of various agro-waste based renewable sources used in the development of edible coatings or films has been depicted in **Figure 1.** These bio-polymer materials have been classified into four groups based on their origin and manufacturing method [20]. Among the biopolymers utilized in the packaging business include natural polysaccharides, proteins, and their derivatives [21, 22]. Polysaccharides (e.g., cellulose, pectin, gum, starch, chitosan) and proteins (e.g., collagen, casein, whey protein, egg protein, gluten) extracted from biomass, polymers synthesized from bio-derived monomers (e.g., polylactic acid (PLA)), and those produced directly by microorganisms (e.g., polyhydroxyalkanoates (PHA), pullulan, curdlan, bacterial cellulose) are examples of such materials based on renewable resources that are commonly used for food packaging applications [21].

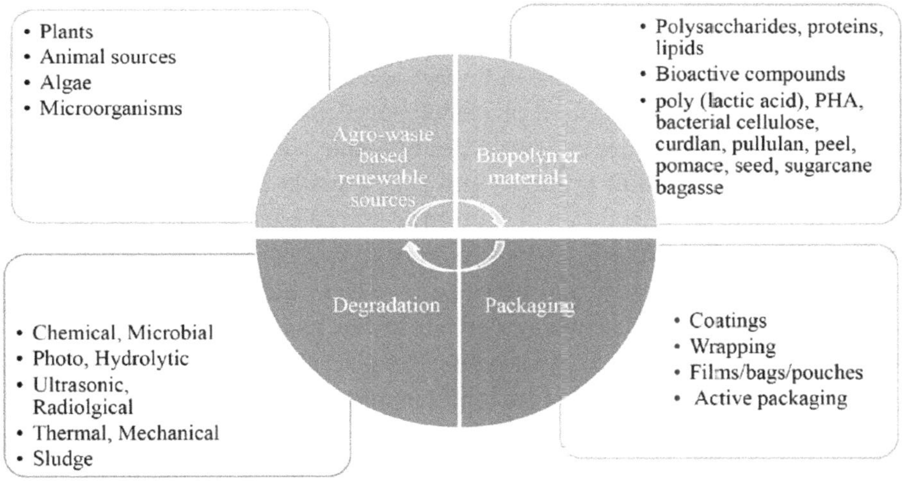

Figure 1. *Life cycle of agro-waste based renewable sources used in the development of edible coatings or films with modifications. (Adopted from Giosafatto et al. [19].)*

These bio-based materials have good barrier properties and can be mass-produced on a large industrial scale for moderately low costs, making them a viable alternative to petroleum-based plastics. However, due to considerable divergence in respect to plastics, such as weak tensile strength, brittleness, thermal instability, and water sensitivity, commercialization of polymers produced from biomass is still limited [23–26]. As a result, various reinforcing materials and chemicals, such as plasticizers (e.g., glycerol, glycol, sorbitol), are combined with edible films and coatings to enhance their quality [27–31]).

In most circumstances, one of the above-mentioned biomaterials can be used alone or in combination with other biopolymers to create an edible coating or film material [16, 17].

PREPARATION OF EDIBLE COATINGS AND FILMS

Edible coatings and films can be liquid, semi-solid or solid matrix that is wrapped around the surface of a food product and can be used as main packaging without providing any sensory or nutritional benefits. These are intended to be tasteless, colorless, non-toxic and unaffected by the sensory properties of the food product [32]. Trinetta [33] reported that when edible coatings and films are applied to cold or hot beverages, edible film dissolves and releases its contents, giving customers comfort portion control, and the elimination of solid waste. Recently, consumer awareness of edible, biodegradable, and environment friendly packaging materials has grown; edible films and coatings are increasingly being employed in the food processing industry for a wide range of food products. A comprehensive list of biopolymers used, their properties, functions and processing methods employed for production of edible coatings and films for various food products has been presented in **Table 1**. The coating method chosen has an impact not only on the preservation effect of the coating generated on the food products, but also on the cost of manufacturing and process efficiency. There are two types of edible packagings: (i) edible coatings applied directly to the food products and (ii) premade films wrapped around the food products. Two processes can be used to create edible coatings [54]. Dry-process methods, such as thermoplastic extrusion, rely on the thermoplastic characteristics of polymers when plasticized and heated above their glass-transition temperature in low-water-content circumstances. Extruded films have the drawback of not being able to cover uneven surfaces. On a commercial scale, extrusion and spraying are the predominant procedures for film creation and coating deposition, respectively. On the other hand, at a lab scale, the casting method for film creation and dipping methods for coating deposition, are simple to use and preferred methods.

Table 1. *List of biopolymers used, their properties, functions and processing methods employed for production of edible coatings and films for various food products.*

Bio-polymers	Properties	Functions in edible coatings and films	Food product	Processing methods	References
Starch Cellulose Pectin Gums Chitosan Agar Alginate Dextran	Thickeners Gellants Stabilizers Coatings	They form base structure of a solid polymer matrix. They control physical changes, microbial growth, nutritional qualities and shelf-life.	Mango, Apple, Tomato, Strawberries, Green chilies, Sausage, Water melon, Plums, Bell pepper, Brinjal, Potatoes, Taro corms	Extrusion Solvent casting, co-extrusion, Dipping	[34–40]

Bio-polymers	Properties	Functions in edible coatings and films	Food product	Processing methods	References
Gelatin Pea protein isolate Casein Collagen Blood protein Fish protein Whey protein	Gellants Thickeners Stabilizers Emulsifiers Foaming	They help in transport of antimicrobials and antioxidants. They control transport of gases (mainly O_2)	Soybean oil Cheese slice, Sausage, Meat slice, Soluble coffee, Walnut kernels, Apple, Blood hake, Beef tenderloins, Pork loins, Salmon fillets Chicken breasts, Rice crisp balls	Solvent casting, Extrusion, Panning, Spraying, Dipping, Compression, Injection Molding, Electrospinning, 3D-printing	[5, 41–48]
Waxes Pullulan esters Chocolates Milk butter Oils (Olive oil and sunflower oil) Paraffin Glycerides	Protectors Coatings	They help to avoid drying or dehydration of the edible films and provide flexibility. They show anti-aging effects.	Strawberry, Fresh-cut apples	Solvent casting, Spreading, Dipping	[49, 50]
Additives (Glycerol, Resins, Polyglycerol Polyricinoleate, Lecithin, Ascorbic acid, Bioactive compounds)	Viscosity resistance Flexibility Antioxidants Stabilizers Anti-fungal Herbicides Fertilizers Antimicrobial Antiinflammatory	They decrease the intermolecular force, melting temperature in the mixture and modify viscosity and rheological properties. They increase the solubility of polymers in water and work as stabilizers as well as protection for the products.	Pear, Mango, Quail eggs, Fresh-cut apples	Solvent casting, Spreading, Dipping	[49, 51–53]

CLASSIFICATION OF EDIBLE COATINGS AND FILMS

Edible coatings and edible films are not the same; edible coatings can be applied directly to the surface of fruits, vegetables, and other food products in liquid form, usually by immersing the product in an edible material solution. Whilst edible films are molded as solid sheets and then utilized to wrap and package the food products [55]. The edible coating and film components are biodegradable and non-toxic. Biopolymer matrixes such as polysaccharides, proteins, lipids, and composite materials are used to create edible materials (**Table 1**). These coatings and films are thin layers created as a coating on a food surface or put (pre-made) between food components. Their goal is to enhance the shelf life of the food product while simultaneously acting as a safety barrier. They can retard moisture migration and the loss of volatile chemicals, as well as inhibit respiration and delay textural changes. In addition, as compared to typical synthetic films, they are good fat and oil barriers and have a high selective gas permeability ratio CO_2/O_2 [56]. They can also serve as carriers for food additives like antioxidants [57] and/or antimicrobial agents [10], as well as improve the product's mechanical integrity and handling properties. For some applications, stand-alone edible films with strong mechanical qualities could replace synthetic packaging films. Composition of biopolymers, their concentration, drainage time, viscosity, and other factors influence the mechanical and barrier properties of edible films and coatings [58].

INCORPORATION OF BIOACTIVE COMPOUNDS INTO EDIBLE COATINGS AND FILMS

Bioactive chemicals are generally secondary metabolites of plants that have both nutritional value and other functions in their metabolism, such as growth stimulant and protection against biotic and abiotic stress [59]. They are widely distributed in nature. Fruits and vegetables [11], leaves, blossoms [2], grains, seeds [60], rhizomes and roots [61], of different sorts of plants are important sources of bioactive components including phenols, proteins, essential oils, terpenoids and flavonoids. Plant-determined bioactive compounds are being viewed as fascinating elements for the creation of biodegradable and bioactive films because of their usefulness and natural origin [7]. Plant extracts and fruit pulps as sources of bioactive compounds or isolated bioactive compounds in film-forming solutions have been demonstrated to have antioxidant and antibacterial effects on the subsequent films, prolonging their utility in bioactive and biodegradable films or packaging [2, 3, 24, 62]. Plant derived naturally bioactive compounds are incorporated directly into agro-based polymers or the encapsulated plant-derived bioactive chemicals are integrated into the biopolymers by spraying during the production process [63]. Some edible coatings obtained from plant-derived bioactive compounds along with their methods of incorporation into the coatings, method of development of the edible coatings and films from these phytochemicals and the functional properties of these coatings and films are enlisted in **Table 2**.

Table 2. *List of some edible coatings obtained from plant-derived bioactive compounds, their methods of incorporation, method of development of the edible coatings and films and their functional properties*

Plant-derived bioactive compounds	Methods for incorporation	Methods of development of edible coatings and films	Functional properties	References
Blackberry pulp	Directly by sprinkling into the film-forming solution	Casting	Increased water vapor permeability and solubility in water; Reduced tensile strength; Antioxidant properties	[8, 12]
Curcumin	Directly into the film forming solution	Casting	Antioxidant and antimicrobial activity	[64]
Cranberry extract	Directly into the film forming solution	Casting	Antioxidant	[65]
Vitamin E (α-tocopherol encapsulated in carboxy methyl cellulose)	Directly into the film forming solution	Solution casting	Antioxidant;	
Decreased water vapor permeability and tensile strength	[66]			
Tea Polyphenols	Directly into the film forming solution	Casting	Antioxidant and antimicrobial properties;	
Improved water barrier properties and tensile strength	[67, 68]			

Plant-derived bioactive compounds	Methods for incorporation	Methods of development of edible coatings and films	Functional properties	References
Babassu	Producing the film using inherently bioactive biopolymeric materials from agricultural by-products	Casting	Antioxidant	[15]
Essential oils	Directly into the film-forming solution	Casting	Improved barrier properties	[2]

PROPERTIES OF EDIBLE COATINGS AND FILMS FROM AGRO-WASTE BASED POLYMERS

In the food processing industry, edible coating/film provides a consistent quality for food items with market safety, nutritional value, and low manufacturing costs [69]. Control of mass transfers, mechanical protection, and sensory appeal are the most significant functions of an edible film or coating. Preventing desiccation of foods, regulating microenvironments of gases around foods, and controlling migration of ingredients and additives in food systems are all part of mass transfer control. Edible coatings on fresh foods can reduce quality changes and quantity losses by modifying and controlling the internal atmosphere of particular foods, which is an alternative to modified atmosphere storage. Even while penetration of oxygen may degrade food quality due to oxidation of the fragrance components in the food, modification of the internal atmosphere by the application of edible coatings can aggravate disorders associated with high carbon dioxide or low oxygen concentration. For fresh items, edible films with high water vapor permeability is also desired to extend the shelf life, while extremely high water vapor permeability is not, as it might cause significant moisture loss in the fruits during storage. The mechanical strength of an edible film must be sufficient to safeguard the packaging's integrity during distribution. The sensory qualities of an edible coating or film are critical aspects in final product approval.

The selection of ingredients is one of the most crucial aspects of making edible films. Polysaccharide-based edible coatings and films, a type of natural macromolecule with a high bioactivity, are often generated from agricultural feedstock or crustacean shell wastes. Polysaccharides that may form gels in water are found all over the plant kingdom. Some of them have been thoroughly explored, such as pectins in higher plants, carrageenans and agarose in algae, algal and bacterial alginates, and xanthans. Mucilages are heteropolysaccharides derived from plant stems, such as cactus stems. Food, cosmetics, pharmaceuticals, and other businesses may benefit from cactus mucilage [70]. The complex polysaccharide is a type of dietary fiber that can absorb adequate water before dissolving and dispersing and generating viscous or gelatinous colloids. The low cost of cactus mucilage as a coating is an important desirable consideration.

On the other hand, proteins such as casein, whey proteins, and maize zein, have been employed as a moisture barrier in edible coatings since they are numerous, abundant, inexpensive, and readily available. Corn zein, soy protein, wheat proteins (gluten, gliadin), peanut protein, gelatin, casein, and milk whey proteins have all been used in the production of protein-based films [5]. Food protein may act as natural vehicles, adapted to carry vital micronutrients (e.g. calcium and phosphate), building blocks (e.g. amino acids), as well as immune system components (e.g. immunoglobulins and lactoferrin) [71]. Furthermore, food proteins can be employed in coating formulations to create environment friendly packagings that are easily degradable and can be applied to a variety of foods such as vegetables, fruits, poultry, and fish items [72]. Protein-based edible films are appealing because, when compared to lipidand polysaccharide-based edible films, they have superior gas barrier properties. Protein films' poor water vapor resistance and lower mechanical strength when compared to synthetic polymers, however, limit their use in food packaging.

Further, despite multifunctional potential of polysaccharides and proteins, the hydrophilic nature of these biopolymers limits their capacity to offer the desired edible film capabilities. Incorporation of hydrophobic chemicals, adjustment of polymer interaction, and production of cross-links are some of the ways to increase the water barrier and mechanical properties of these films. Lipid molecules used in edible coatings include neutral lipids of glycerides, which are esters of glycerol and fatty acids, and waxes, which are esters of long-chain monohydric alcohols and fatty acids. They are used to give hydrophobicity to food coatings [73]. Although protein films have good oxygen barrier and mechanical properties at low and intermediate relative humidity due to their large number of polar groups and extensive polymer inter-chain, the hydrophilic nature of proteins prevents edible protein films from acting as an effective water vapor barrier. However, lipid films have limited water vapor permeability due to their hydrophobic nature, but they are extremely brittle because of their monomeric structure. Furthermore, lipids generate opaque films or coatings and are susceptible to oxidation. These characteristics may affect the organoleptic characteristics of food and reduce their marketability. Natural antioxidants and antibacterial agents have also been added to the edible film to help prevent autoxidation of high-fat foods and boost oil resistance in fried foods [74, 75]. During storage, edible materials acts as barrier against moisture and gases from fresh produce, slowing enzymatic oxidation and protecting the food from browning and texture softening. These may also have the capacity to preserve natural volatile flavor compounds and prevent color components from discoloration [76]. Edible coatings and films aid to preserve phytochemical (phenolic, antioxidants, color) and physicochemical (total soluble solids, weight loss, pH, and respiration rate) attributes in fresh and minimally processed fruits and vegetables over time [77].

PHYSIOCHEMICAL AND MORPHOLOGICAL CHARACTERIZATION OF EDIBLE COATINGS AND FILMS

The main attributes and techniques for characterization of edible coatings and films are given below:

Physicochemical characterization

- Mechanical properties
- Solubility
- Color and transparency
- Thermal properties
- Microscopy
- Barrier properties

Performance evaluation

- Moisture loss, color, film thickness, microbial test
- Sensory properties
- Barrier properties (O_2 permeability, CO_2 permeability, water vapor permeability)

Morphological characterization

- Atomic force microscopy
- Transmission electron microscopy
- Scanning electron microscopy
- Universal tensile machine
- Fourier transform infrared spectroscopy
- Thermal methods
- Differential scanning calorimetry

- Thermogravimetric analysis
- Differential thermal analysis
- Dynamic mechanical analysis
- X-Ray diffraction
- Nuclear magnetic resonance spectroscopy analysis
- Dynamic light scattering
- Rheological analysis
- Zeta potential analysis

APPLICATION OF EDIBLE COATINGS AND FILMS

Edible coatings and films produced from a variety of biopolymers can effectively preserve the nutritional and organoleptic qualities of various foods (**Figure 2**).

Edible coatings and films are known as eco-friendly packaging materials, as they replace synthetic or plastic packaging materials and lower the post-harvest losses of fruits and vegetables [78–81]. The edible coating extends shelf life [76, 82–84], prevent microbiological contamination [85], minimizes lipid oxidation [86], and lowers their degradation effect [87].

Biopolymers-based edible coatings operate as barrier layers against gas diffusion, fragrance alterations, water migration, and various volatile exchange [88, 89]. Because of their great selective permeability to oxygen and carbon dioxide, polysaccharides have mostly been employed for food wrapping. The majority of these low-cost films are made from cellulose and its derivatives, such as ethers and esters, starch, pectins, and gums, which are used in food preservation. Fresh fruit products such as tomatoes, cherries, fresh beans, strawberries, mangoes, and bananas have all been coated with cellulose-based edible coatings to prevent quality loss. Chitosan is a polysaccharide that is commonly used to prevent post-harvest deterioration in fresh fruits and vegetables. Chitosan is made up of chitin, which is found in nature just next to cellulose in quantity [90]. Tahir et al. [91] investigated the efficiency of a gum arabic edible coating for increasing total antioxidant content in strawberry fruits during cold storage, with an increase in anthocyanin and phenolic contents. Furthermore, edible coatings containing chitosan and essential oils of oregano or thyme can inhibit the growth of spoilage and pathogen microorganisms while also improving the sensory quality of peeled shrimp [92]. Edible coatings are good transporter of functional ingredients and additives such color, flavor, minerals, vitamins, and antioxidant agents to improve the nutritional value, durability, and functionality of foods [77]. Ebrahimi and Rastegar [93] reported that a guar-based edible coating coated with *A. vera* and Spirulina platensis was effective in preserving the ascorbic acid content, total phenol, and antioxidant activity of mango maintained at room temperature. In addition, the coating extended the shelf life of mango fruit by reducing respiration and weight loss. Active films offer promising approach for slowly delivering the functional additives to the food surface, which could help to prevent food spoilage. Active packaging – a novel offshoot of the family of edible films-is quite encouraging as it can be carrier for a wide range of food additives, such as vitamins, antioxidants, minerals, colorants, fragrances, and antibacterial agents for the packaged food products [94].

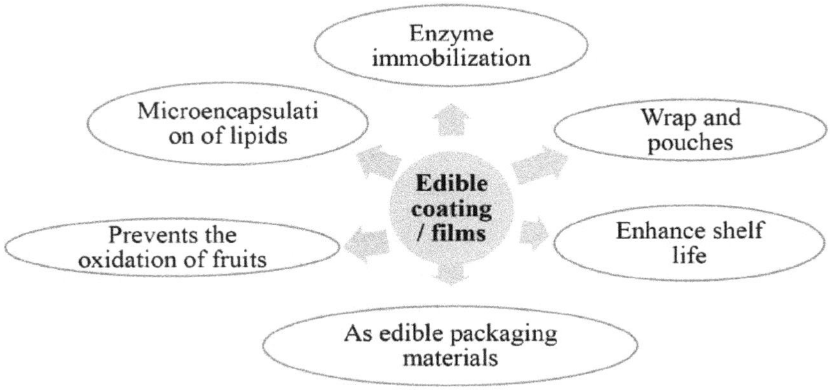

Figure 2. *Application of edible coating and films.*

CONCLUSIONS AND FUTURE PROSPECTS

Starch, pectin, collagen, sodium alginate, gelatin, chitin, whey protein, chitosan, soy protein, gluten proteins and lipids are all examples of agro-industrial leftovers based biopolymers that have been widely used in the manufacturing of environment friendly food coatings and films. However, limited mechanical strengths and moisture barrier properties of most biopolymers-based edible coatings and films are the notable drawbacks. Combining agroindustrial leftovers based biopolymers with plant-derived bioactive substances (vitamins, carotenoids, phenolic compounds and phytochemicals, among others) permit the formation of bioactive films with antioxidants, antibacterial action. It is feasible to improve the mechanical and moisture barrier and physical properties of films by combining proteins (e.g., milk proteins, soy protein, collagen, and gelatin) with polysaccharides (e.g., starches, alginates, cellulose, and chitosan) or other polymers and hydrophobic compounds (lipids). Crosslinking procedures, on the other hand, could be a fascinating process that take into consideration chemical, enzymatic, and physical processes to produce biodegradable packaging materials with improved qualities from agroindustrial wastes.

ABSTRACT

Mostly, food packaging employs synthetic materials obtained from nonrenewable sources. These packaging materials are based on petrochemicals and cause substantial environmental problems by producing massive amounts of nonbiodegradable solid wastes. Edible coatings and films are considered as the potential solution to these problems of non-biodegradable packaging solid wastes for maintaining food-environment interactions, retaining food quality, and extending shelf life. In addition, edible coatings and films offer prevention from microbial spoilage of packed foods by controlling moisture and gas barrier characteristics. Increasing environmental concerns and consumer demands for high-quality eco-friendly packaging have fueled the advancement of innovative packaging technologies, for instance, the development of biodegradable films from renewable agricultural and food processing industry wastes. Therefore, the current chapter presents the application of edible coatings and films as an alternative to conventional packaging, emphasizing the fundamental characterization that these biodegradable packaging should hold for specific applications such as food preservation and shelf life enhancement. The primary employed components (e.g., biopolymers, bioactive, and additives components), manufacturing processes (for edible films or coatings), and their application to specific foods have all been given special consideration in this chapter. Besides, a future vision for the use of edible films and coatings as quality indicators for perishable foods is presented.

REFERENCES

1. Zhao G, Lyu X, Lee J, Cui X, Chen WN. Biodegradable and transparent cellulose film prepared eco-friendly from durian rind for packaging application. Food Packaging and Shelf Life. 2019; 21:100345.
2. Mahcene Z, Khelil A, Hasni S, Akman PK, Bozkurt F, Birech K, Goudjil MB, Tornuk F. Development and characterization of sodium alginate based active edible films incorporated with essential oils of some medicinal plants. International Journal of Biological Macromolecules. 2020; 145:124-132.
3. Zahan KA, Azizul NM, Mustapha M, Tong WY, Rahman AMS, Sahuri IS. Application of bacterial cellulose film as a biodegradable and antimicrobial packaging material. Materials Today: Proceedings. 2020; 31:83-88.
4. Gouveia TIA, Biernacki K, Castro MCR, Gonçalves MP, Souza HKS. A new approach to develop biodegradable films based on thermoplastic pectin. Food Hydrocolloids. 2019; 97:105175.

5. Álvarez-Castillo E, Felix M, Bengoechea C, Guerrero A. (2021). Proteins from Agri-Food Industrial Biowastes or Co-Products and Their Applications as Green Materials. Foods. 2021; 10:981. DOI: 10.3390/ foods10050981

6. Fakhouri FM, Costa D, Yamashita F, Martelli SM, Jesus RC, Alganer K, Collares-Queiroz FP, Innocentini-Mei LH. (2013). Comparative study of processing methods for starch/gelatin films. Carbohydrate Polymers. 2013; 95:681-689.

7. Jafarzadeh S, Jafari SM, Salehabadi A, Nafchi AM, Uthaya Kumar US, Khalil HPSA. (2020). Biodegradable green packaging with antimicrobial functions based on the bioactive compounds from tropical plants and their by-products. Trends in Food Science and Technology. 2020. 100:262-277.

8. Nogueira GF, Soares CT, Cavasini R, Fakhouri FM, de Oliveira RA. Bioactive films of arrowroot starch and blackberry pulp: Physical, mechanical and barrier properties and stability to pH and sterilization. Food Chemistry. 2019(a); 275:417-425.

9. Riaz A, Lagnika C, Luo H, Dai Z, Nie M, Hashim MM, Liu C, Song J, Li D. (2020) Chitosan-based biodegradable active food packaging film containing Chinese chive (*Allium tuberosum*) root extract for food application. International Journal of Biological Macromolecules. 2020; 150:595-604.

10. Fu Y, Dudley EG. (2021). Antimicrobial-coated films as food packaging: A review. Comprehensive Review in Food Science and Food Safety. 2021; 1-34. DOI: 10.1111/1541-4337.12769

11. Brito TB, Carrajola JF, Gonçalves ECBA, Martelli-Tosi M, Ferreira MSL. Fruit and vegetable residues flours with different granulometry range as raw material for pectin-enriched biodegradable film preparation. Food Research International. 2019; 121:412-421.

12. Nogueira GF, Fakhouri FM, de Oliveira RA. Incorporation of spray dried and freeze dried blackberry particles in edible films: Morphology, stability to pH, sterilization and biodegradation. Food Packaging and Shelf Life. 2019(b); 20:100313.

13. Zarandona I, Barba C, Guerrero P, de la Caba K, Maté J. Development of chitosan films containing-cyclodextrin inclusion complex for controlled release of bioactives. Food Hydrocolloids. 2020; 104:105720

14. Maniglia BC, Domingos JR, de Paula RL, Tapia-Blácido DR. (2014). Development of bioactive edible film from turmeric dye solvent extraction residue. LWT:Food Science and Technology. 2014; 56, 269-277.

15. Maniglia BC, Tessaro L, Lucas AA, Tapia-Blácido DR. Bioactive films based on babassu mesocarp flour and starch. Food Hydrocolloids. 2017; 70:383-391.

16. Shit SC, Shah PM. Edible polymers: Challenges and opportunities. Journal of Polymers. 2014; 427259:1-13.

17. Pop OL, Pop CR, Dufrechou M, Vodnar DC, Socaci SA, Dulf FV, et al. Edible films and coatings functionalization by probiotic incorporation: A Review. Polymers. 2020; 12(12):1-15.

18. Mohee R, Unmar GD, Mudhoo A, Khadoo P. Biodegradability of biodegradable/ degradable plastic materials under aerobic and anaerobic conditions. Waste Management. 2008; 28:1624-1629.

19. Giosafatto CVL, Antonio Fusco A, Al-Asmar A, Mariniello L. (2020). Microbial Transglutaminase as a Tool to Improve the Features of HydrocolloidBased Bioplastics. International Journal of Molecular Science. 2020; 21:3656. DOI:10.3390/ijms21103656

20. DeGruson ML. Biobased polymer packaging. In Reference Module in Food Science; Smithers, G., Ed.; Elsevier Inc.: Amsterdam, The Netherlands. 2016.

21. Gruji´c R, Vujadinovi´c D, Savanovi´c D. Biopolymers as food packaging materials. In Advances in Applications of Industrial Biomaterials; Pellicer E, Nikolic D, Sort J, Baró M, Zivic F, Grujovic N, Gruji´c R, Pelemis S. Eds.; Springer International Publishing: Cham, Switzerland, 2017; 139-160. ISBN 9783319627663.

22. Pellicer E, Nikolic D, Sort J, Baró MD, Zivic F, Grujovic N, Gruji´c R, Pelemis S. Advances in Applications of Industrial Biomaterials, 1st ed.; Springer International Publishing: New York, NY, USA, 2017; ISBN 9783319627670.

23. Jabeen N, Majid I, Nayik GA. (2015). Bioplastics and food packaging: A review. Cogent Food and Agriculture. 2015; 1:1117749.

24. Chen H, Wang J, Cheng Y, Wang C, Liu H, Bian H, ..., Han W. (2019). Application of protein-based films and coatings for food packaging: A review. Polymers. 2019; 11(12):2039.

25. Milani JM, Tirgarian B. (2020). An overview of edible protein-based packaging: Main sources, advantages, drawbacks, recent progressions and food applications. Journal of Packaging Technology and Research. 2020; 4:103-115.

26. Zhong Y, Godwin P, Jin Y, Xiao H. Biodegradable polymers and greenbased antimicrobial packaging materials: A mini-review. Advanced Industrial and Engineering Polymer Research. 2020; 3:27-35.

27. Borges JA, Romani VP, Cortez-Vega WR, Martins VG. Influence of different starch sources and plasticizers on properties of biodegradable films. International Food Research Journal. 2015; 22(6).

28. Jeyasubramanian K, Balachander R. Starch bioplastic film as an alternative food-packaging material. Journal of Achievements in Materials and Manufacturing Engineering. 2016; 75:78-84.

29. Shendurse A. Milk protein based edible films and coatings—Preparation, properties and food applications. Journal of Nutritional Health and Food Engineering. 2018; 8:219-226.

30. Zakaria NH, Muhammad N, Sandu AV, Abdullah MMAB. Effect of mixing temperature on characteristics of thermoplastic potato starch film. IOP Conference Series: Materials Science and Engineering. 2018; 374:012083.

31. Rahman R. Bioplastics for food packaging: A review. International Journal of Current Microbiology and Applied Sciences. 2019; 8:2311-2321.

32. Viana RM, Sá NMSM, Barros MO, Borges MDF, Azeredo HMC. Nanofibrillated bacterial cellulose and pectin edible films added with fruit purees. Carbohydrate Polymers, 2018; 196:27-32.

33. Trinetta V. Edible packaging. Reference Module in Food Sciences, 2016; 1-4.

34. Aly SS, Mohamed EN, Abdou ES. Research Article Effect of Edible Coating on Extending the Shelf Life and Quality of Fresh Cut Taro. American Journal of Food Technology. 2017; 12:24-131.

35. Calderón-Castro A, Vega-García MO, de Jesús Zazueta-Morales J, Fitch-Vargas PR, Carrillo-López A, Gutiérrez-Dorado R, ..., Aguilar-Palazuelos E. (2018). Effect of extrusion process on the functional properties of high amylose corn starch edible films and its application in mango (*Mangifera indica* L.) cv. Tommy Atkins. Journal of Food Science and Technology. 2018; 55(3): 905-914.

36. Licciardello F, Lombardo S, Rizzo V, Pitino I, Pandino G, Strano MG, Muratore G, Restuccia C, Mauromicale G. Integrated agronomical and technological approach for the quality maintenance of ready-to-fry potato sticks during refrigerated storage. Postharvest Biology and Technology. 2018; 136:23-30.

37. Fitch-Vargas, PR, Aguilar-Palazuelos E, Vega-García MO, Zazueta-Morales JJ, Calderón-Castro A, Montoya-Rodríguez A, et al. Effect of a corn starch coating obtained by the combination of extrusion process and casting technique on the postharvest quality of tomato. Revista Mexicana de Ingeniera Quimica, 2019. 18(3):789-801.

38. Zhang W, Li X, Jiang W. Development of antioxidant chitosan film with banana peels extract and its application as coating in maintaining the storage quality of apple. International Journal of Biological Macromolecules. 2019.

39. Hilbig J, Hartlieb K, Herrmann K, Weiss J, Gibis M. Influence of calcium on white efflorescence formation on dry fermented sausages with co-extruded alginate casings. Food Research International. 2020; 131(109012):1-8.

40. Abral H, Pratama AB, Handayani D, Mahardika M, Aminah I, Sandrawati N, ..., Ilyas RA. Antimicrobial Edible Film Prepared from Bacterial Cellulose Nanofibers/Starch/Chitosan for a Food Packaging Alternative. International Journal of Polymer Science, 2021. DOI: 10.1155/2021/6641284

41. Antoniewski MN, Barringer SA, Knipe CL, Zerby HN. Effect of a gelatin coating on the shelf life of fresh meat. Journal of Food Science. 2007; 72(6):382-387.

42. Chiu Po-en, Lih-shiuh L. Antimicrobial Activities of Tapioca Starch / Decolorized Hsian-Tsao Leaf Gum Coatings Containing Green Tea Extracts in Fruit-Based Salads, Romaine Hearts and Pork Slices. International Journal of Food Microbiology. 2010; 139:23-30.

43. Carrión-Granda X, Fernández-Pan I, Rovira J, Maté J I. (2018). Effect of antimicrobial edible coatings and modified atmosphere packaging on the microbiological quality of cold stored hake (merluccius merluccius) fillets. Journal of Food Quality2018; 1-12.

44. Fadini AL, Rocha FS, Alvim ID, Sadahira MS, Queiroz MB, Alves RMV, et al. (2013). Mechanical Properties and Water Vapour Permeability of Hydrolysed Collagen-Cocoa Butter Edible Films Plasticised with Sucrose. Food Hydrocolloids. 30:625-631.

45. Feng Z, Wu G, Liu C, Li D, Jiang B, Zhang X. Edible coating based on whey protein isolate nanofibrils for antioxidation and inhibition of product browning. Food Hydrocolloids, 2018; 79:179-188.

46. Grosso, AL, Asensio CM, Nepote V, Grosso NR. Quality preservation of walnut kernels using edible coatings. Grasas Y Aceites. 2018; 69(4):1-9.

47. Puscaselu R, Gutt G, Amariei S. Biopolymer-based films enriched with *Stevia rebaudiana* used for the development of edible and soluble packaging. Coatings. 2019; 9(6).

48. Huntrakul K, Yoksan R, Sane A, Harnkarnsujarit N (2020). Effects of pea protein on properties of cassava starch edible films produced by blownfilm extrusion for oil packaging. Food Packaging and Shelf Life. 2020; 24.

49. Khan, MKI, Cakmak H, Tavman S, Schutyser M, Schroën K. Anti-browning and barrier properties of edible coatings prepared with electrospraying. Innovative Food Science and Emerging Technologies. 2014; 25:9-13.

50. Niu B, Shao P, Chen H, Sun P. Structural and physiochemical characterization of novel hydrophobic packaging films based on pullulan derivatives for fruits preservation. Carbohydrate Polymers, 2019; 208:276-284.

51. Dubey PK, Shukla RN, Srivastava G, Mishra AA, Pandey, A. Study on Quality Parameters and Storage Stability of Mango Coated with Developed Nanocomposite Edible Film. International Journal of Current Microbiology and Applied Sciences. 2019; 8(4):2899-2935.

52. Yang K, Dang H, Liu L, Hu X, Li X, Ma Z, et al. (2019). Effect of syringic acid incorporation on the physical, mechanical, structural and antibacterial properties of chitosan film for quail eggs preservation. International Journal of Biological Macromolecules. 2019; 141:876-884.

53. Rodríguez GM, Sibaja JC, Espitia PJP, Otoni CG. (2020). Antioxidant active packaging based on papaya edible films incorporated with Moringa oleifera and ascorbic acid for food preservation. Food Hydrocolloids, 2020; 103

54. Antonio GC, Alves DG, Azoubel PM, Murr FEX, Park KJ. Influence of osmotic dehydration and high temperature short time processes on dried sweet potato (*Ipomoea batatas Lam.*). Journal of Food Engineering. 2008; 84(3):375-382.

55. Aguirre-Joya JA, De Leon-Zapata MA, Alvarez-Perez OB, Torres-León C, Nieto-Oropeza DE, Ventura-Sobrevilla JM, ..., Aguilar CN. (2018). Basic and applied concepts of edible packaging for foods. In *Food packaging and preservation*. Academic Press. 2018; pp:1-61. DOI: 10.3390/foods10050981

56. Falguera V, Quintero JP, Jimenez A, Munoz JA, Ibarz A (2011) Edible films and coatings: structures, active functions and trends in their use. Trends in Food Science and Technology 22(6): 292-303.

57. Song Y, Liu L, Shen H, You J, Luo Y. Effect of sodium alginate-based edible coatings containing different antioxidants on quality and shelf life of refrigerated bream (*Megalobrama amblycephala*). Food Control 2011; 22(3-4): 608-615.

58. Lin D, Zhao Y. Innovations in the development and application of edible coatings for fresh and minimally processed fruits and vegetables. Comprehensive Reviews in Food Science and Food Safety, 2007; 6(3):60-75.

59. Banožić M, Babić J, Jokić S. (2020). Recent advances in extraction of bioactive compounds from tobacco industrial waste-a review. Industrial Crops and Products. 2020; 144:112009.

60. Rodsamran P, Sothornvit R. (2018). Bioactive coconut protein concentrate films incorporated with antioxidant extract of mature coconut water. Food Hydrocolloids. 2018; 79:243-252.

61. Schaefer EW, Pavoni JMF, Luchese CL, Faccin DJL, Tessaro IC. Influence of turmeric incorporation on physicochemical, antimicrobial and mechanical properties of the cornstarch and chitosan films. International Journal of Biological Macromolecules. 2020; 148:342-350.

62. Shirahigue LD, Ceccato-Antonini SR. Agro-industrial wastes as sources of bioactive compounds for food and fermentation industries. Ciência Rural. 2020; 50.

63. Nogueira GF, de Oliveira RA, Velasco JI, Fakhouri FM. Methods of Incorporating Plant-Derived Bioactive Compounds into Films Made with Agro-Based Polymers for Application as Food Packaging: A Brief Review. Polymers. 2020; 12:2518. DOI:10.3390/polym12112518

64. Nieto-Suaza L, Acevedo-Guevara L, Sánchez LT, Pinzón MI, Villa CC. (2019). Characterization of Aloe vera-banana starch composite films reinforced with curcumin-loaded starch nanoparticles. Food Structure. 2019; 22:100131.

65. Wang S, Xia P, Wang S, Liang J, Sun Y, Yue P, Gao X. Packaging films formulated with gelatin and anthocyanins nanocomplexes: Physical properties, antioxidant activity and its application for olive oil protection. Food Hydrocolloids. 2019; 96:617-624.

66. Mirzaei-Mohkam A, Garavand F, Dehnad D, Keramat J, Nasirpour A. Physical, mechanical, thermal and structural characteristics of nanoencapsulated vitamin E loaded carboxymethyl cellulose films. Progress in Organic Coatings. 2020; 138:105383.

67. Farhan A, Hani NM. Active edible films based on semi-refined

68. κ-carrageenan: Antioxidant and color properties and application in chicken breast packaging. Food Packaging and Shelf Life. 2020; 24:100476. Wu H, Lei Y, Zhu R, Zhao M, Lu J, Xiao D, Jiao C, Zhang Z, Shen G, Li S. Preparation and characterization of bioactive edible packaging films based on pomelo peel flours incorporating tea polyphenol. Food Hydrocolloids. 2019; 90:41-49.

69. Bhardwaj A, Alam T, Talwar N. Recent advances in active packaging of agrifood products: A review. Journal of Postharvest Technology, 2019; 7(1):33-62.

70. Del-Valle V, Hernández-Muñoz P, Guarda A, Galotto MJ. Galotto Development of a cactus-mucilage edible coating (*Opuntia ficus indica*) and its application to extend strawberry (*Fragaria ananassa*) shelf-life. Food Chemistry. 2005; 91:751-756.

71. Livney YD. Milk proteins as vehicles for bioactives. Current Opinion in Colloid and Interface Science. 2010; 15(1-2):73-83.

72. Rodriguez-Turienzo L, Cobos A, Moreno V, Caride A, Vieites JM, et al. Whey protein-based coatings on frozen Atlantic salmon (Salmo salar): Influence of the plasticiser and the moment of coating on quality preservation. Food Chemistry. 2011; 128(1):187-194.

73. Aydin F, Kahve HI, Ardic M. Lipid Based Edible Films. Journal of Scientific and Engineering Research. 2017; 4(9):86-92.

74. Daraei GA, Mirzaei HO, Maghsudlo Y, Kashaninejad M, Jafari SM. Production of low fat frenchfries with single and multi-layer hydrocolloid coatings. Journal of Food Science and Technology. 2012; 1(8):660-669.

75. Ganiari S, Choulitoudi E, Oreopoulou V. Edible and active films and coatings as carriers of natural antioxidants for lipid food. Trends in Food Science and Technology, 2017; 68:70-82.

76. Sapper M, Amparo C. Starch-based coatings for preservation of fruits and vegetables. Coatings. 2018; 8(152):2-20.

77. Sharma P, Shehin VP, Kaur N, Vyas P. Application of edible coatings on fresh and minimally processed vegetables: A review. International Journal of Vegetable Science, 2019; 25(3):295-314.

78. Campos-Requena VH, Rivas BL, Pérez MA, Garrido-Miranda KA, Pereira ED. Polymer/clay nanocomposite films as active packaging material: Modeling of antimicrobial release. European Polymer Journal. 2015; 71:461-475.

79. Medina JC, Gutiérrez TJ, Goyanes S, Bernal C, Famá L. Biodegradability and plasticizing effect of yerba mate extract on cassava starch edible films. Carbohydrate Polymers. 2016; 151:150-159.

80. Bashir K, Jan K, Aggarwal M. Thermo-rheological and functional properties of gamma-irradiated whole wheat flour. International Journal of Food Science and Technology. 2017; 52(4):927-935.

81. Moghadam M, Salami M, Mohammadian M, Khodadadi M, Emam-Djomeh Z. Development of antioxidant edible films based on mung bean protein enriched with pomegranate peel. Food Hydrocolloids. 2020. 105735.

82. Grosso AL, Asensio CM, Grosso NR, Nepote V. Increase of walnuts' shelf life using a walnut flour protein-based edible coating. LWT:Food Science and Technology. 2020; 118.

83. Hasan KSM, Ferrentino G, Scampicchio M. Nanoemulsion as advanced edible coatings to preserve the quality of fresh-cut fruits and vegetables: A review. International Journal of Food Science and Technology. 2020; 55:1-10.

84. De Pilli T. Development of a vegetable oil and egg proteins edible film to replace preservatives and primary packaging of sweet baked goods. Food Control. 2020; 114.

85. Bajpai S, Navin-Chand, Lodhi, R. Water sorption properties and antimicrobial action of zinc oxide nano particles loaded sago starch film. Journal of Microbiology, Biotechnology and Food Sciences. 2020; 2(4):2368-2387.

86. Kazemian-Bazkiaee F, Ebrahimi A, Hosseini SM, Shojaee-Aliabadi S, Farhoodi M, Rahmatzadeh B, et al. Evaluating the protective effect of edible coatings on lipid oxidation, fatty acid composition, aflatoxins levels of roasted peanut kernels. Journal of Food Measurement and Characterization. 2020; 14(2):1025-1038.

87. Dong C, Wang B, Li F, Zhong Q, Xia X, Kong B. Effects of edible chitosan coating on Harbin red sausage storage stability at room temperature. Meat Science. 2020; 2020159.

88. Sabbah M, Di Pierro P, Dell'Olmo E, Arciello, A, Porta R. Improved shelf-life of Nabulsi cheese wrapped with hydrocolloid films. Food Hydrocolloids. 2019; 96:29-35.

89. Kõrge K, Bajić1 M, Likozar B, Novak U. Active chitosan-chestnut extract films used for packaging and storage of fresh pasta. 2020. DOI: 10.1111/IJFS.14569.

90. Harish, KV, Prashanth, Tharanathan RN. Chitin/chitosan: modifications and their unlimited application potential-an overview. Trends in Food Science and Technology. 2007; 18(3):117-131.

91. Tahir HE, Xiaobo Z, Jiyong S, Mahunu GK, Zhai X, Mariod AA. Quality and postharvest-shelf life of cold-stored strawberry fruit as affected by gum arabic (Acacia senegal) edible coating. Journal of Food Biochemistry. 2018; 42(3):12527.

92. Carrión-Granda X, Fernández-Pan I, Jaime I, Rovira J, Maté JI. Improvement of the microbiological quality of ready-to-eat peeled shrimps (*Penaeus vannamei*) by the use of chitosan coatings.

93. International Journal of Food Microbiology, 2016; 232:144-149. Ebrahimi F, Rastegar S. Preservation of mango fruit with guar-based edible coatings enriched with Spirulina platensis and Aloe-vera extract during storage at ambient temperature. Scientia Horticulturae. 2020; 265.

94. Yong H, Liu J. (2021). Active packaging films and edible coatings based on polyphenol-rich propolis extract: A review. Comprehensive Review in Food Science and Food Safety. 2021; 20:2106-2145.

Section 4

COMPUTATIONAL APPLICATIONS

Chapter 7

COMPUTATIONAL APPLICATIONS FOR THE EVALUATION AND SIMULATION OF THE THERMAL TREATMENT OF CANNED FOODS

INTRODUCTION

Heat treatment is a process of utmost importance to ensure food safety. If the product, a canned food, for example, does not receive an adequate heat treatment, it might cause intoxication and even death of the consumer [1]. The treatment must be designed correctly to guarantee efficient results, reducing the negative impact on the food, caused by the use of high temperatures [2]. For this, in order to optimize the process, it is necessary to know the thermal properties of the food, the kinetics of the changes in its quality, in addition to the quantitative and qualitative characteristics of the microbial load and/or enzymes [3].

To evaluate and simulate a thermal treatment, there are computational tools that use techniques that have been designed to evaluate the time of a process, or its F-value and/or simulate a thermal process.

In order to evaluate a thermal process there are two groups of methods, the Formula Methods and the General Methods. The **Figure 1** shows that there are several Formula methods to evaluate a heat treatment. The classic Formula Methods are those of Charles Olin Ball and Charles Raymond Stumbo, which are called Ball's

Formula Method [4–6], and Stumbo's Formula Method [7–10], respectively. There are other Formula Methods such as the one proposed by Kan-Ichi Hayakawa and the Formula Method developed by Quang Tuan Pham. These Formula Methods are called Hayakawa's Formula Method [11–13], and Pham's Formula Method [9, 13–15].

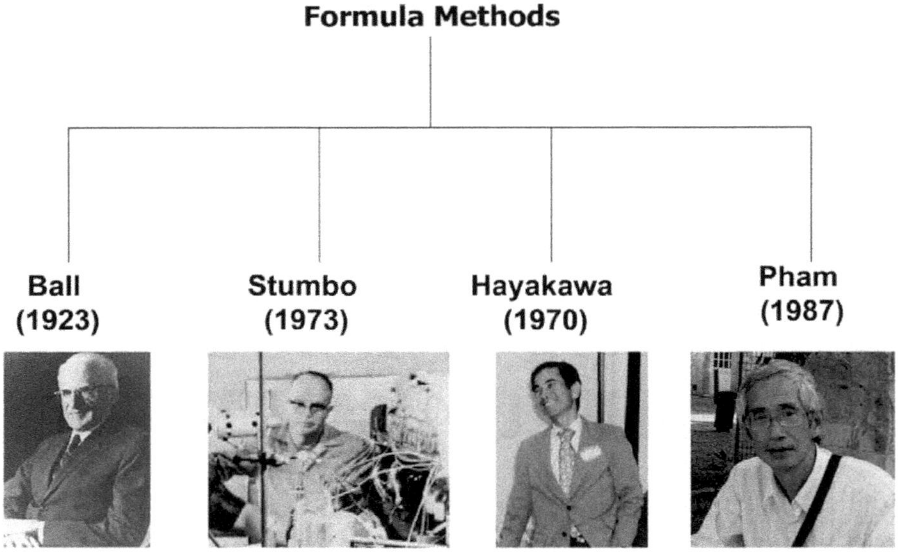

Figure 1. *Various formula methods to evaluate a thermal process of packaged foods.*

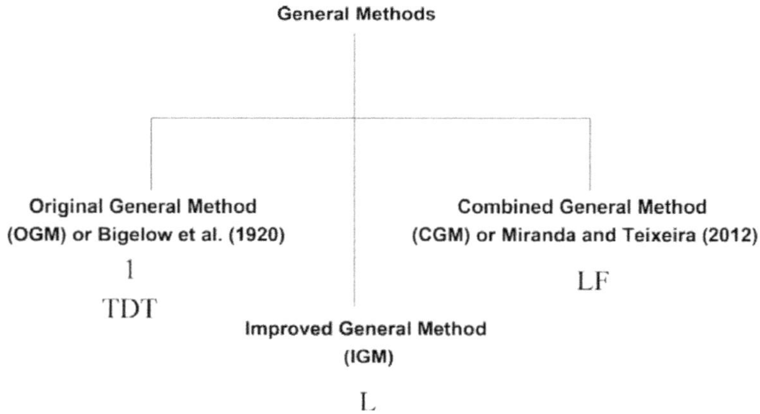

Figure 2. *Various general methods to evaluate a thermal process of packaged foods. TDT = thermal death time, L = lethal rate, and LF = LF-value.*

General methods can also be used to evaluate a packaged food. There are three General Methods (**Figure 2**). The Original General Method (OGM) was plated by Willard Dell Bigelow and his team in 1920 [15, 16]. The Improved General Method (IGM) was proposed by Charles Olin Ball after 1923 [15, 17–20].

The Combined General Method (CGM) was proposed by William R. MirandaZamora, and Arthur A. Teixeira in 2012, as its name says it is a combination of the two previous General Methods [2, 15].

Both General Methods and Formula Methods can be solved using software or computer applications [21–23].

The simulation of the heat treatment of packaged foods can be done using the finite difference method or using the finite element method [24–29].

GENERAL AND TECHNICAL ASPECTS OF THE HEAT TREATMENT OF CANNED FOODS

Knowing the acidity of the food is crucial to determine the severity of the heat treatment. If the food has a pH < 4.6, as in the case of apple nectar, canned mango, pickles, citrus fruit juices, jam, sauerkraut, some sauces, etc.; in this case, the acidic medium will inhibit the proliferation of sporulated bacteria such as *C. botulinum* and, therefore, severe heat treatment (<100°C) will not be required. In contrast, if the food has low acidity (pH > 4.6), such as canned asparagus, milk, among other meat products, seafood and canned vegetables, it is essential to use high temperatures >100°C [8, 14], which guarantee the sterilization of the food, increasing its shelf life. There are particular cases such as, for example, although milk has an almost neutral pH, being susceptible to microbiological spoilage, a mild pasteurization is applied to preserve its nutritional characteristics, in addition to the fact that milk is a product with a short shelf life; however, in many cases a high temperatureshort time (HTST) pasteurization is applied and, being for a few seconds, the impact on the quality of the product is considerably avoided [2, 15].

Thermal treatment is based on two aspects: the biological (or microbiological) and the physical. The biological aspect refers to the microorganism, as the reference microorganism or target microorganism. The important thing is the number of decimal log reductions, n, [30–33], which is defined as:

$$n = \log\left(\frac{N_0}{N}\right) \tag{1}$$

where N_0 = the number of spores of the target microorganism, and N = the number of spores of the target microorganism that remains after heat treatment.

For example, if we start with a load of 1000 spores of *C. sporogenes*, and after heat treatment it is reduced to 1 spore. Therefore, the number of decimal log reductions will be 3 according to Eq. (1). In order to destroy the spores of *C. botulinum*, 12 decimal log reductions are needed [2, 15, 34–36]. The F_0 value is the standard used worldwide to quantify the F-value at a reference temperature. The reference temperature used for commercial sterilization is 121.1°C on the Celsius scale, or 250°F on the Fahrenheit scale. The F_0 value is the minutes that are necessary to evaluate the lethal effect of heat at the reference temperature of 121.1°C = 250°F [2, 15, 16, 21].

Other important values to determine are the D value and the z value, which are determined in the laboratory, using different methods. The D-value and z-value are derived from the thermal death or destruction curves and the thermal resistance curve, respectively [2]. The D-value is defined as the time required to destroy 90% of the initial microbial load, or go through a logarithmic cycle. The smaller the D value the faster the destruction rate [15]. The z-value is defined as the variation in degrees Celsius or degrees Fahrenheit required to reduce 90% of the D-value, or for the D-value to go through a logarithmic cycle [2, 15, 16].

One of the methods to determine the z-value and the decimal reduction time or D-value is the thermoresistometer [37–39]. The D-value and z-value can also be determined to the nutrients by means of a thermoresistometer [40–42]. The D value or decimal reduction time, is related to the number decimal log reductions through Eq. (2). Furthermore, the D-value and the z-value are characteristic of each spore or vegetative cell of the microorganism, or nutrient [43–45].

$$F_T^z = n \cdot \log\left(\frac{N_0}{N}\right) = n \cdot D \tag{2}$$

where F_T^z = F-value that depends on the temperature and the z-value of the microorganism. Generally, for low acid canned preserves a D value of 0.21 minutes and 12 decimal log reductions are taken, which gives a value of 2.52 minutes using Eq. (2). For handling in a food safety and process plant, it takes about F_T^z =3 minutes [2, 15]. The physical aspect has to do with recording the temperature history using temperature sensors [46–50].

Cans and/or packaging have an important role in heat transfer efficiency and, therefore, they should be properly selected, considering their physical, mechanical, thermal and even electrical and optical characteristics. According to Berk [31], the most commonly used material is tinplate, especially because of its low cost. Aluminum is also widely used in the manufacture of cans for alcoholic and non-alcoholic beverages, and although it is more ductile and lighter than tinplate, it has a higher cost [18]. Glass is also used for the packaging of beverages and canned food, being characterized by its impermeability, rigidity, thermal resistance and transparency, obtaining attractive containers; however, they are very fragile and have a high weight [29]. Considering its thermal properties at 20°C, it has a thermal conductivity (k) of 0.75 W m^1 K^1, a specific heat (C_p) of 800 J kg^1 K and a thermal diffusivity (α) of 0.35 x 106 m^2 s^1, with a great difference compared to aluminum, whose k, C_p and α values are 230 W m^1 K^1, 900 J kg^1 K and 95 x 106 m^2 s^1, respectively. It is important to mention that these values depend on other characteristics, such as thickness, which, the higher the thickness, the better the thermal resistance, with a lower heat transfer rate [2, 16]. To avoid environmental impact, thinner materials are used, as in the case of tinplate, but like other metallic materials, they still maintain optimum qualities [34].

HEAT PENETRATION STUDIES OF CANNED FOODS

Heat penetration testing or studies is done by placing containers with temperature sensors in the coldest zone of the autoclave or retort. From the heat penetration tests it is interesting to determine the heat penetration factors f and j [51, 52].

The EVATMi-ZA v 1.0 software includes the determination of the heat penetration parameters f and j [1, 2] (heating, f_h and j_h [53–55] or cooling, f_c and j_c). The heat penetration parameters for heating include the determination of the delay factor $j_{\_CUT}$, based on the CUT value "come-up time" [56]. The come-up time is the time it takes for the autoclave or retort to reach process temperature. The 0.58 CUT is the new origin when using the Charles Olin Ball model or Ball's Formula Method [57–60]. **Figure 3** shows the behavior of steam within a canned food. The trend of temperature versus retort temperature or autoclave temperature is linear on semi logarithmic paper [61–64].

From **Figure 3** we deduce:

$$\frac{1}{f_h} = \frac{\log\left(j_h/j_{h\_CUT}\right)}{0.58CUT} \tag{3}$$

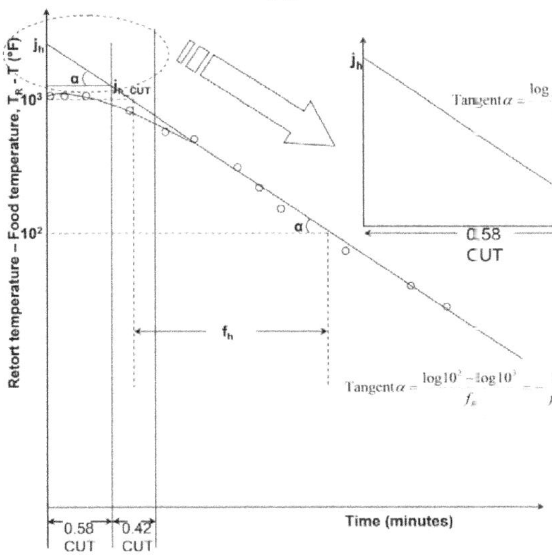

Figure 3. *Simple heating heat penetration curve.*

Therefore, f_h as a function of CUT, j and $j_{\_CUT}$ is:

$$f_h = \frac{0.58CUT}{\log\left(j_h/j_{h\_CUT}\right)} \tag{4}$$

The EVATMi-ZA v 1.0 software includes the determination of the broken curve penetration parameters (f_h, f_{h2}, x_{bh}, x_{bh_CUT}, j_h, j_{h_CUT}). There are many canned foods that exhibit a broken curve [55, 65–70].

4. HEAT TREATMENT EVALUATION GENERAL METHOD

Willard Dell Bigelow in 1920 presented a method for calculating the F-value of packaged foods, which was essentially graphical, it was determined by weighing (using scissors and analytical balance), counting squares (using graph paper) or planimetry (using planimeter) [71–74]. Initially, they constructed a graph in Cartesian coordinates, the curve of rise and fall of the temperature (heat penetration) at the slowest heating point (critical point) of the product during sterilization [75–80]. The thermal resistance of the bacteria was represented by the thermal destruction time curve (TDT-curve) obtained by plotting the time required to destroy a high percentage of spores from a population versus the degree temperature [81–83]. From the TDT-curve, the values of "thermal death or destruction time" (TDT) were calculated for each time of the heat penetration curve. This is known as the Original General Method (OGM).

To use the Original General Method, the Improved General Method and the Combined General Method it is necessary to calculate 1/TDT (min^{-1}), L (lethal rate) and LF (min) respectively. For which the following expressions will be used respectively [5, 15]:

$$\frac{1}{TDT} = \frac{10^{\frac{T-T_{ref}}{z}}}{\left(F^{z}_{T_{ref}}\right)_{Required}} \tag{5}$$

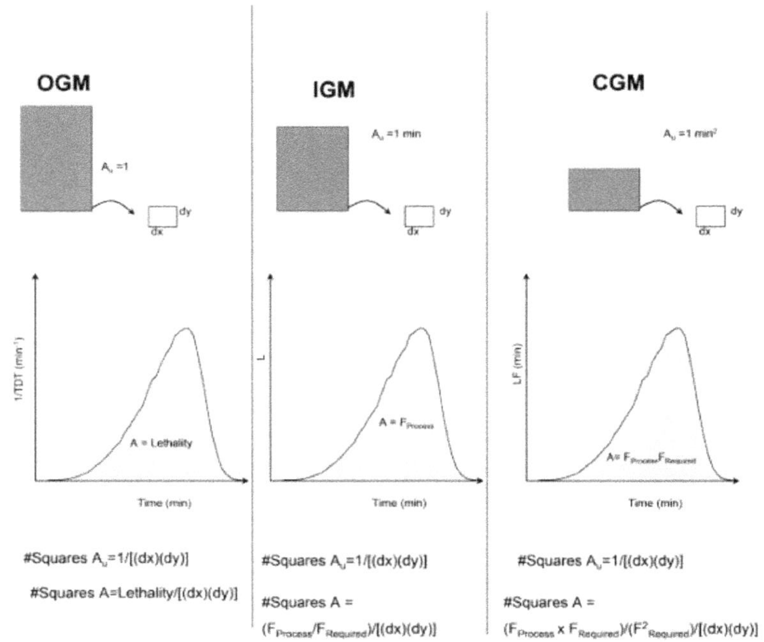

Figure 4. *General methods: Original general method (OGM), improved general method (IGM), and combined general method (CGM) using in a practical way the counting of squares on graph*

paper.

$$L = 10^{\frac{T-T_{ref}}{z}} \tag{6}$$

$$LF = \left(F^{z}_{T_{ref}}\right)_{Required} \times 10^{\frac{T-T_{ref}}{z}} \tag{7}$$

Figure 4 shows how to solve using General Methods using the counting of squares technique in a practical way. Also, it can be solved using numerical techniques such as Simpson's rule, the rectangular rule, or the trapezoidal rule. The EVATMi-ZA v 1.0 software can solve the General methods using numerical techniques (rectangular, trapezoidal and/or Simpson) [15].

CALCULATION AND PREDICTION OF HEAT TREATMENT BY BALL'S METHOD

Charles Olin Ball developed and published his Formula Method in 1923 [84–90]. Olin Ball had already participated in research with Willard Dell Bigelow in 1920 [86]. The success of the Ball's Formula Method is that it appears in a Bulletin that goes directly to the canning industry [91]. Ball uses a hyperbola to represent the curvilinear part at the beginning of cooling. Ball does not use a hyperbolic function [91–93]. Ball's Formula Method is the favorite of the food industry. Ball's Formula Method is ninety-eight years old, it has crossed the threshold of time, although its handling is not well understood [91, 93]. Ball's Formula Method uses the heat penetration factors of heating. The Formula Method has two variants, simple curve and broken heating curve [94–96]. Formula Methods, such as Ball's Formula Method allow you to predict the F-value of the process, or calculate the process time. There are two cases to solve, or the time, or the F value. The Formula Methods are preferred to simulate or predict the process time, knowing the heat penetration parameters [97, 98]. The EVATMi-ZA v 1.0 software can solve time or F-value cases using Ball's Formula Method [99, 100].

HEAT TREATMENT MODELING AND SIMULATION

Although heat treatments are of utmost importance due to their role in food preservation and also because they confer specific characteristics to food [101], the heat used during the process causes the degradation of nutrients, significantly affecting their quality; therefore, it is essential to minimize such damage, but without affecting the desired sterility of the products. For this purpose, it is necessary to develop a heat treatment that involves precise operating conditions in terms of temperature and time, achieving a minimal but efficient process [102], which guarantees optimum food quality [103]. Although currently it is still a challenge, thanks to technological advances, there are mathematical modeling and computer simulation techniques [104] that allow predicting the quality of food during processing and storage, in addition to providing the opportunity to optimize the process [105]. To achieve an ideal design, it is essential to have knowledge of the transport phenomena [106], such as mass, heat (by conduction, convection and radiation) and momentum, involved in the process [107], principles such as, for example, for solid foods, heat and moisture transfer, which are generally modeled by Fourier's and Fick's law, respectively [108] and for fluids, the continuity and Navier–Stokes Equations [109]. Likewise, the other physical mechanisms involved in the respective thermal process must be understood in order to achieve better results.

As it is known, preventing the deterioration of the organoleptic and nutritional characteristics of food [110], improves food safety, an important objective in the industry [108] and which is difficult to fulfill, due to the complex and dynamic nature of food [111] and thermal processing, requiring knowledge not only in engineering, but also in chemistry and microbiology [103]; therefore, as mentioned, numerical solutions are required [101] that, unlike traditional analytics, allow a fast and intelligent management of the large database obtained [112], helping to estimate the behavior of the food during thermal processing. It is necessary to mention that the pioneering study on the subject was by Datta & Teixeira [113], who performed a numerical modeling of the natural convection heating process of a liquid food packaged in cylindrical cans, successfully predicting the thermal (TP) and velocity (VP) profiles [114, 115].

By predicting and optimizing the process, modeling also helps to reduce time and costs, due to the reduction of experiments [116], which are very high under normal conditions [117], having as a disadvantage the obtaining of results in long periods [118]. Basically, one could, for example, after determining the effect of pasteurization temperature on the microbial load in different areas of a food, generate a mathematical model to help predict what the microbial spoilage would be if the processing temperature increases, without neglecting its influence on food quality. Likewise, a correct modeling enhanced with simulation, it would be possible to experiment with different changes in the variables, besides acquiring other advantages such as having control, and a broad and concise vision of the process [119].

For the execution of the computational techniques, first of all, the transfer phenomena that govern the thermal process, mainly heat, are represented by partial differential equations (PDE) [120] to be subsequently converted into a discrete model [107], which will be solved with some numerical method, such as finite differences, finite element and finite volume [108] or also called computational fluid dynamics (CFD) which is the most used [121], since it allows effectively designing a process or optimizing an existing one [122], through the development of three-dimensional models of the system and a numerical solution that describes it with high accuracy and realism. Similarly, it should be noted that, thanks to artificial intelligence, there are other modeling and optimization techniques, such as artificial neural networks and genetic algorithms, which are based on human intelligence and evolution, respectively [103].

CFD has been used since the 1950s, and has developed rapidly up to the present [117], since it is characterized by providing, through numerical algorithms, an easy resolution of the multiple physical phenomena involved in the process [120], which, in addition to heat, mass and momentum transfer (or fluid mechanics), also includes phase changes and chemical reactions [123]. Regarding thermal processing of canned foods, in which thermal processing is more difficult, because the temperature change is affected by the complex characteristics of the product [124], but also, the shape [125], type, size and orientation of the package [126]. CFD has been widely employed to solve challenges such as TP and VP monitoring [109], determination of the slowest heating zone (SHZ), slowest cooling zone (SCZ) [118] and even the kinetics of microbial inactivation and nutrient degradation. Specifically, in addition to sterilization [119] and pasteurization [127], it has covered a wide range of thermal treatments such as cooking [101] baking, drying [103] and cooling [128], and there is even information on its use in non-thermal processes such as microwave heating, ohmic heating, among others, which confirms its feasibility, versatility and suitability for food processing.

Since the since the development of CFD, there have been several commercial software, having in the early 1970s, a great boom and continuous improvement until today [127], emphasizing greater ease of use [129]. Of the extensive list of computer codes, some are FIDAP [116], CFX, FLUENT, PHOENICS [10], ANSYS, ANDINA-F, CFD++ [120], FLOW-3D, STAR-CD, CFD-ACE+ [128] and MSC Marc [105], of which most are still operational or have had some changes; for example, FLUENT and CFX are currently owned by ANSYS inc [129, 130], a leading developer of advanced engineering software. ANSYS offers several types of analysis and concerning heat treatments, it includes the three forms of heat transfer, phase change, internal heat source, contact thermal resistance, among other evaluations [131]. It is based on the creation of a geometry, which is divided into a finite number of units to form a computational geometry, then the governing PDEs are discretized, solved with numerical methods [122] and finally, the results are interpreted by the analyst. CFD simulation with ANSYS, has been applied to different packaged products such as in the pasteurization of beer [102] and water [115], in solid–liquid mixtures such as peas in water [132] and carrot-orange soup [133], food models such as waxy corn starch [126] and sucrose solution [134], in potato refrigeration [135], and even in the improvement of equipment processing parameters, such as in those of an industrial meat dryer [136] and a hydrofluidization freezing chamber [137]. There are also studies on the use of other software in solving the equations of energy, mass and moment, to determine the thermal behavior. A research deals with the effect of using 3.5% cornstarch packaging with an immobile can and rotating continuously at 146 rpm, using FIDAP 7.6 as software [138]. Furthermore, the same authors carried out a similar experiment, but evaluating the effect of intermittent axial agitation (0–146 rpm), and two retort temperatures (111 and 131°C) [139]. In another study, LabVIEW 8.5 was used to compare freezing results of guava pulp packed in stacked boxes, buckets, and unstacked drums, 34, 20, and 200 L, respectively [140].

OPTIMIZATION OF HEAT TREATMENT

Until now, process optimization is essential to determine the best parameters that help to obtain the ideal results, in less time and with a significant reduction in costs. Considering the food industry, this task is much more difficult since there are multiple variables that intervene in the quality of the product [141] or process.

Likewise, if it focuses mainly on the thermal processing of packaged foods and its importance in food safety, its optimization and especially its control are a challenge [142], due to the excessive use of high temperatures and for prolonged periods of time, and for Therefore, the improvement of the treatment conditions is essential to maintain the maximum characteristics of the food [143] and, consequently, a better acceptance by the consumer.

Regarding modeling and simulation, optimization, being related, for its application requires computational modeling and prediction techniques and that, due to advances in hardware, software and engineering related to processing thermal, it is becoming easier and faster to find the best solution [144]. In addition, one must have knowledge of heat transfer, quality change and microbial reduction, the three axes on which heat treatment is based.

Some techniques are the parametrization of the control vector, the principle of the continuous minimum, the super-simple optimization, the dynamic optimization, the neural network [145], the genetic algorithms, the simulated annealing of multiple initiation, among others. It should be noted that local optimization techniques have been used for thermal processing, which are the oldest and are hardly used, and global optimization techniques that are becoming increasingly popular [146]. This is due to the fact that, with the traditional ones, only the influence of a factor (independent variable) on a response (dependent variable) could be evaluated, therefore, as a general result of the processing was not obtained, the effect that the other variables had. This is problematic considering the complexity of the food, its dynamics [147] and also the characteristics of the container [145], including all the changes caused during heat treatment. For this, in contrast, global techniques [143], such as TOPSIS (Technique for Order of Preference by Similarity to Ideal Solution), focus on multiobjective optimization. In an investigation, it was applied in pressure pasteurization of green soybean tofu, managing to optimize the process with a thermal denaturation of 85°C and 10 min, and a high pressure homogenization of 80 MPa, for 4 cycles, achieving an increase in hardness, capacity of retention of water, proteins, fat and yield, in 155.7, 34.48, 30.31, 29.11 and 21.42%, respectively [148].

CONCLUSIONS

In this chapter we explore general and technical aspects of the heat treatment of canned foods, which become the basis or foundation. Heat penetration studies are the requirement for evaluation of heat treatment either by the General Method or by the Ball's Formula Method. Finally, in the final part, we review the modeling and simulation of the heat treatment, in order to achieve the optimization of the heat treatment. Thanks to the computational advances, it has been possible to improve the optimization techniques and even more the global ones, which are ideal to face the complexity of the thermal processing of packaged foods; however, there are still certain limitations such as the relative delay in executing the experimental design, due to the number of dependent and independent variables, with their respective levels (values) and replicas (repetitions).

ABSTRACT

Throughout this chapter we will explore the computational applications that can help us in the evaluation, calculation and simulation of the thermal treatment of canned foods. Although some basic principles of microbial death kinetics will be recalled, the course is basically focused on the exploration and use of computational applications to evaluate and simulate the heat treatment of low-acid foods, considering *C. botulinum* as the reference microorganism. I hope that this book chapter will be useful for you and that you will be able to explore all the contents that are planned to be developed: General and technical aspects of the heat treatment of canned foods, heat penetration studies of canned foods, heat treatment evaluation General method, calculation and prediction of heat treatment by Ball's Method, heat treatment modeling and simulation, and optimization of heat treatment.

REFERENCES

1. Ling B, Tang J, Kong F, Mitcham EJ, Wang S. Kinetics of food quality changes during thermal processing: A review. Food and Bioprocess Technology. 2014; 8(2), 343–358. DOI: 10.1007/s11947-014-1398-3

2. Tucker GS, Featherstone S, MirandaZamora WR. Fundamentos del procesamiento térmico de alimentos. Madrid: AMV; 2020. 516 p.

3. van Boekel M, Fogliano V, Pellegrini N, Stanton C, Scholz G, Lalljie S, Somoza V, Knorr D, Jasti PR, Eisenbrand G. A review on the beneficial aspects of food processing. Molecular Nutrition & Food Research. 2010; 54(9), 1215–1247. DOI:10.1002/ mnfr.200900608

4. Stier RF. Technical and quality management of canning. Swainson's Handbook of Technical and Quality Management for the Food Manufacturing Sector. 2019; 505–527. DOI:10.1016/b978-1-78242-2754.00020-4

5. Pillai SD, Shayanfar S. Aseptic packaging of foods and its combination with electron beam processing. Electron Beam Pasteurization and Complementary Food Processing Technologies. 2015; 83–93. DOI:10.1533/ 9781782421085.2.83

6. Hayakawa K, Ball CO. Charts for calculating average temperature of thermally conductive food in a cylindrical can during heat processing. Canadian Institute of Food Technology Journal. 1969; 2(1), 12–19. DOI:10.1016/ s0008-3860(69)74336-7

7. Sablani S., Shayya WH. Computerization of Stumbo's method of thermal process calculations using neural networks. Journal of Food Engineering. 2001; 47(3), 233–240. DOI: 10.1016/s0260-8774(00)00121-7

8. Serment-Moreno V, Welti-Chanes J. Sterilization of foods. Encyclopedia of Food and Health. 2016; 175–180. DOI: 10.1016/b978-0-12-384947-2.00663-2

9. Afaghi M, Ramaswamy, HS, Prasher SO. Thermal process calculations using artificial neural network models. Food Research International. 2001; 34(1), 55–65. DOI: 10.1016/s0963-9969(00)00132-0

10. Mittal GS, Zhang J. Prediction of food thermal process evaluation parameters using neural networks. International Journal of Food Microbiology. 2002; 79(3), 153–159. DOI:10.1016/s0168-1605(02)00109-5

11. Noronha J, Hendrickx M, Van Loey A, Tobback P. New semi-empirical approach to handle time-variable boundary conditions during sterilisation of non-conductive heating foods. Journal of Food Engineering. 1995; 24 (2), 249–268. DOI:10.1016/0260-8774 (94)p2646-m

12. Hayakawa K, Giannoni-Succar EB, Huang F, Zhou L. Use of the empirical temperature response function for modified Duhamel's theorem application. Journal of Food Engineering. 1997; 34(3), 331–353. DOI: 10.1016/s0260-8774(97)00084-8

13. Rattan NS. Heating behavior and quality changes in canned potatoes subjected to agitation processing [thesis]. Montreal, Canada: McGill University; 2012.

14. Stoforos NG. Thermal process design. Food Control. 1995; 6(2), 81–94. DOI:10.1016/0956-7135(95)98911-j

15. Miranda-Zamora WR, Teixeira AA. Principios matemáticos del proceso térmico de alimentos. Madrid: AMV; 2012. 560 p.

16. Heldman DR, Hartel RW. Principles of food processing. Gaithersburg, Maryland: Aspen Publishers, Inc; 1997. 288.

17. Friso D. A new mathematical model for food thermal process prediction. Modelling and Simulation in Engineering. 2013; 1–8. DOI:10.1155/ 2013/569473

18. Simpson R, Almonacid S, Teixeira A. Bigelow's General method revisited: Development of a New Calculation Technique. Journal of Food Science. 2003; 68(4), 1324–1333. DOI: 10.1111/j.1365-2621.2003.tb09646.x

19. Körmendy L, Zsarnóczay G, Mihályi V. A new, modified acid phosphatase assay for determining the extent of heat treatment in canned hams. Food Chemistry. 1992; 44(5), 367–375. DOI:10.1016/0308-8146(92) 90270-c

20. Simpson R, Figueroa I, Teixeira A. Optimum on-line correction of process deviations in batch retorts through simulation. Food Control. 2006; 17(8), 665–675. DOI:10.1016/j. foodcont.2005.06.004

21. Miranda Zamora WR, Sanchez Chero MJ, Sanchez Chero JA. Software for the determination of the time and the F value in the thermal processing of packaged foods using the Modified Ball Method. In: Ahram T., Karwowski W., Vergnano A., Leali F., Taiar R. (eds) Intelligent Human Systems Integration 2020. IHSI 2020. Advances in Intelligent Systems and Computing. 2020; vol 1131. Springer, Cham. DOI: 10.1007/978-3-030-39512-4_78

22. Zamora WRM, Villarreyes SSC, Povis NLL, More LAV, Chero MJS, Panca CMA, Morales MVS. A new mathematical solution for packaged food thermal processing. In: Mrugalska B., Trzcielinski S., Karwowski W., Di Nicolantonio M., Rossi E. (eds) Advances in Manufacturing, Production Management and Process Control. AHFE 2020. Advances in Intelligent Systems and Computing. 2020; vol 1216. Springer, Cham. DOI:10.1007/978-3030-51981-0_49

23. Zamora WRM, Chero MJS, TimanáAlvarez M, Seminario-Morales V, NiñoCarmona C, Leyva N, More LAV, Ticona-Carrizales L, Ygnacio A. Program in Visual Basic Language: A Simplified Procedure for Thermal Treatment Evaluation of Packaged Foods. In: Russo D., Ahram T., Karwowski W., Di Bucchianico G., Taiar R. (eds) Intelligent Human Systems Integration 2021. IHSI 2021. Advances in Intelligent Systems and Computing. 2021; vol 1322. Springer, Cham. DOI:10.1007/978-3-030-680176_71

24. Özişik MN, Orlande HRB, Colaço MJ, Cotta RM. Finite difference methods in heat transfer. 2nd Edition. Boca Raton, FL: CRC Press; 2017. 600.

25. Welt BA, Teixeira AA, Chau KV, Balaban MO, Hintenlang DE. Explicit finite difference methods for heat transfer simulation and thermal process design. Journal of Food Science. 1997; 62 (2), 230–236. DOI:10.1111/j.1365-2621.1997.tb03974.x

26. Gosz MR. Finite Element Method. Boca Raton, FL: CRC Press; 2006. 400

27. Banga JR, Alonso AA, Gallardo JM, Perez-Martin RI. Mathematical modelling and simulation of the thermal processing of anisotropic and nonhomogeneous conduction-heated canned foods: Application to canned tuna. Journal of Food Engineering. 1993; 18(4), 369–387. DOI:10.1016/0260-8774 (93)90053-m

28. Tucker GS. Development and use of numerical techniques for improved thermal process calculations and control. Food Control. 1991; 2(1), 15–19. DOI:10.1016/0956-7135(91)90113-b

29. Teixeira AA, Dixon JR, Zabradnik JW, Zinsmeister GE. Computer optimization of nutrient retention in the thermal processing of conduction heated foods. Food Techology. 1969; 23, 137–142.

30. Sekhon AS, Singh A, Michael M. Short communication: Decimal log reductions of Salmonella Senftenberg 775 W and other Salmonella serovars in nonfat milk and powder. Journal of Dairy Science. 2020; 103, 8, 6894–6899. DOI:10.3168/jds.2019-17844

31. Berk Z. Thermal processing. Food Process Engineering and Technology. 2018; 399–420. DOI:10.1016/b978-012-812018-7.00017-8

32. Sant'Ana AS, Rosenthal A, Massaguer PR. Heat resistance and the effects of continuous pasteurization on the inactivation of Byssochlamys fulva ascospores in clarified apple juice. Journal of Applied Microbiology. 2009; 107(1), 197–209. DOI:10.1111/j.1365-2672.2009.04195.x

33. Rachon G. Survival of pathogens in low moisture foods [thesis]. Whiteknights, United Kingdom: University of Reading; 2017.

34. Tucker G, Featherstone S. Essentials of thermal processing. Second edition. Chichester, UK: Wiley-Blackwell; 2021. 352.

35. Diao MM, André S, Membré J-M. Meta-analysis of D-values of proteolytic *Clostridium botulinum* and its surrogate strain *Clostridium sporogenes* PA 3679. International Journal of Food Microbiology. 2014; 174, 23–30. DOI: 10.1016/j.ijfoodmicro.2013.12.029

36. Rosnes JT, Fernandez PS, Periago PM, Shara T. Microorganisms of relevance in thermally processed foods. In: Valdramidis, V., Van Impe, J.F.M. (Eds.), Progress on Quantitative Approaches of Thermal Food Processing. Nova Science Publishers, New York; 2012. p. 1–37.

37. Condón S, Arrizubieta MJ, Sala FJ. Microbial heat resistance determinations by the multipoint system with the thermoresistometer TRSC Improvement of this methodology. Journal of Microbiological Methods. 1993; 18(4), 357–366. DOI:10.1016/ 0167-7012(93)90017-c

38. André S, Leguerinel I, Palop A, Desriac N, Planchon S, Mafart P. Convergence of Bigelow and Arrhenius models over a wide range of heating temperatures. International Journal of Food Microbiology. 2019; 291, 173–180. DOI:10.1016/j. ijfoodmicro.2018.11.019

39. Garre A, González-Tejedor GA, Aznar A, Fernández PS, Egea JA. Mathematical modelling of the stress resistance induced in Listeria monocytogenes during dynamic, mild heat treatments. Food Microbiology. 2019; 84, 103238. DOI:10.1016/j. fm.2019.06.002

40. Al Fata N, Georgé S, André S, Renard CMGC. Determination of reaction orders for ascorbic acid degradation during sterilization using a new experimental device: The thermoresistometer Mastia ®. LWT-Food Science and Technology. 2017; 85, 487–492. DOI:10.1016/j.lwt.2016.08.043

41. Blasco R, Esteve MJ, Frígola A, Rodrigo M. Ascorbic acid degradation kinetics in mushrooms in a hightemperature short-time process controlled by a thermoresistometer. LWT-Food Science and Technology. 2004; 37(2), 171–175. DOI:10.1016/j. lwt.2003.08.003

42. Al Fata N, Georgé S, Dlalah N, Renard CMGC. Influence of partial pressure of oxygen on ascorbic acid degradation at canning temperature. Innovative Food Science & Emerging Technologies. 2013; 49, 215–221. DOI: 10.1016/j. ifset.2017.11.007

43. Gabriel AA, Ubana MA. Decimal reduction times of Salmonella Typhimurium in guinataang kuhol: An indigenous Filipino dish. LWT-Food Science and Technology. 2007; 40(6), 1108–1111. DOI:10.1016/j. lwt.2006.06.003

44. Membré J-M, Diao M, Thorin C, Cordier G, Zuber F, André S. Risk assessment of proteolytic *Clostridium botulinum* in canned foie gras. International Journal of Food Microbiology. 2015; 210, 62–72. DOI: 10.1016/j.ijfoodmicro.2015.06.002

45. Wei X, Lau SK, Chaves BD, Danao M-GC, Agarwal S, Subbiah J. Effect of water activity on the thermal inactivation kinetics of Salmonella in milk powders. Journal of Dairy Science. 2020; 103, 8, 6904–6917. DOI:10.3168/ jds.2020-18298

46. Mulla R, Dunnill CW. Single material thermocouples from graphite traces: Fabricating extremely simple and low cost thermal sensors. Carbon Trends. 2021; 4, 100077. DOI:10.1016/j. cartre.2021.100077

47. Fryer PJ, Simmons MJH, Cox PW, Mehauden K, Hansriwijit S, Challou F, Bakalis S. Temperature Integrators as tools to validate thermal processes in food manufacturing. Procedia Food Science. 2011; 1, 1272–1277. DOI: 10.1016/j. profoo.2011.09.188

48. Tarzan-Lorente M, Ceravalls J, Bosch J, Cama JMG, Pardo A. Electronic system for controlling the food cooking process. Procedia Chemistry. 2009; 1 (1), 489–492. DOI:10.1016/j. proche.2009.07.122

49. Gil AG, Ochoa González OA, Cardona Sepúlveda LF, Alvarado Torres PN. Venting stage experimental study of food sterilization process in a vertical retort using temperature distribution tests and energy balances. Case Studies in Thermal Engineering. 2020; 100736. DOI:10.1016/j. csite.2020.100736

50. Sullivan JJ. Wireless data loggers to study heat penetration in retorted foods. In-Pack Processed Foods. 2008; 116– 130. DOI:10.1533/9781845694692.2.116

51. Stumbo CR. Thermobacteriology in food processing. Second edition. New York: Academic Press; 1973. 329.

52. Etzel MR, Willmore P, Ingham BH. Heat penetration and thermocouple location in home canning. Food Science & Nutrition. 2014; 3(1), 25–31. DOI: 10.1002/ fsn3.185

53. Ranganathan K, Rangaswamy S, Subramanian V, Shanmugam N. Modelling of drying kinetics and heat penetration studies on carrot. International Journal of Engineering and Technical Research. 2015; 3(5), 371–376.

54. Awuah GB, Khurana A, Weddig LM, Balestrini CG. A comparative study of heat penetration data using remote sensors and needle or rod-in-tube thermocouples. Journal of Food Process Engineering. 2007; 30(4), 458–471. DOI:10.1111/ j.1745-4530.2007.00106.x

55. Smout C, Ávila I, Van Loey AML, Hendrickx, MEG, Silva C. Influence of rotational speed on the statistical variability of heat penetration parameters and on the non-uniformity of lethality in retort processing. Journal of Food Engineering. 2000; 45(2), 93– 102. DOI:10.1016/s0260-8774(00) 00045-5

56. Simpson R, Almonacid S, Nuñez H, Urtubia A, Teixeira AA. Is there a need for the come-up time correction factor in Ball's Formula Method? A critical analysis. Food Engineering Reviews. 2012; 4(2), 107–113. DOI:10.1007/ s12393-012-9049-9

57. Radrigan R. Computer simulation of thermal processing for food. Heat Transfer Phenomena and Applications. 2012; 183–202.DOI:10.5772/51815

58. Dixon WR, Watts EG, King JA, Fu X, Wicker L. Shelf-stable sustainable shrimp thermally processed with reciprocal agitation. Frontiers in Sustainable Food Systems. 2020; 4, 1–12 DOI:10.3389/fsufs.2020.569790

59. Adepoju MA, Omitoyin BO, Mohan CO, Zynudheen AA. Heat penetration attributes of milkfish (Chanos chanos) thermal processed in flexible pouches: a comparative study between steam application and water immersion. Food Science & Nutrition. 2016; 5(3), 521–524. DOI:10.1002/ fsn3.426

60. Berry MR. Prediction of Come-Up time correction factors for batch-type agitating and still retorts and the influence on thermal process calculations. Journal of Food Science. 1983; 48(4), 1293–1299. DOI:10.1111/ j.1365-2621.1983. tb09214.x

61. Datta AK. On the theoretical basis of the asymptotic semilogarithmic heat penetration curves used in food processing. Journal of Food Engineering. 1990; 12(3), 177–190. DOI:10.1016/ 0260-8774(90)90009-w

62. Körmendy I, Körmendy L. Considerations for calculating heat inactivation processes when semilogarithmic thermal inactivation models are non-linear. Journal of Food Engineering. 1997; 34(1), 33–40. DOI: 10.1016/s0260-8774(97)00071-x

63. Körmendy I, Körmendy L, Ferenczy A. Thermal inactivation kinetics of mixed microbial populations. A hypothesis paper. Journal of Food Engineering. 1998; 38(4), 439–453. DOI: 10.1016/s0260-8774(98)00119-8

64. Downing DL. Heat penetration determinations and thermal process calculations. A Complete Course in Canning and Related Processes. 1996; 39–102. DOI:10.1533/9781845696207.39

65. Reynaga W. Estudio del tratamiento térmico de enlatado de pechuga de pollo (*Gallus gallus*) en trozos y desmenuzado [thesis]. La Molina, Lima: Universidad Nacional Agraria La Molina; 2014.

66. Yang WH, Rao MA. Numerical study of parameters affecting broken heating curve. Journal of Food Engineering. 1998; 37(1), 43–61. doi: 10.1016/s0260-8774(98)00070-3

67. Berry MR, Bush RC. Establishing thermal processes for products with broken-heating curves from data taken at other retort and initial temperatures. Journal of Food Science. 1987; 52(4), 958–961. DOI:10.1111/j.1365-2621.1987. tb14252.x

68. Denys S, Noronha J, Stoforos NG, Hendrickx M, Tobback P. Evaluation of process deviations, consisting of drops in rotational speed, during thermal processing of foods in rotary water cascading retorts. Journal of Food Engineering. 1996; 30(3–4), 327–338. DOI:10.1016/s0260-8774(96)00057-x

69. Wiese KL, Wiese, KF. A comparison of numerical techniques to calculate broken line heating factors of a thermal process. Journal of Food Processing and Preservation. 1992; 16(5), 301–312. DOI: 10.1111/j.1745-4549.1992.tb00211.x

70. Llave YA, Hagiwara T, Sakiyama T. Artificial neural network model for prediction of cold spot temperature in retort sterilization of starch-based foods. Journal of Food Engineering. 2012; 109(3), 553–560. DOI:10.1016/j. jfoodeng.2011.10.024

71. Fellows PJ. Heat sterilisation. Food Processing Technology. 2017; 581–622. DOI:10.1016/b978-0-08-1005224.00012-2

72. Kramer A, Twigg BA. Principles and instrumentation for the physical measurement of food quality with special reference to fruit and vegetable products. Advances in Food Research. 1960; 9, 153–220. DOI:10.1016/s0065-2628(08)60276-1

73. Stumbo CR. The General Method. Thermobacteriology in Food Processing. 1973; 143–151. DOI:10.1016/b978-0-12-675352-3.50019-x

74. Biran A. Geometric properties of areas and volumes. Geometry for Naval Architects. 2019; 121–194. DOI:10.1016/ b978-0-08-100328-2.00012-2

75. Rinaldi M, Chiavaro E, Massini R. Real-time estimation of slowest heating point temperature and residual cooking time by coupling multipoint temperature measurement and mathematical modelling: Application to meat cooking automation. Food Control. 2012; 23(2), 412–418. DOI:10.1016/j. foodcont.2011.08.009

76. Huang XJ, Hanzawa T, Sakai N. The characteristics of the slowest heating point in a canned food in oil. Nippon Shokuhin Kogyo Gakkaishi. 1992; 39(1), 1–7. DOI:10.3136/nskkk1962.39.1

77. Hanzawa T, Wang QH, Suzuki M, Sakai N. Numerical analysis of slowest heating or cooling point in a canned food in oil. Journal of Chemical Engineering of Japan. 1998; 31(3), 451–455.DOI:10.1252/jcej.31.451

78. Wang Q-Z, Hanzawa T, Sakai N. Estimating heating time for sterilizing process in a canned liquid foods with particles. Nippon Shokuhin Kagaku Kogaku Kaishi. 1998; 45(11), 676–682. DOI:10.3136/nskkk.45.676

79. Assan MYA, Watanabe H, Mihori T. Temperature distribution at the surface of cans in an industrial scale static retort during saturated steam sterilization. Food Science and Technology Research. 2000; 6(3), 196–200. DOI:10.3136/ fstr.6.196

80. Yasui T, Esselen WB, Fukazawa T, Hashimoto Y. Processing studies on canned corned beef and canned luncheon meat. Agricultural and Biological Chemistry. 1961; 25(8), 632– 636. DOI:10.1271/bbb1961.25.632

81. Hayakawa K. Selective review of research results related to thermal process lethality estimation. Japan Journal of Food Engineering. 2001; 2(2), 47–52. DOI:10.11301/jsfe2000.2.47
82. Yamamoto Y, Ono N, Higashi K, Yoshii H. Studies on growth inhibition of food spoilage microorganisms for low salt foods. Part VIII. Effects of adipic acid on growth and thermal resistance of spores of anaerobic sporeforming bacteria. Nippon Shokuhin Kogyo Gakkaishi. 1989; 36(7), 551–556. DOI: 10.3136/nskkk1962.36.7_551
83. Nakae T, Nakanishi T. Application of steady state conduction system to microbiology Part V. Journal of the Agricultural Chemical Society of Japan. 1967; 41(9), 465–469. DOI:10.1271/ nogeikagaku1924.41.9_465
84. Gould WA. Understanding our past. Fundamentals of Food Processing and Technology. 1997; 9–20. DOI:10.1533/ 9781845696092.9
85. Robertson GL. History of food packaging. Reference Module in Food Science. 2019; 1–49. DOI:10.1016/ b978-0-08-100596-5.22535-3
86. Hayakawa K, Ball CO. A note on theoretical heating curve of a cylindrical can of thermally conductive food. Canadian Institute of Food Technology Journal. 1968; 1(2), 54–60. DOI:10.1016/ s0008-3860(68)74464-0
87. Bermudez A, Martinez A. A state constrained optimal control problem related to the sterilization of canned foods. Automatica. 1994; 30(2), 319– 329. DOI:10.1016/0005-1098(94) 90033-7
88. Heldman DR. Introduction. Food Preservation Process Design. 2011; 1–18. DOI:10.1016/b978-0-12-3724861.00001-4
89. Mitchell EL. A review of aseptic processing. Advances in Food Research. 1988; 1–37. DOI:10.1016/s0065-2628 (08)60284-0
90. Mohamed IO. Computer simulation of food sterilization using an alternating direction implicit finite difference method. Journal of Food Engineering. 2003; 60(3), 301–306. DOI:10.1016/ s0260-8774(03)00051-7
91. Miranda-Zamora WR, Ludeña AL, Tapia DA, Bazán JF. Herramientas computacionales aplicadas a la evaluación de tratamientos térmicos de los alimentos envasados usando el método de Ball. Piura: UNP; 2010. 94.
92. Miranda WR. Note: hyperbolic function or equation of the hyperbola? Research, March 2016, 11. DOI: 10.13140/RG.2.1.1239.5287
93. Miranda-Zamora WR, Heldman DR. Diseño de procesos térmicos y alta presión de alimentos. Madrid: AMV; 2018. 634 p.

94. Miranda-Zamora WR, Tucker GS. Procedimientos del tratamiento térmico de alimentos. Madrid: AMV; 2017. 290 p.

95. Miranda-Zamora WR. Manual de tratamiento térmico y envasado de alimentos. Madrid: AMV; 2017. 622 p.

96. Miranda-Zamora WR, Stoforos NG. Procesamiento térmico de alimentos teoría, práctica y cálculos. Madrid: AMV; 2016. 330 p.

97. Miranda-Zamora WR, Vignolo TG, Leyva NL. Ingeniería del tratamiento térmico de alimentos. Piura: UNP; 2012. 268 p.

98. Hayakawa K. Selective review of research results related to thermal process lethality estimation I. Lethality estimation and heat transfer. Japan Journal of Food Engineering. 2001; 2(1), 1–9. DOI:10.11301/jsfe2000.2.1

99. Ball CO, Olson FCW. Sterilization in food technology theory, practice, and calculations. New York: McGraw-Hill; 1957. 654 p.

100. Ball CO. Mathematical solution of problems on thermal processing of canned food. Berkeley: University of California Press; 1928. 245 p.

101. Trystram G. Modelling of food and food processes. Journal of Food Engineering. 2012;110:269-277. DOI: 10.1016/j.jfoodeng.2011.05.001

102. Augusto PED, Pinheiro TF, Cristianini M. Using Computational Fluid-Dynamics (CFD) for the evaluation of beer pasteurization: effect of orientation of cans. Ciência e Tecnologia de Alimentos. 2010;30(4): 980-986. DOI: 10.1590/S0101-20612010000400022

103. Singh A, Singh AP, Ramaswamy HS. Computational techniques used in heat transfer studies on canned liquid-particulate mixtures. Trends in Food Science & Technology. 2015;43: 83-103. DOI: 10.1016/j. tifs.2015.02.001

104. Lemus-Mondaca RA, VegaGálvez A, Moraga NO. Computational Simulation and Developments Applied to Food Thermal Processing. Food Engineering Reviews. 2011;3:121-135. DOI: 10.1007/s12393-011-9040-x

105. Martins RC. Simple finite volumes and finite elements procedures for food quality and safety simulations. Journal of Food Engineering. 2006;73:327-338. DOI: 10.1016/j.jfoodeng.2005.01.033

106. Ho QT, Carmeliet J, Datta AK, Defraeye T, Delele MA, Herremans E, Opara L, Ramon H, Tijskens E, van der Sman R, Liedekerke PV, Verboven P, Nicolaï, BM. Multiscale modeling in food engineering. Journal of Food Engineering. 2013;114:289-291. DOI: 10.1016/j.jfoodeng.2012.08.019

107. Erdogdu F, Sarghini F, Marra F. Mathematical Modeling for Virtualization in Food Processing. Food Engineering Reviews. 2017;9:295-313. DOI: 10.1007/s12393-017-9161-y

108. Wang L, Sun DW. Recent developments in numerical modelling of heating and cooling processes in the food industry—a review. Trends in Food Science & Technology. 2003; 14: 408-423. DOI: 10.1016/S0924-2244(03) 00151-1

109. Augusto PED, Cristianini M. Numerical Simulation of Packed Liquid Food Thermal Process Using Computational Fluid Dynamics (CFD). International Journal of Food Engineering. 2011;7(4):16. DOI: 10.2202/1556-3758.2418′

110. Norton T, Sun DW. Computational fluid dynamics (CFD) -an effective and efficient design and analysis tool for the food industry: A review. Trends in Food Science & Technology. 2006;17: 600-620. DOI: 10.1016/j. tifs.2006.05.004

111. Datta AK. Status of Physics-Based : Models in the Design of Food Products, Processes, and Equipment. Comprehensive Reviews in Food Science and Food Safety. 2008;7: 121-129. DOI: 10.1111/ j.1541-4337.2007.00030.x

112. Verboven P, Defraeye T, Datta AK, Nocolai B. Digital twins of food process operations: the next step for food process models? Current Opinion in Food Science. 2020;35:79-87. DOI: 10.1016/j.cofs.2020.03.002

113. Datta AK, Teixeira AA. Numerically predicted transient temperature and velocity profiles during natural convection heating of canned liquid foods. Journal of Food Science. 1988;53:191-195. DOI: 10.1111/j.1365-2621.1988.tb10206.x

114. Boz Z, Erdogdu Z. Evaluation of two-dimensional approach for computational modelling of heat and momentum transfer in liquid containing horizontal cans and experimental validation. Foods and Bioproducts Processing. 2013;91:37-45. DOI: 10.1016/ j.fbp.2012.08.005

115. Lee MG, Yoon WB. Developing an effective method to determine the deviation of F value upon the location of a still can during convection heating using CFD and subzones. Journal of Food Process Engineering 2014;37: 493-505. DOI: 10.1111/jfpe.12107

116. Farazbakht F, Zamindar N, Hojjatoleslamy M, Toghraie D. Numerical simulation of transient heat transfer for tomato paste in semi rigid aluminum container. Journal of Food Measurement and Characterization. 2017;11:479-487. DOI 10.1007/s11694-016-9415-z

117. Malekjani N, Jafari SM. Simulation of food drying processes by Computational Fluid Dynamics (CFD); recent advances and approaches. Trends in Food Science & Technology. 2018;78: 206-223. DOI: 10.1016/j.tifs.2018.06.006

118. Serami MS, Ramezan Y, Khashehchi M. CFD simulation and experimental validation of in-container thermal processing in Fesenjan stew. Food Science & Nutrition. 2020;9:1079-1087. DOI: 10.1002/fsn3.2083

119. Fadiji T, Coetzee CJ, Berry TM, Ambaw A, Opara UL. The efficacy of finite element analysis (FEA) as a design tool for food packaging: A review. Biosystems Engineering. 2018;174: 20-40. DOI: 10.1016/j.biosystemseng.2018.06.015

120. Norton T, Tiwari B, Sun DW. Computational Fluid Dynamics in the Design and Analysis of Thermal Processes: A Review of Recent Advances. Critical Reviews in Food Science and Nutrition. 2013;53(3): 251-275. DOI: 10.1080/10408398.2010.518256

121. Park HWP, Yoon WB. Computational Fluid Dynamics (CFD) Modelling and Application for Sterilization of Foods: A Review. Processes. 2018;6:62. DOI: 10.3390/ pr6060062

122. Boz Z, Erdogdu F, Tutar M. Effects of mesh refinement, time step size and numerical scheme on the computational modeling of temperature evolution during natural-convection heating. Journal of Food Engineering. 2014;123: 8-16. DOI: 10.1016/j.jfoodeng.2013. 09.008

123. Zhao CJ, Han JW, Yang XT, Qian JP, Fan BL. A review of computational fluid dynamics for forced-air cooling process. Applied Energy. 2016;168:314-331. DOI: 10.1016/j.apenergy.2016.01.101

124. Erdogdu F, Karatas O, Sarghini F. A short update on heat transfer modelling for computational food processing in conventional and innovative processing. Current Opinion in Food Science. 2018;23:113-119. DOI: 10.1016/j.cofs.2018.10.003

125. Shafiekhani S, Zamindar N, Hojatoleslami M, Toghraie D. Numerical simulation of transient temperature profiles for canned apple puree in semirigid aluminum based packaging during pasteurization. Journal of Food Science and Technology. 2016;53:2770-2778

126. Rinaldi M, Malavasi M, Cordioli M, Barbanti D. Investigation of influence of container geometry and starch concentration on thermal treated inpackage food models by means of Computational Fluid Dynamics (CFD). Food and Bioproducts Processing. 2018; 108:1-11. DOI: 10.1016/j.fbp.2017.12.003

127. Kuriakose R, Anandharamakrishnan C. Computational fluid dynamics (CFD) applications in spray drying of food products. Trends in Food Science & Technology. 2010;21:383-398. DOI: 10.1016/j.tifs.2010.04.009

128. Xia B, Sun DW. Applications of computational fluid dynamics (CFD) in the food industry: a review. Computers and Electronic in Agriculture. 2002;34 (1–3):5-24. DOI: 10.1016/S0168-1699 (01)00177-6

129. Nicolaï BM, Verboven P, Scheerlinck N. Modelling and simulation of thermal processes. In: Richardson P, editor. Thermal technologies in food processing. 1st ed. Cambridge: Woodhead Publishing; 2001. p. 91-112. DOI: 10.1533/9781855736610.2.91

130. Lyczkowski WR. The Rise of the First Commercial CFD Codes: PHOENICS, FLUENT, FIDAP, CFX, FLOW-3D, and STAR-CD. In: Memoir P, editor. The History of Multiphase Science and Computational Fluid Dynamics. 1st ed. Cham: Springer; 2018. P. 185-187. DOI: 10.1007/978-3-319-66502-3_14

131. Li B, Kang Z, Ma H. Research on the Meat Food Vacuum Cooling Model based on ANSYS Simulation. Revista Ibérica de Sistemas e Tecnologias de Informação. 2016;E6: 184-196.

132. Kızıltas S, Erdogdu F, Palazoglu TK. Simulation of heat transfer for solid–liquid food mixtures in cans and model validation under pasteurization conditions. Journal of Food Engineering. 2010;97:449-456. DOI:10.1016/j.jfoodeng.2009.10.042

133. Ghani AGA, Farid MM, Chen XD, Richards P. Thermal sterilization of canned food in a 3-D pouch using computational ¯uid dynamics. Journal of Food Engineering. 2001;48(2):147-156. DOI: 10.1016/S0260-8774(00)00150-3

134. Siriwattanayotin S, Yoovidhya T, Meepadung T, Ruenglertpanyakul W: Simulation of sterilization of canned liquid food using sucrose degradation as an indicator. Journal of Food Engineering. 2006;73(4):307-312. DOI: 10.1016/j.jfoodeng.2004.08.008

135. Chourasia MK, Goswaki TK. CFD simulation of effects of operating parameters and product on heat transfer and moisture loss in the stack of bagged potatoes. Journal of Food Engineering. 2007;80:947-960. DOI: 10.1016/j.jfoodeng.2006.07.015

136. Mirade PS. Prediction of the air velocity field in modern meat dryers using unsteady computational fluid dynamics (CFD) models. Journal of Food Engineering. 2003;60:41-48. DOI: 10.1016/S0260-8774(03)00009-8

137. Stebel M, Smolka J, Palacz M, Adamczyk W, Piechnik E. Numerical investigation of the fluid flow distribution for the hydrofluidisation food freezing method. International Journal of Thermal Sciences. 2020;151: 106284. DOI: 10.1016/j.ijthermalsci.2020.106284

138. Tattiyakul J, Rao MA, Datta AK. Simulation of heat transfer to a canned corn starch dispersion subjected to axial rotation. Chemical Engineering and Processing. 2001;40(4):391-399. DOI: 10.1016/S0255-2701(01)00116-7

139. Tattiyakul J, Rao MA, Datta AK. Heat transfer to a canned corn starch dispersion under intermittent agitation. Journal of Food Engineering. 2002;54 (4):321-329. DOI: 10.1016/S0260-8774 (01)00218-7

140. Okita WM, Reno MJ, Peres AP, Resende JV. Heat transfer computational fluid dynamics in the air blast freezing of guava pulp in large containers. Brazilian Journal of Chemical Engineering. 2013;30(4): 811-824. DOI: 10.1590/ S0104-66322013000400013

141. Malekjani N, Jafari SM. Food process modeling and optimization by response surface methodology (RSM). In: Sevda S, Singh A, editors. Mathematical and Statistical Applications in Food Engineering. 1st ed. Florida: CRC Press; 2020. p. 181-203. DOI: 10.1201/9780429436963-13

142. Simpson R, Ramirez C, Jiménez D, Almonacid S, Nuñez H, Angulo A: Simultaneous multi-product sterilization: Revisited, explored, and optimized. 2019;241:149-158. DOI: 10.1016/j.jfoodeng.2018.08.007

143. Alonso AA, Pitarch JL, Antelo LT, Vilas C. Event-based dynamic optimization for food thermal processing: High-quality food production under raw material variability. Food and Bioproducts Processing. 2021;127: 162-173. DOI: 10.1016/j.fbp.2021.02.013

144. Abakarov A, Nuñez M. Thermal food processing optimization: Algorithms and software. Journal of Food Engineering. 2013;115(4):428-442. DOI: 10.1016/j.jfoodeng.2012.02.013

145. Holdsworth SD, Simpson R. Thermal processing of packaged foods. 3rd edition. Cham: Springer; 2016. 516 p. DOI: 10.1007/978-3-319-24904-9

146. Miri T, Tsoukala A, Bakalis S, Pistikopoulos EN, Rustem B, Fryer PJ. Global optimization of process conditions in batch thermal sterilization of food. Journal of Food Engineering. 2008;87(4):485-494. DOI: 10.1016/j. jfoodeng.2007.12.032

147. Banga JR, Balsa-Canto E, Alonso AA. Quality and Safety Models and Optimization as Part of ComputerIntegrated Manufacturing. Comprehensive Reviews in Food Science and Food Safety. 2008;7: 168-174. DOI: 10.1111/j.1541-4337.2007.00023.x

148. Li J, Wang K, Gao Y, Ma C, Sun D, Hussain MA, Qayum A, Jiang Z, Hou J. Effect of thermal treatment and pressure on the characteristics of green soybean tofu and the optimization conditions of tofu processing by TOPSIS analysis. LWT – Food Science and Technology. 2021;136(1):110314. DOI: 10.1016/j.lwt.2020.110314

Chapter 8

ENHANCEMENT OF HEAT TRANSFER USING TAYLOR VORTICES IN THERMAL PROCESSING FOR FOOD PROCESS INTENSIFICATION

INTRODUCTION

In manufacturing processes, including those specific to the food industry, sustainable development is necessary because there is a limit on the energy and other resources. To achieve this goal, chemical industries have considered process intensification (PI) that might result in a paradigm shift. Although the definition of PI is still under discussion, a dramatic reduction in the process size is one of the common goals. One of the methods to achieve size reduction is the enhancement of transport rates, such as momentum, heat, and mass. For example, Harvey [1] successfully showed that, in the ester saponification process, the apparatus size was reduced by one-tenth compared with a traditional batch reactor, using an oscillatory baffled reactor exhibiting an excellent mixing performance. Therefore, PI technologies would bring about innovation in all the manufacturing processes. In fact, the introduction of

PI technologies has promoted various processes, for example, biopharmaceutical processes [2, 3]. The concept of PI should also be applied to food processing to establish energy/resource-saving processing. However, PI has not gained significant attention in the food industry. Boom et al. [4] analyzed three reasons for this: 1) food processing is largely based on traditional methods; 2) processing costs represent a small fraction of the total cost of food production, with the raw material representing the major portion of the total cost in most cases; and 3) the requirement of absolute food safety is a necessary obstacle to processing innovation. However, we should consider the transition from traditional food processing to novel processing by leveraging PI technologies, considering the environment, energy, and increasing population.

Few researchers have attempted to accomplish food process intensification by controlling fluid (liquid food) motion to enhance the mixing and heat/mass transfer. For example, Müller et al. [5] proposed a novel UV-C treatment device for juices based on the Dean vortex technology. Dean vortex flow occurs in a coiled tube owing to centrifugal instability [6]. They successfully showed that the Dean vortex flow promoted the inactivation of microorganisms because the fluid element is more frequently transported to the irradiation region through convective motion. Zhang et al. [7] successfully demonstrated the efficient manufacturing method of *Fuzhu* (also known as *Yuba*) through Rayleigh–Bénard convection. Rayleigh– Bénard convection is the flow between horizontal planes whose temperature at the lower plane is higher than that at the upper plane [8]. The driving cause underlying the Rayleigh–Bénard convection is the local distribution of the fluid density. Therefore, the novel concept based on fluid engineering has the potential for innovation in food processing.

In this chapter, aiming toward food process intensification, the application of a unique vortex flow between rotating cylinders (Taylor–Couette flow) to thermal processing is introduced.

TAYLOR–COUETTE FLOW

Taylor [9] first discovered and analyzed the unique vortex flow generated between cylinders with the inner cylinder rotating. This flow experiences several transitions with an increase in the rotational speed of the inner cylinder. The flow dynamics are characterized by the Reynolds number [$Re = \rho R_i \omega d/\eta$] in the circumferential direction, where ρ, R_i, ω, d, and η are the fluid density, inner cylinder radius, rotational speed of the inner cylinder, gap width, and fluid viscosity, respectively. At a relatively low Re, a Couette flow is observed with no pressure gradient in the flow direction. When Re exceeds a critical Re (Re_{cr}), toroidal vortices appear to be counter-rotating, spaced regularly along the axis, as shown in **Figure 1.** These vortices are called Taylor vortex flows. In addition, each vortex cell is called a Taylor cell. The value of Re_{cr} that depends on the radius ratio R_i/R_o, was theoretically derived by Taylor [9]. After the initial transition, as Re increases, the Taylor vortex flow cascadingly transitions to a singly periodic wavy vortex flow, quasiperiodic wavy vortex flow, and weakly turbulent wavy vortex flow [10, 11]. Finally, the flow develops into a fully turbulent vortex flow. The dynamics of the Taylor– Couette flow are interesting from the viewpoint of not only fluid mechanics, but also process engineering because this flow system has few advantageous characteristics as a reactor. First, mixing and heat/mass transfer are enhanced by the toroidal motion within the Taylor cells. Second, each Taylor cell is extruded through a single file without breakdown when a small axial flow is added. In addition, mass transfer between the Taylor cells is prevented by an inward boundary where an inward secondary flow is formed. This implies that the axial dispersion that is a negative factor for uniform processing, is suppressed, while local mixing and heat/mass transfer are enhanced. Therefore, the continuous and uniform production is possible using this flow system as a reactor. Since Kataoka et al. [12] reported the excellent performance of Taylor–Couette flow in a chemical reactor, Taylor– Couette flow has been applied to various chemical processes, such as emulsion polymerization [13], photocatalytic reaction [14], particle synthesis [15], reverse osmosis [16], particle classification [17], and gas absorption [18]. According to these studies, the Taylor–Couette flow reactor has the potential to effectively intensify the processes compared with a traditional reactor, such as stirred tank reactor. In fact, among the chemical engineering communities, it is well-known that the Taylor–Couette reactor enables PI. Furthermore, few researchers have suggested that the Taylor–Couette flow apparatus is suitable for processes with shear-sensitive materials, such as food and bio-processes because the local strong shear force is absent. Haut et al. [19] applied the Taylor–Couette flow to the cultivation of CHO cells and reported the possibility of the Taylor–Couette flow-based bioreactor. To the best of our knowledge, the research conducted by Giordano et al. [20] is the first attempt of applying the Taylor–Couette flow in food processes. They showed that fructose–glucose isomerization could be efficiently conducted using a Taylor–Couette flow reactor. Subsequently, few researchers reported the excellent performance of Taylor–Couette flow in the non-thermal inactivation of bacteria in juice [21–23].

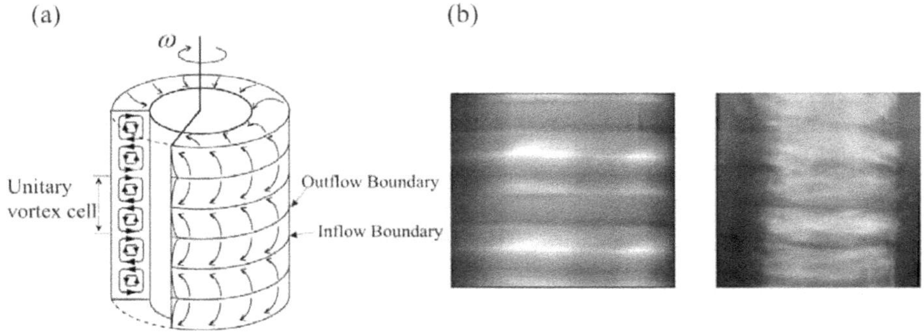

Figure 1. *Taylor–Couette flow: (a) schematic picture and (b) flow visualization. The left and right figures show the laminar Taylor vortex flow and wavy vortex flow, respectively. The black band corresponds to inflow boundaries.*

Although the application of Taylor–Couette flow to food processes is rather limited to non-thermal processing, efficient heat transfer by Taylor vortices should be utilized in thermal processing. Few research groups including the author have been trying to intensify thermal food processing based on Taylor–Couette flow. The intensification of starch processing and heat sterilization is introduced in this chapter.

FOOD PROCESS INTENSIFICATION USING TAYLOR–COUETTE FLOW

Intensification of starch processing

Starch is typically a biopolymer that consists of 25% amylose (linear structure) and 75% amylopectin (branched structure). Detailed information on starch from the viewpoint of chemistry is reviewed in other articles [24]. Starch processing is frequently encountered in the manufacturing process of various types of food, such as beer, beverages, bread, and sauce. From a practical viewpoint, one of the most important types of starch processing is starch hydrolysis that comprises gelatinization, liquefaction, and saccharification. Enzymatic hydrolysis is described in this chapter because it is more prevalent than the other starch modifications, such as thermal and chemical treatment [25, 26]. In the starch hydrolysis process, the fluid viscosity intricately changes, as shown in **Figure 2**.

A significant increase in the viscosity was observed during gelatinization. Further, when enzyme (α-amylase) is added, the viscosity decreases as starch chains are broken down into glucose, maltose, maltotriose, and few higher oligomers. This intricate viscosity change is not favorable to the food engineers because the key operation is different between gelatinization and enzymatic liquefaction/saccharification processes. During gelatinization, heat transfer from the heated surface due to starch suspension and mass transfer between starch grains and water are required. In liquefaction/saccharification, highly efficient mixing of gelatinized starch and a small amount of enzyme is the most important operation. Therefore, individual apparatuses must be used. Consequently, the total size of starch hydrolysis process becomes large, as Baruque et al. indicated [28].

To make the total size compact, Baks et al. proposed the simultaneous and continuous processing of gelatinization and liquefaction/saccharification using an extruder [29, 30]. As shown in their studies, even at a high concentration of starch (600 g/L), gelatinization was completely conducted using the extruder. However, a high shear force was applied to the starch suspension in the extruder. This high shear force induces inactivation of the enzyme (α-amylase) [31, 32]. Therefore, other apparatuses such as stirred vessels are necessary for liquefaction/ saccharification after gelatinization [30]. Paolucci-Jeanjean et al. [33] proposed a unique membrane reactor to conduct enzymatic hydrolysis using only one apparatus. However, the starch concentration was limited to 150 g/L because of the absence of mechanical agitation.

To intensify starch hydrolysis using a single apparatus, Masuda et al., Hubacz et al., and Matsumoto et al. applied a Taylor–Couette flow reactor for continuous starch hydrolysis [27, 34–39]. The features of the Taylor–Couette flow are described in the previous section. Taylor–Couette flow enhances not only mixing, but also heat/ mass transfer. Therefore, it is expected that both gelatinization, where heat/mass transfer is necessary and liquefaction/saccharification, where mixing is necessary, are intensified using a single Taylor–Couette flow reactor.

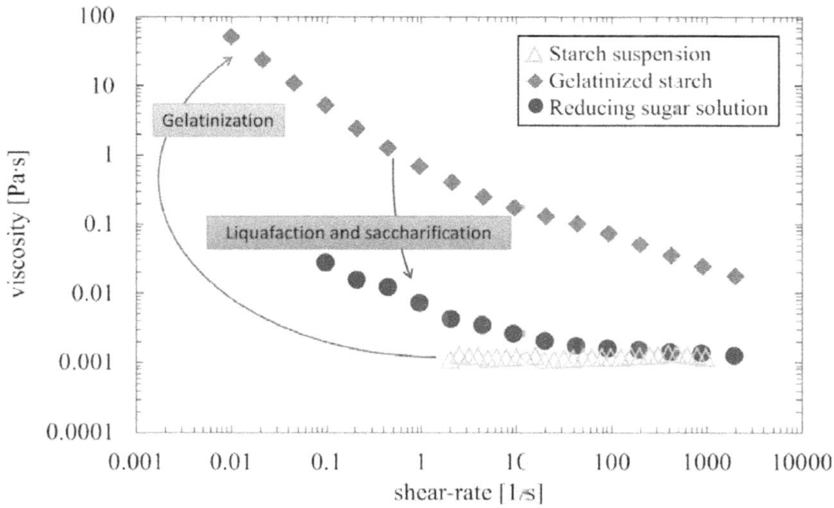

Figure 2. *Viscosity change at various shear rates during starch gelatinization/liquefaction/ saccharification [27].*

As an example, a Taylor–Couette flow reactor utilized by Masuda et al. [35] is shown in **Figure 3**. The reactor consisted of a rotating inner cylinder, a stationary outer cylinder, and two divided water jackets. A starch suspension was introduced into the inlet. The enzyme was continuously fed using a syringe pump from the port set in the middle of the reactor. Therefore, the first and second half parts of the reactor were regarded as corresponding to the gelatinization and liquefaction/ saccharification processes, respectively. High-temperature water was pumped in the first water jacket to promote gelatinization. Furthermore, moderate temperature water was pumped into the second water jacket to avoid the thermal deactivation of α-amylase.

The effects of Taylor vortices on starch gelatinization and hydrolysis were experimentally and numerically investigated in detail. **Figure 4** shows the impact of Taylor vortex flow on the degree of starch gelatinization (DSG). A high value of DSG was obtained when Taylor vortices were formed because the Taylor vortex flow enhanced the heat transfer from the heating surface. It should be noted that microscopic mass transfer around the starch granules was not considered in their simulation [36].

However, ascertaining whether Taylor vortices are formed within the reactor is not straightforward because the reactor is enwrapped in water jackets made of stainless steel. Therefore, to simulate the fluid flow in the reactor during starch gelatinization, Hubacz et al. [36] empirically established a mathematical model to describe the change in the rheological properties in response to gelatinization, as follows:

$$\eta = \frac{(0.0013DSG + 0.0031)}{1312^{n-1}}\dot{\gamma}^{n-1}, \tag{1}$$

where n [–] and γ_ [1/s] are the rheological model parameter and shear rate, respectively. **Figure 5** shows the axial velocity distribution near the inlet when the initial concentration of starch, C_0, is 50 g/L at the following values of ω: (a) 10 and (b) 22 rad/s. As clearly shown in **Figure 5**, at ω = 10 rad/s, it is confirmed that there are no Taylor vortices, except near the inlet because of the lower centrifugal force. Therefore, the rheological model is reasonably advantageous for the practical design of the starch gelatinization process based on Taylor–Couette flow. In addition, as Van Den Einde et al. [40] indicated, starch granule degradation by thermomechanical treatment should also be considered. Hubacz et al. [36] confirmed that, as shown in **Figure 6**, there was no mechanical destruction of starch granules, and thermal degradation was not visible. Therefore, Taylor–Couette flow is suitable for the intensification of starch gelatinization owing to the efficient heat transfer without violent shear force.

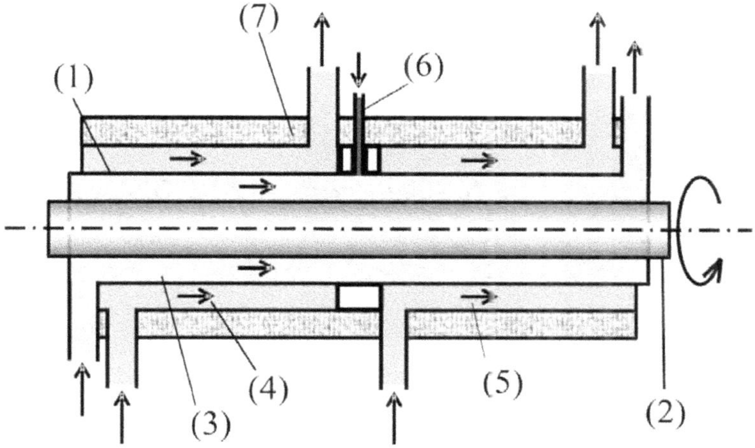

Figure 3. *Taylor–Couette flow reactor used for starch processing [35]: (1) stationary outer cylinder, (2) rotational inner cylinder, (3) starch suspension, (4) hot water for gelatinization, (5) variable temperature water for enzymatic reaction, (6) enzyme injection port, (7) insulator.*

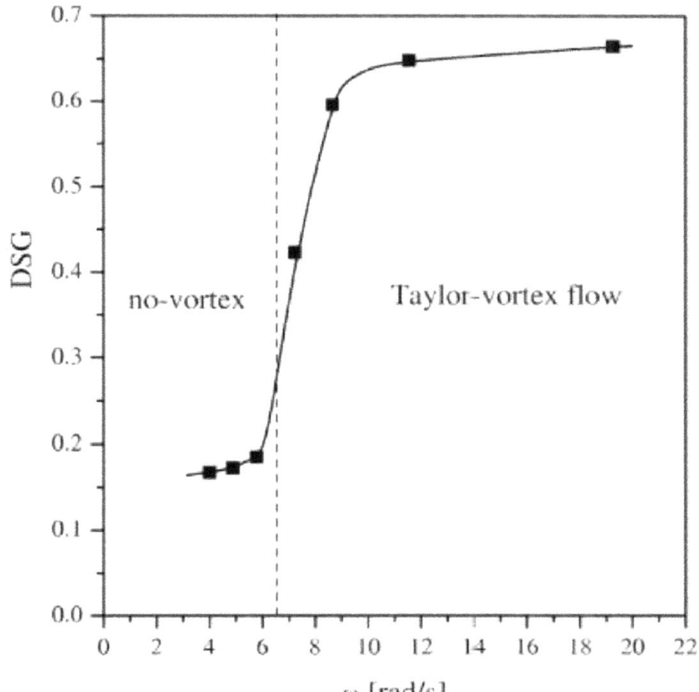

Figure 4. *Dependence of DSG (degree of starch gelatinization), obtained via two-dimensional simulation, on the rotational speed of inner cylinder (ω) [36]. The water jacket temperature,* $\mathbf{T}_{hj}$*, was assumed to be 65°C.*

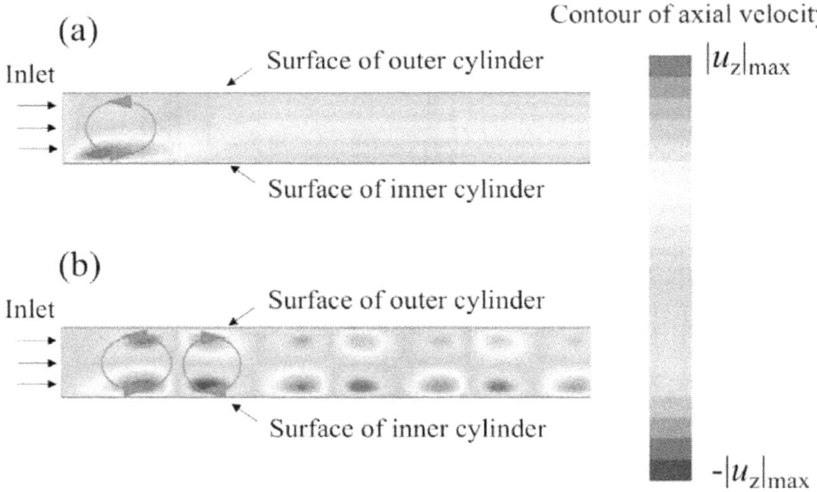

Figure 5. *Velocity distribution near the inlet during gelatinization at* $\mathbf{C}_0$ *= 50 g/L, ω = (a) 10 rad/s and (b) 22 rad/s. The circles in the figures denote vortex motion.*

The Taylor–Couette flow reactor intensifies starch gelatinization and liquefaction/saccharification. **Figure 7** shows that the relationship between the concentration of reducing sugar and effective Reynolds number at C_0 = 50, 150, and 300 g/L for the axial velocity *u* of 0.024 cm/s. It is noted that the flow condition was evaluated by the effective Reynolds number Re_{eff} because the apparent viscosity spatially changes due to the shear-thinning property of starch suspension. The detailed procedure for defining and calculating Re_{eff} is described in a paper by Masuda et al. [41]. As clearly shown in **Figure 7**, a higher yield of reducing sugar is obtained through the operation above Re_{cr} (dashed line in the figure) in all cases of C_0. Remarkably, starch is continuously and efficiently hydrolyzed even at relatively high concentrations of the starch suspension. The maximum yield is comparable to that obtained using a stirred batch reactor. Therefore, the conversion from batch to continuous is possible for food process intensification. However, a slight decrease in the yield was observed at higher Re_{eff} values. This decrease is explained by the axial dispersion and destabilization of the vortex structure during the enzymatic reaction [27]. At higher *Re*, a wavy motion is observed (called the wavy Taylor vortex flow). The wavy vortex flow significantly enhances mixing and heat/mass transfer within Taylor cells; furthermore, this also leads to axial dispersion through by-pass flow [42]. According to Richter et al. [43, 44], Taylor vortices can be stabilized and immobilized by a ribbed inner cylinder, as shown in **Figure 8**.

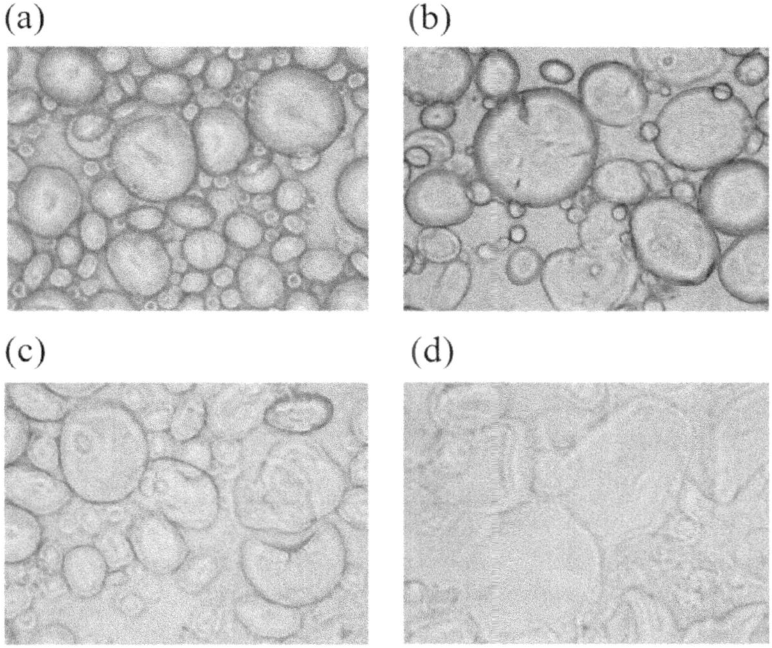

Figure 6. *Structure of starch observed using a light microscope: (a) native starch, gelatinized starch after treatment at (b) u = 0.099 cm/s, ω = 11.56 rad/s, T_{hj} = 60°C, (c) u = 0.099 cm/s, ω = 19.56 rad/s, T_{hj} = 65°C and (d) u = 0.099 cm/s, ω = 19.56 rad/s, T_{hj} = 85°C [36].*

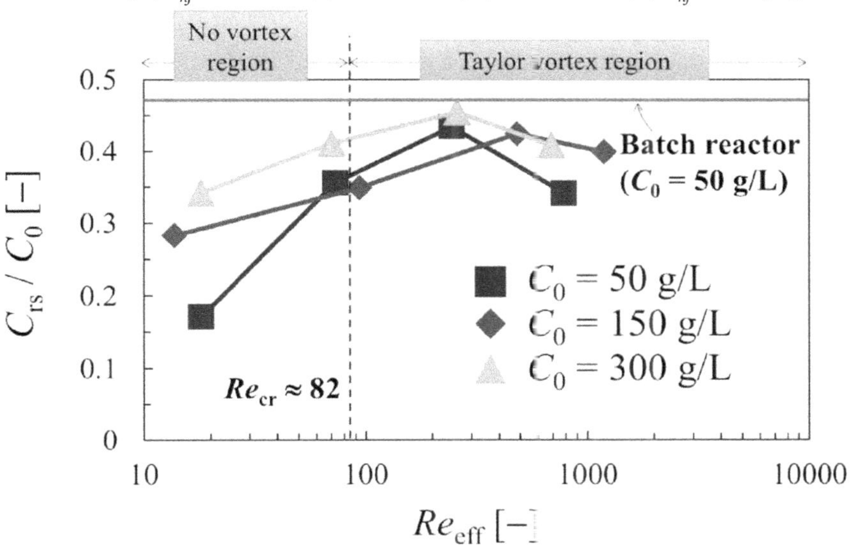

Figure 7. *Relation between the yield of reducing sugar (**C_{rs}/C_0**) and effective Reynolds number (**Re_{eff}**) at **C_0** = 50, 150, 300 g/L, **u** = 0.024 cm/s, **T_{hj}** = 45°C [39]. The dashed line denotes the critical **Re** (**Re_{cr}**) where Taylor vortices are fully formed.*

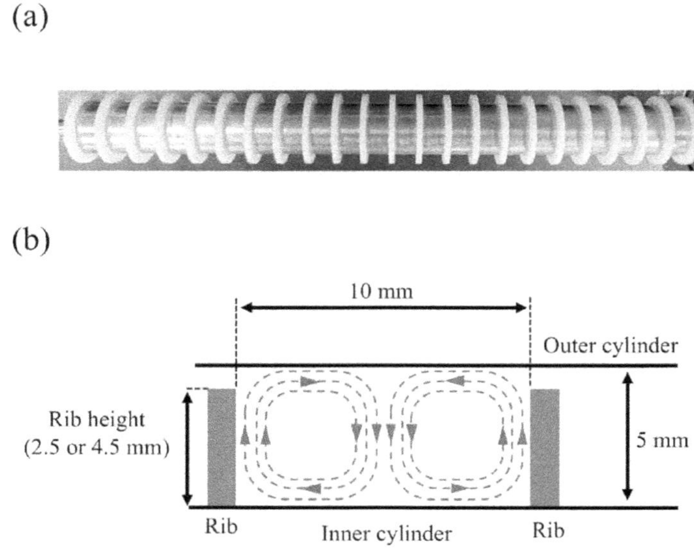

Figure 8. *Ribbed inner cylinder system [27]: (a) picture and (b) cross-sectional view of a pair of Taylor vortices between ribs.*

Consequently, the axial dispersion was suppressed even at a higher *Re*. Masuda et al. [27] successfully showed that the decrease in the yield at a higher Re_{eff} is suppressed by the equipment of ribs in the inner cylinder, as shown in **Figure 9**. Furthermore, as shown in **Figure 10**, the yield of small saccharides (glucose, maltose, and maltotriose) was significantly enhanced by utilizing the ribbed inner cylinder. This is because the ribbed inner cylinder enables the enhancement of mixing, while the axial dispersion is suppressed at a higher *Re*. Finally, the effect of the axial velocity on the reducing sugar yield at C_0 = 150 g/L is shown in **Figure 11**.

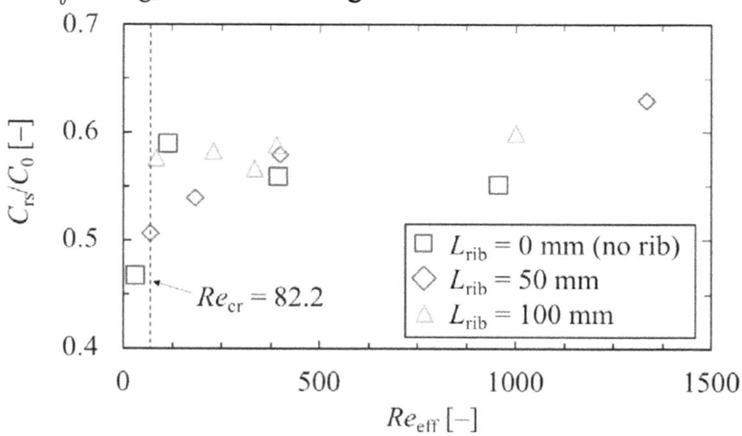

Figure 9. *Effect of* $\mathbf{Re}_{eff}$ *on* $\mathbf{C}_{rs}/\mathbf{C}_0$ *with three types of cylinders (*$\mathbf{L}_{rib}$ *= 0, 50, 100 mm) at* **u** *= 0.024 cm/s in starch hydrolysis experiments [27].* $\mathbf{L}_{rib}$ *refers to the length of the ribbed section from the outlet.*

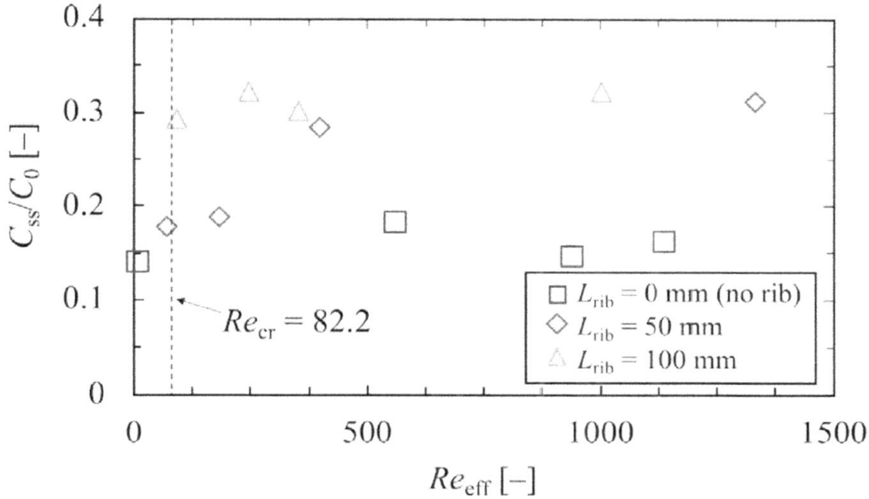

Figure 10. *Effect of* C_{ss}/C_0 *on* Re_{eff} *with three types of cylinders* (L_{rib} *= 0, 50, 100 mm) at* **u** *= 0.024 cm/s [27].* C_{ss} *refers to small saccharide concentration.*

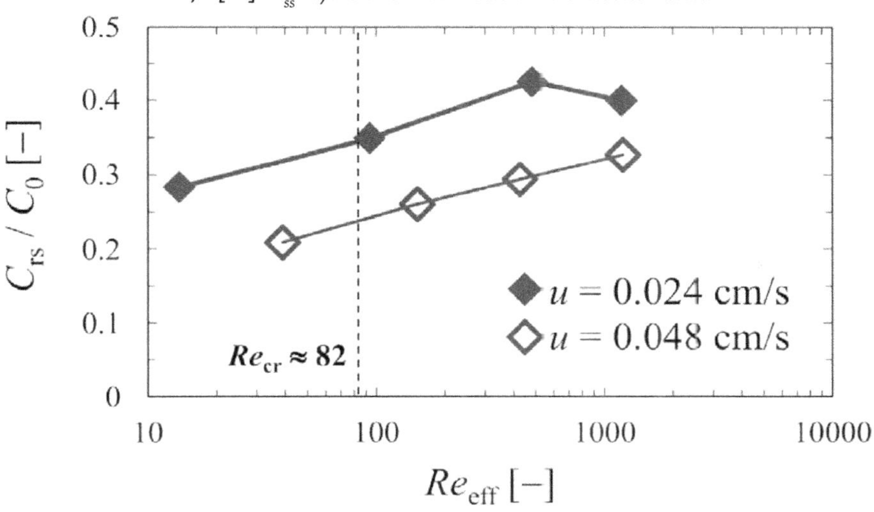

Figure 11. *Effect of axial velocity on* C_{rs}/C_0 *at* C_0 *= 150 g/L [39].*

At a higher axial velocity (u = 0.048 cm/s), the yield monotonically increases with Re_{eff} without a decrease at a higher Re_{eff}. Masuda et al. [39] explained that the transition from laminar Taylor vortex flow to wavy Taylor vortex flow occurs at a higher Re_{eff} than at a lower u because the axial flow enhances the stability of the Taylor vortex flow [45]. This should be investigated from the viewpoint of fluid mechanics. Nevertheless, the Taylor–Couette flow reactor promotes innovation in starch processing, for example, dramatic size reduction, high efficiency, and energy saving.

Intensification of heat sterilization processing

Heat sterilization is important for human health. Although novel technologies such as ultraviolet, ultrasonic, high-pressure, and cold plasma have been utilized [46], thermal sterilization plays a major role in the food industry. Recently, ohmic heating has recently been applied to heat sterilization processes [47]. However, the principle of scale-up for industries is under consideration. A traditional heat sterilizer, including a double-pipe, plate, and scrapped surface heat exchanger, faces problems such as clogging and high-pressure loss in the case of highly viscous liquid food. Therefore, heat sterilizers should be utilized for food process intensification. We consider the functions of an ideal sterilizer as follows:

1. High heat transfer performance in rapid heating;
2. Low shear force to avoid mechanical degradation of nutritional component;
3. Low pressure loss for saving energy.

These three functions are satisfied by adequately controlling the motion of liquid food. For example, chaotic advection and Dean vortex flow enable efficient and continuous heat sterilization [48, 49]. Taylor–Couette flow also offers a novel heat sterilization process. As described in the previous section, the Taylor–Couette flow offers efficient and moderate heat transfer. In addition, the shear-thinning properties of many liquid foods should be considered. Another advantage is that a lower power is required for pumping because the apparent viscosity decreases owing to the rotation of the inner cylinder. The Taylor–Couette flow sterilizer has the potential for food process intensification. Masuda et al. [50–52] numerically investigated the performance of a Taylor–Couette flow sterilizer. They assumed the sterilization process of highly viscous liquid food such as mayonnaise or ketchup, including the thermal destruction of the spores of *Clostridium botulinum* and the retention of thiamine.

Figure 12 shows the computational domain used in [51]. To eliminate the effect of back flows through Taylor vortex flow at the outlet, an extended section is imposed where the inner cylinder is stationary. This attempt does not affect the simulation results. They have solved the conservation equations of mass, momentum, heat, and chemical species, as follows [51]:

$$\nabla . \mathrm{u} = 0 \tag{2}$$

$$(\mathbf{u} \cdot \nabla)\mathbf{u} = -\frac{\nabla p}{\rho} + \frac{1}{\rho}\nabla \cdot (2\eta \mathbf{D}) - \mathbf{g}\alpha(T - T_{\mathrm{ref}}) \tag{3}$$

$$\mathbf{u} \cdot \nabla T = \frac{\lambda}{\rho C_{\mathrm{p}}} \nabla^2 T, \tag{4}$$

$$\mathrm{u}.\nabla \mathrm{C} = \nabla.(\mathrm{D_c} \nabla \mathrm{C}) + \mathrm{S}, \tag{5}$$

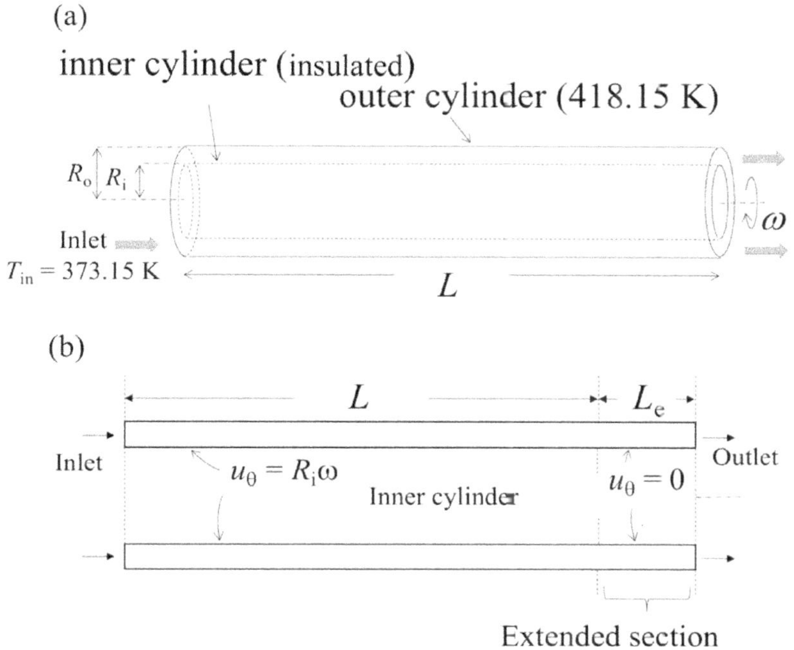

Figure 12. *Computational domain: (a) three-dimensional view without an extended section, (b) cross-sectional view with an extended section [51].*

where **u** is the velocity, p is the pressure, ρ is the density, η is the viscosity depending on the shear rate, **D** (= (∇**u** + ∇**u**^{T}) /2) is the rate of deformation tensor, **g** is the gravitational acceleration, α is the coefficient of volume expansion, T is the temperature, T_{ref} is the reference temperature, λ is the thermal conductivity, C_p is the specific heat capacity, C is the concentration, D_c is the diffusion coefficient, and S is the scalar source term. Because this simulation was assumed to be in a steady state, the time derivative term is omitted in Eqs. (2)–(5). It was assumed that the model fluid had a moderate shear-thinning property. According to Horak and Kessler [53], the thermal destruction of thiamine is followed by a second-order reaction model. The decrease in thiamine concentration due to destruction was included in the sink term, S, as shown in Eq. (5). Detailed information on the numerical procedure is described in [51]. The simulation code was validated.

Figure 13 shows the temperature distribution with the velocity vectors near the inlet at various values of Re_{eff}. In the case of Re_{eff} = 101.1 and 172.6 (**Figure 13(c)** and **(d)**), Taylor vortices were fully developed near the inlet, and consequently, heat transfer from the surface of the outer cylinder was significantly enhanced. This enhancement of the heat transfer is clearly confirmed from the bulk temperature distribution along the axis, as shown in **Figure 14**.

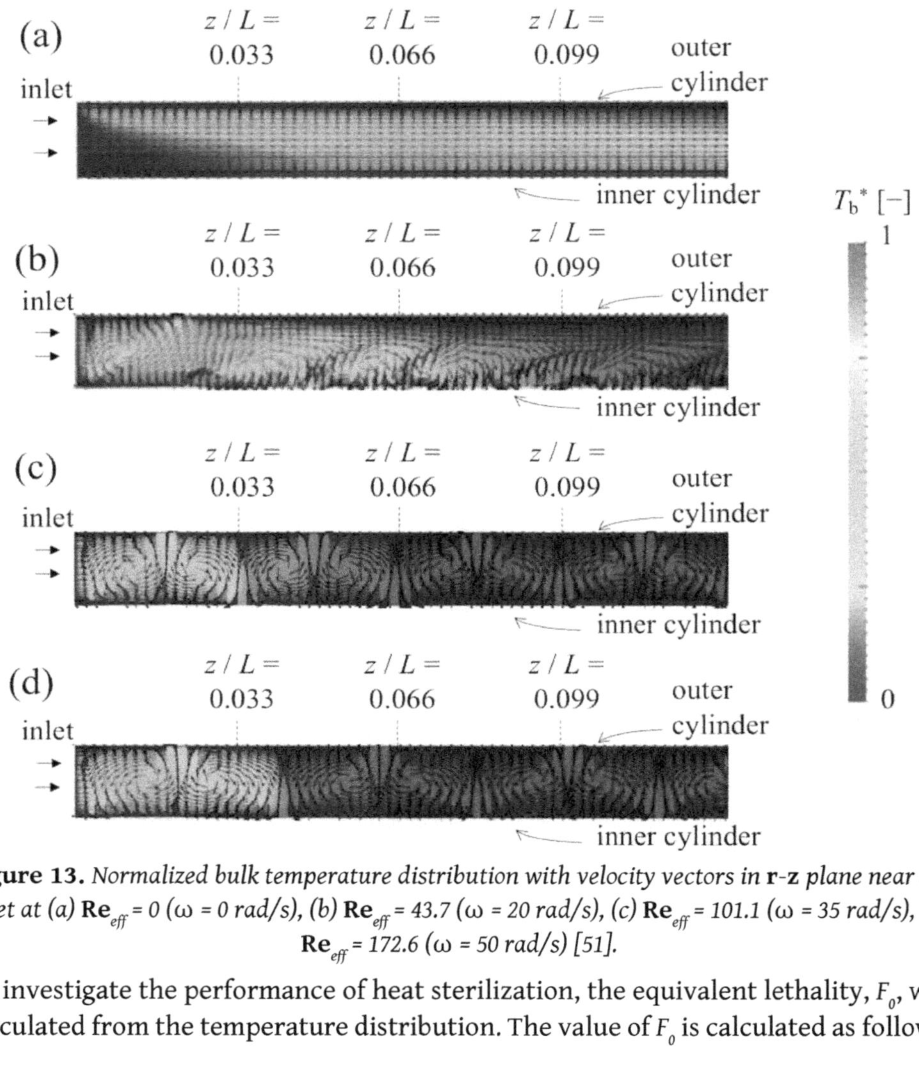

Figure 13. *Normalized bulk temperature distribution with velocity vectors in* **r**-**z** *plane near the inlet at (a)* $\mathbf{Re}_{eff}$ *= 0 (ω = 0 rad/s), (b)* $\mathbf{Re}_{eff}$ *= 43.7 (ω = 20 rad/s), (c)* $\mathbf{Re}_{eff}$ *= 101.1 (ω = 35 rad/s), (d)* $\mathbf{Re}_{eff}$ *= 172.6 (ω = 50 rad/s) [51].*

To investigate the performance of heat sterilization, the equivalent lethality, F_0, was calculated from the temperature distribution. The value of F_0 is calculated as follows:

$$F_0 = \int_0^t \exp\left[\frac{E_a}{R}\left(\frac{1}{394.25} - \frac{1}{T(t)}\right)\right] dt, \tag{6}$$

where E_a is the activation energy for the destruction of *Clostridium botulinum*, and R is the gas constant. Finally, the local value of F_0 at an arbitrary axial position z is calculated as follows:

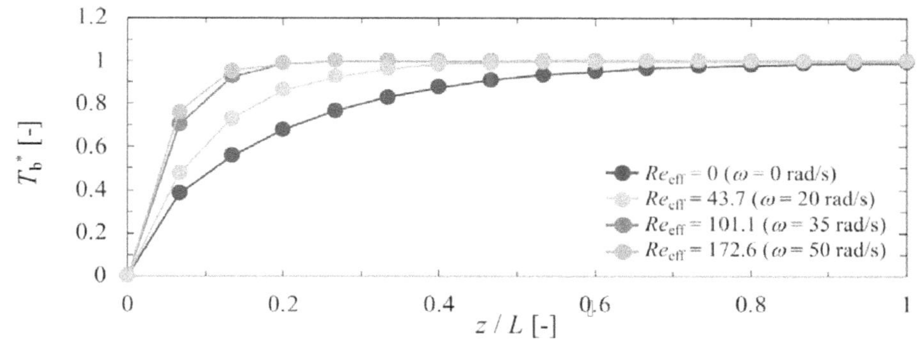

Figure 14. *Normalized bulk temperature distribution along the axis [51].*

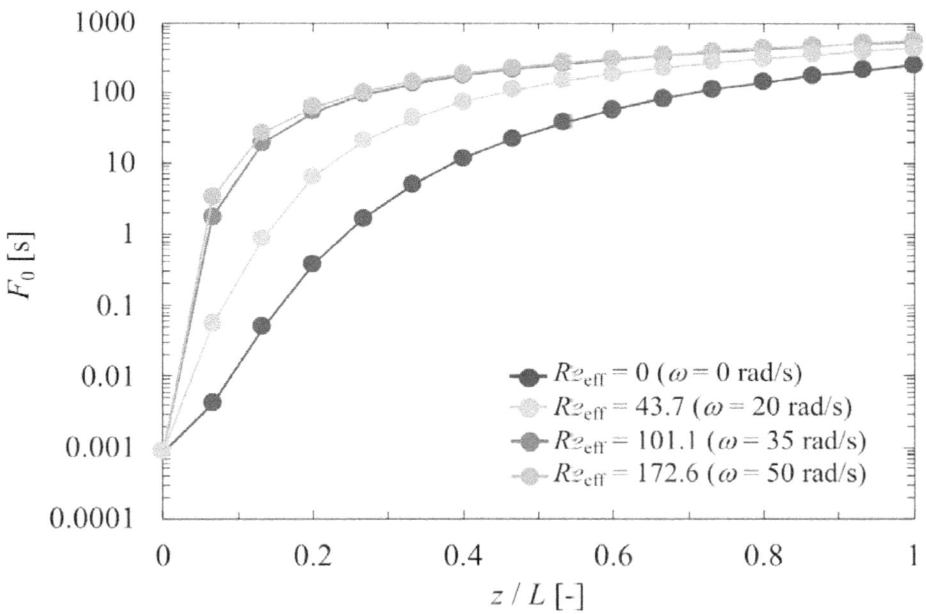

Figure 15. *Equivalent lethality distribution along the axis [51].*

Figure 15 shows the axial distribution of F_0 along the axis. The significant increase in F_0 (higher than F_0 = 500 s) is observed under the condition at which Taylor vortices are developed (Re_{eff} = 101.1 and 172.6), as shown in **Figure 15**. This result indicates that Taylor–Couette flow has the potential to intensify the heat sterilization process. **Figure 16** shows the retention performance of thiamine during the sterilization process. Comparing the result at Re_{eff} = 101.1 with that at Re_{eff} = 172.6, it is confirmed from **Figure 15** that there is no clear difference in F_0. In addition, a clear difference in the thermal destruction of thiamine is not observed in **Figure 16**. Nevertheless, Ilo and Berghofer [54] indicated the mechanical destruction of thiamine by shear force.

Therefore, the operation at Re_{eff} = 101.1, is preferable because of the lower shear force. It is valuable to investigate the effect of shear force on thiamine destruction in the future.

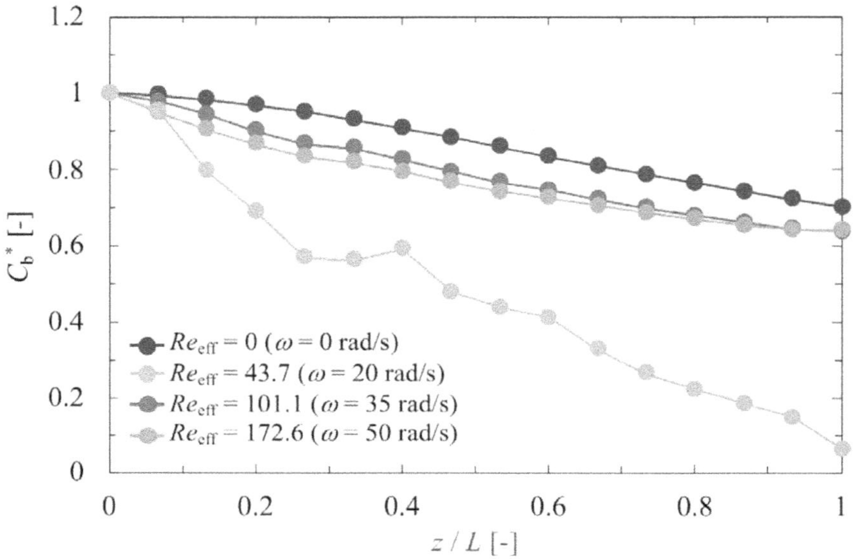

Figure 16. *Normalized thiamine concentration distribution along the axis [51].*

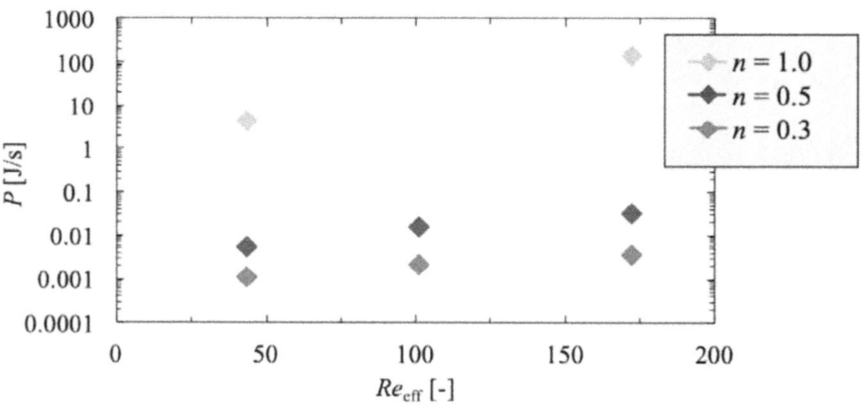

Figure 17. *Effect of power consumption on rheological properties in Taylor–Couette flow sterilizer.*

Finally, the characteristics of energy consumption that are important for practical applications, are shown in **Figure 17**. In **Figure 17**, the energy consumption was calculated from the shear stress at the surface of the inner cylinder, as follows:

$$P = \omega R_{\mathrm{i}} \iint \tau_{\mathrm{r}\theta} \mathrm{d}A, \tag{8}$$

where $\tau_{r\theta}$ is the component of the shear stress tensor at the surface of the inner cylinder, and dA is the differential surface of the inner cylinder. It is noted that the value of n indicates the strength of the shear-thinning property. For Newtonian fluids, n corresponds to 1. Remarkably, **Figure 17** shows that the power consumption significantly decreases with an increase in the shear-thinning property because the apparent viscosity decreases owing to the shear force generated by the rotation of the inner cylinder. Therefore, the Taylor–Couette flow sterilizer enables energy-saving sterilization processing of liquid foods with shear-thinning properties.

CONCLUSIONS

In this chapter, novel food processing utilizing Taylor–Couette flow was introduced for food process intensification. As examples, starch processing and heat sterilization processes were specifically selected. With respect to starch processing, continuous and efficient gelatinization/liquefaction/saccharification were successfully conducted even in the case of high-concentration starch suspension. In addition, no clear thermal degradation of the starch granules was observed. Therefore, in the future, Taylor–Couette flow could be practically utilized in industries. In heat sterilization processing, enhancement of heat transfer by Taylor–Couette flow significantly improved the thermal destruction of *Clostridium botulinum*. Actually, the sufficient value of F_0 (higher than F_0 = 500 s) was obtained due to Taylor vortices. Based on the lethality, thermal destruction of nutritional components such as thiamine and mechanical destruction by shear force, the optimum operational conditions were proposed.

Taylor–Couette flow has the potential to intensify other processes as well. For example, an appropriate mixing performance of Taylor vortices would facilitate the manufacturing of sophisticated emulsions, such as multiple emulsions. Furthermore, other fluid techniques, such as chaotic advection, could incorporate novel processing. This chapter provides all food engineers with new insights into food process intensification.

ABSTRACT

We are witnessing a transition from the traditional to novel processing technologies in the food industry to address the issues regarding energy, environment, food, and water resources. This chapter first introduces the concept of food process intensification based on vortex technologies to all food engineers/ researchers. Thereafter, the novel processing methods for starch gelatinization/ hydrolysis and heat sterilization based on Taylor–Couette flow are reviewed. In fluid mechanics communities, the Taylor–Couette flow is well-known as a flow between coaxial cylinders with the inner cylinder rotating. Recently, this unique flow has been applied in food processing. In starch processing, enhanced heat transfer through Taylor vortex flow significantly improves gelatinization. In addition, effective and moderate mixing leads to an increase in the reducing sugar yield. In sterilization processing, the enhanced heat transfer also intensifies the thermal destruction of *Clostridium botulinum*. However, a moderate heat transfer should be ensured because excessive heat transfer also induces thermal destruction of the nutritional components. The Taylor–Couette flow is only an example considered here. There are various flows that intensify the heat/mass transfer and mixing in food processing. It is expected that this chapter will stimulate the development of food processing based on fluid technologies, toward food process intensification.

REFERENCES

1. Harvey AP, Mackley MR, Stonestreet P. Operation and optimization of an oscillatory flow continuous reactor. Industrial & Engineering Chemistry Research. 2001; 40:5371-5377. DOI: 10.1021/ie0011223
2. Shallan AI, Priest C. Microfluidic process intensification for synthesis and formulation in the pharmaceutical industry. Chemical Engineering and Processing-Process Intensification. 2019;142:107559. DOI: 10.1016/j.cep.2019.107559
3. Satyawali Y, Vanbroekhoven K, Dejonghe W. Process intensification: The future for enzymatic processes?. Biochemical Engineering Journal. 2017; 121:196-223. DOI: 10.1016/j. bej.2017.01.016
4. Boom RM, van der Goot AJ, Janssen AEM, Schroën KGPH. Food process intensification for better sustain-ability. In: Proceedings of 11th International Congress on Engineering and Food; 22-26 May 2011; Athnes; FMS1276
5. Müller A, R. Stahl M, Graef V, M.A.P. Franz C, Huch M. UV-C treatment of juices to inactivate microorganisms using Dean vortex technology. Journal of Food Engineering. 2011;107:268-275. DOI: 10.1016/j.jfoodeng.2011.05.026
6. Dean WR, Hurst JM. Note on the motion of fluid in a curved pipe. Mathematika. 1959;6:77-85. DOI: 10.1112/S0025579300001947
7. Zhang J, Peng X, Guo S. Protein-lipid film (fuzhu) prepared from soymilk: Effects of soymilk convection on its formation, composition, and quality. LWT – Food Science and Technology. 2021;141:110909. DOI: 10.1016/j. lwt.2021.110909

8. Bodenschatz E, Pesch W, Ahlers G. Recent developments in Rayleigh-Bénard convection. Annual Review of Fluid Mechanics. 2000;32:709-778. DOI: 10.1146/annurev.fluid.32.1.709

9. Taylor GI. VIII. Stability of a viscous liquid contained between two rotating cylinders. Proceedings of the Royal Society A. 1923;223:289-343. DOI: 10.1098/rsta.1923.0008

10. Di Prima RC, Swinney HL. Instabilities and transition in flow between concentric rotating cylinder. In: Swinney HL, Gollub JP, editors. Hydrodynamic Instabilities and the Transition to Turbulence. Berlin: Springer; 1981. p. 139-180

11. Kataoka K. Taylor vortices and instabilities in circular Couette flows. In: Cheremisinoff, editor. Encyclopedia of Fluid Mechanics. Houston: Gulf Publications; 1986. p. 236-274

12. Kataoka K, Doi H, Hongo T, Futagawa M. Ideal plug-flow properties of Taylor vortex flow. Journal of Chemical Engineering of Japan. 1975;8: 472-476. DOI:

13. Kataoka K, Ohmura N, Kouzu M, Simamura Y, Okubo M. Emulsion polymerization of styrene in a continuous Taylor vortex flow reactor. Chemical Engineering Science. 1995;50: 1409-1416. DOI: 10.1016/0009-2509 (94)00515-S

14. Sczechowski JG, Koval CA, Noble RD. A Taylor vortex reactor for heterogeneous photocatalysis. Chemical Engineering Science. 1995;50: 3163-3173. DOI: 10.1016/0009-2509(95)00176-6

15. Ogihara T, Matsuda G, Yanagawa T, Ogata N, Fujita K, Nomura M. Continuous synthesis of monodispersed silica particles using Couette-Taylor vortex flow. Journal of the Ceramic Society of Japan. 1995; 103:151-154. DOI: 10.2109/jcersj.103.151 10.2109/jcersj.103.151

16. Lee S, Lueptow RM. Rotating reverse osmosis: a dynamic model for flux and rejection. Journal of Membrane Science. 2001;192:129-143. DOI: 10.1016/S0376-7388(01)00493-8

17. Saomoto K, Horie T, Kumagai N, Takigawa T, Noui-Mehidi MN, Ohmura N. Dispersion of floating particles in a Taylor vortex flow reactor. Journal of Chemical Engineering of Japan. 2010;43:319-325. DOI: 10.1252/jcej.09We07

18. Masuda H, Zheng W. Horie T, Ohmura N. Enhancement of gas hold-up with a Taylor vortex flow system equipped with ribs. Journal of Chemical Engineering of Japan. 2013;46:27-32. DOI: 10.1252/jcej.12we067

19. Haut B, Ben Amor H, Coulon L, Jacquet A, Halloin V. Hydrodynamics and mass transfer in a Couette–Taylor bioreactor for the culture of animal cells. Chemical Engineering Science. 2003;58: 777-784. DOI: 10.1016/S0009-2509(02) 00607-3

20. Giordano RLC, Giordano RC, Cooney CL. Performance of a continuous Taylor–Couette–Poiseuille vortex flow enzymic reactor with suspended particles. Process Biochemistry. 2000;35:1093-1101. DOI: 10.1016/S0032-9592(00)00143-6

21. Forney LJ, Pierson JA, Ye Z. Juice Irradiation with Taylor-Couette flow: UV inactivation of *Escherichia coli*. Journal of Food Protection. 2004;67: 2410-2415. DOI: 10.4315/0362-028X67.11.2410

22. Milly PJ, Toledo RT, Kerr WL, Armstead D. Hydrodynamic cavitation: characterization of a novel design with energy considerations for the inactivation of *Saccharomyces cerevisiae* in apple juice. Journal of Food Science. 2008;73:M298-M303. DOI: 10.1111/ j.1750-3841.2008.00827.x

23. Orlowska M, Koutchma T, Kostrzynska M, Tang J, Defelice C. Evaluation of mixing flow conditions to inactivate *Escherichia coli* in opaque liquids using pilot-scale Taylor–Couette UV unit. Journal of Food Engineering. 2014;120:100-109. DOI: 10.1016/j. jfoodeng.2013.07.020

24. Ellis RP, Cochrane MP, Dale MFB, Duffus CM, Lynn A, Morrison IM, Prentice RDM, Swanston JS, Tiller SA. Starch production and industrial use. Journal of the Science of Food and Agriculture. 1998;77:289-311. DOI: 10.1002/(SICI)1097-0010(199807)77:3 <289::AID-JSFA38>3.0.CO;2-D

25. Tester RF, Debon SJJ. Annealing of starch — a review. International Journal of Biological Macromolecules. 2000;27: 1-12. DOI: 10.1016/S0141-8130(99) 00121-X

26. Tharanathan RN. Starch — Value addition by modification. Critical Reviews in Food Science and Nutrition. 2005;45:371-384. DOI: 10.1080/ 10408390590967702

27. Matsumoto M, Masuda H, Hubacz R, Horie T, Iyota Y, Shimoyamada M, Ohmura N. Enzymatic starch hydrolysis performance of Taylor-Couette flow reactor with ribbed inner cylinder. Chemical Engineering Science. 2021;231: 116270. DOI: 10.1016/j.ces.2020.116270

28. Baruque EA, Baruque MDA, Sant'Anna GL. Babassu coconut starch liquefaction: An industrial scale approach to improve conversion yield. Bioresource Technology. 2000;75:49-55. DOI: 10.1016/S0960-8524(00)00026-2

29. Baks T, Ngene IS, van Soest JJG, Janssen AEM, Boom RM. Comparison of methods to determine the degree of gelatinization for both high and low starch concentrations. Carbohydrate Polymers. 2007;67:481-490. DOI: 10.1016/j. carbpol.2006.06.016

30. Baks T, Kappen FHJ, Janssen AEM, Boom RM. Towards an optimal process for gelatinisation and hydrolysis of highly concentrated starch–water mixtures with *alpha*-amylase from *B. licheniformis*. Journal of Cereal Science. 2008; 47:214-225. DOI: 10.1016/j. jcs.2007.03.011

31. Özbek B, Yüceer S. α-Amylase inactivation during wheat starch hydrolysis process. Process Biochemistry. 2001;37:87-95. DOI: 10.1016/S0032-9592(01)00175-3

32. van der Veen ME, Van Iersel DGV, van der Goot AJ. Shear-induced inactivation of α-amylase in a plain shear field. Biotechnology Progress. 2004;20:1140-1145. DOI: 10.1021/ bp049976w

33. Paolucci-Jeanjean D, Belleville MP, Rios GM, Zakhia N. Kinetics of continuous starch hydrolysis in a membrane reactor. Biochemical Engineering Journal. 2000;6:233-238. DOI: 10.1016/S1369-703X(00)00092-9

34. Hubacz R, Buczyńska M. Starch gelatinisation in Couette-Taylor flow apparatus. Chemical and Process Engineering. 2011;4:257-279. DOI: 10.2478/v10176-011-0021-7

35. Masuda H, Horie T, Hubacz R, Ohmura N. Process intensification of continuous starch hydrolysis with a Couette–Taylor flow reactor. Chemical Engineering Researh and Design. 2013; 91:2259-2264. DOI: 10.1016/j. cherd.2013.08.026

36. Hubacz R, Ohmura N, Dłuska E. Intensification of Starch Processing Using Apparatus with Couette–Taylor Flow. Journal of Food Process Engineering. 2013;36:774-785. DOI: 10.1111/jfpe.12046

37. Masuda H, Horie T, Hubacz R, Ohmura N. Numerical and experimental investigation of continuous starch hydrolysis with a Couette–Taylor flow reactor. In: Proceedings The 24th International Symposium on Transport Phenomena (ISTP 24) ; 1-5 November 2013; Yamaguchi. Japan; 2013. p. 402-407

38. Hubacz R, Masuda H, Horie T, Ohmura N. Thermal treatment of starch slurry in Couette-Taylor flow apparatus. Chemical and Process Engineering. 2017;38:345-361. DOI: 10.1515/cpe2017-0027

39. Masuda H, Horie T, Hubacz R, Ohmura N, Shimoyamada M. Process development of starch hydrolysis using mixing characteristics of Taylor vortices. Bioscience, Biotechnology and Biochemistry. 2017;81:1-7. DOI: 10.1080/09168451.2017.1282806

40. Van Den Einde RM, Van Der Goot AJ, Boom RM. Understanding molecular weight reduction of starch during heating-shearing processes. Journal of Food Science. 2008;68: 2396-2404. DOI: 10.1111/j.1365-2621.2003.tb07036.x

41. Masuda H, Horie T, Hubacz R, Ohta M, Ohmura N. Prediction of onset of Taylor-Couette instability for shearthinning fluids. Rheologica Acta. 2017; 56:73-84. DOI: 10.1007/s00397-0160987-7

42. Lueptow RM, Docter A, Min K. Stability of axial flow in an annulus with a rotating inner cylinder. Physics of Fluids. 1992;4:2446-2455. DOI: 10.1063/ 1.858485

43. Richter O, Hoffmann H, KraushaarCzarnetzki B. Effect of the rotor shape on the mixing characteristics of a continuous flow Taylor-vortex reactor. Chemical Engineering Science. 2008;63: 3504-3513. DOI: 10.1016/j. ces.2008.04.003

44. Richter O, Menges M, KraushaarCzarnetzki B. Investigation of mixing in a rotor shape modified Taylor-vortex reactor by the means of a chemical test reaction. Chemical Engineering Science. 2009;64:2384-2391. DOI: 10.1016/j. ces.2009.02.015

45. Wereley ST, Lueptow RM. Velocity field for Taylor–Couette flow with an axial flow. Physics of Fluids. 1999;11: 3637-3649. DOI: 10.1063/1.870228

46. Li X, Farid M. A review on recent development in non-conventional food sterilization technologies. Journal of Food Engineering. DOI: 10.1016/j. jfoodeng.2016.02.026

47. Tiravibulsin C, Lorjaroenphon Y, Udompijitkul P, Kamonpatana P. Sterilization of coconut milk in flexible packages via ohmic-assisted thermal sterilizer. LWT. 2021;147:111552. DOI: 10.1016/j.lwt.2021.111552

48. Kelder JDH, Ptasinski KJ, Kerkhof PJAM. Power-law foods in continuous coiled sterilisers. Chemical Engineering Science. 2002;57: 4605-4615. DOI: 10.1016/ S0009-2509(02)00321-4

49. Tian S, Barigou M. Using chaotic advection to enhance the continuous heat-hold-cool sterilisation process. Innovatice Food Science & Emerging Technologies. 2016;34:352-366. DOI: 10.1016/j.ifset.2016.03.007

50. Masuda H, Horie T, Ohmura N, Shimoyamada M. Intensification of heat sterilization process for liquid foods using Taylor–Couette flow system. Chemical Engineering Transactions. 2017;57:1753-1758. DOI: 10.3303/ CET1757293

51. Masuda H, Hubacz R, Shimoyamada M, Ohmura N. Numerical simulation of sterilization process for shear-thinning food in Taylor–Couette flow system. Chemical Engineering & Technology. 2019; 42:859-866. DOI: 10.1002/ ceat.201800600

52. Masuda H, Hubacz R, Ohmura N, Shimoyamada M. Effect of rheological properties of liquid foods on heat sterilization process in Taylor–Couette flow apparatus. Chemical Engineering Transactions. 2019;75:31-36. DOI: 10.3303/CET1975006

53. Horak F, Kessler H. Thermische Thiaminschädigung—Eine Reaktion 2. Ordnung. Zeitschrift Lebensmittal Untersuchung-und Forschung A. 1981; 173:1-6. DOI: 10.1007/BF01042831

54. Ilo S, Berghofer E. Kinetics of thermomechanical destruction of thiamin during extrusion cooking. Journal of Food Science. 1998;63: 312-316. DOI: 10.1111/ j.1365-2621.1998. tb15732.x

Chapter 9

THOUGHTS FOR FOODS: IMAGING TECHNOLOGY OPPORTUNITIES FOR MONITORING AND MEASURING FOOD QUALITY

INTRODUCTION

The quality and safety of human food is the concern of all the elements of this system from the consumer, producers, and manufacturers of food, food control organizations, food safety organizations, and the market niche requirements. When agricultural and food products do not meet the quality standards and safety criteria, consumers lose faith in producers leading to the loss of these products' competitiveness in the market, and then significant economic loss. Although some systems are proposed to achieve food safety and quality by achieving a set of conditions that fall under the so-called Good Manufacturing Practices (GMP) and Hazard Analysis and Critical Control Point (HACCP) which represents the best way to achieve food security through all production steps. Unfortunately, with all these requirements for GMP and HACCP systems and others, they are not sufficient to ensure the production of safe food free of contaminants and defects, so it has become necessary to introduce modern technologies to quality inspect and detect blemishes and contamination. Therefore, the focus was on the development of non-destructive, modern, fast, reliable, and applicable methods that meet the needs of both food manufacturers and producers, as well as the desires of the consumer. In the present scientific climate, an acceleration in the growth of image processing technology has been observed among rapidly growing technologies. As well, image technology forms a core research area not only in engineering and computer science disciplines but also in the agricultural and food sectors. Image analysis is a study technique that aims to quantify the characteristics of each part of the image, both concerning their location and their filling. In fact, the densities are analytically observed as a function of the position concerning a reference point. The image is observed in its most basic units which are pixels (PICture ELement). They can have a square or rectangular shape and can take one or more values depending on the type of acquisition that is made, whether mono or multispectral. Even images taken with common cameras have more than one piece of information on each pixel. In fact, in this case, each pixel will have three numerical information with numbers ranging from 0 to 255, meaning by 0 the absence of color and with 255 the maximum intensity for that band of the electromagnetic spectrum. Image analysis is of interest in many areas of study, from the medical to the criminological, from the building to the agricultural sector. This diffusion is due to the fact that it is a very objective type of analysis that is based on a certain source (the image) and bases a series of calculations on it with a rigorous statistical approach. In the agri-food sector, image analysis is very successful because the appearance of a food product has a whole series of qualitative information that is difficult to parameterize by classical methods. Sensory approaches always remain invaluable tools of judgment, especially if conducted on a representative number of subjects and if carried out with an adequate plan of relief data and statistical adaptations. In addition, as far as the visual component is concerned, human vision, as already indicated, is limited to wavelengths between 390 and 700 nm, with greater sensitivity around 550 nm. The human eye is also influenced by the brightness of the background (simultaneous contrast effect) and tends to overestimate or underestimate information at the boundary between objects of different intensities (Mach Band effect) [1]. According to the widely accepted Retinex theory [2], there are three systems of independent cones consisting of receptors that read in three different wavelength regions of the visible spectrum.

Image analysis techniques, applied in the food field, show the following main advantages: objectivity, continuity over time, and rapid decision-making. An image is a representation of a two-dimensional or three-dimensional reality according to the independent spatial coordinates of an object. This is a one-plane transposition of the object's descriptive information, placed in exact positions concerning a reference point. In this context, [3] defined the image as a two-dimensional function, f (x, y), where x and y are the spatial coordinates and the amplitude of function, f at any pair of coordinates (x, y) is called the intensity or gray level of the image at that point as depicted in **Figure 1**. Consequently, if x,y, and the amplitude values of f are finite and discrete quantities, the image is defined as a digital image. So, the digital image is composed of a finite number of elements called pixels, each of which has a particular location and value. The images are generated by the combination of an energy source (electromagnetic, but also ultrasound) and the reflection of the energy emitted by the object.

A digital image is an ordered set of numbers each representing the intensity of reflection of the electromagnetic band of which the camera is sensitive. So, the digital image is composed of a finite number of elements called pixels, each of which has a particular location and value. Most digital cameras are equipped with sensors that allow the reading of individual pixels arranged in an arrangement called the Bayer mosaic [4] as shown in **Figure 2A**, consisting of blocks of four pixels, two green, one red, and one blue. Because each pixel is sensitive only to its own color, the end result is an image with scattered red, green, and blue dots. To achieve gradual tones and smooth photography, the processor or editing software must subsequently debayer **Figure 2B**. Because each pixel is filtered to record only one of the three colors, the data for each pixel cannot fully specify each of the red, green, and blue values on its own. To obtain a color image, various demosaicing algorithms can be used to interpolate a set of complete red, green, and blue values for each pixel. These algorithms use the surrounding pixels of the corresponding colors to estimate values for a particular pixel. Different algorithms that require various amounts of computing power result in final images of variable quality. This can be done inside the camera, producing a JPEG or TIFF image, or outside the camera using raw data directly from the sensor. Multispectral cameras are equipped with as many sensors as there are bands of the spectrum from which you want to get the reflected information.

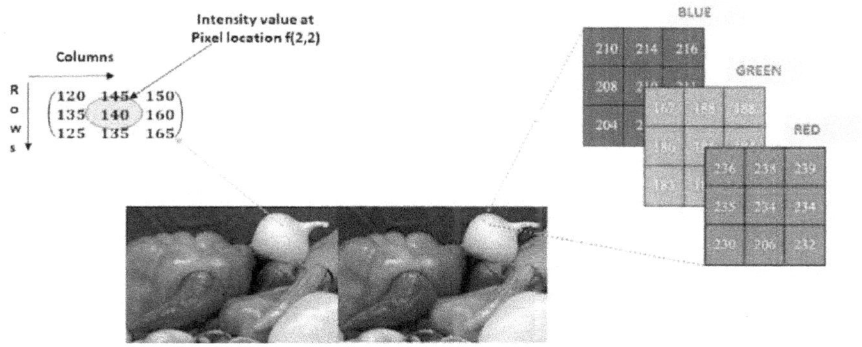

Figure 1. *Illustrates gray and color images: an array of pixels intensity and color values.*

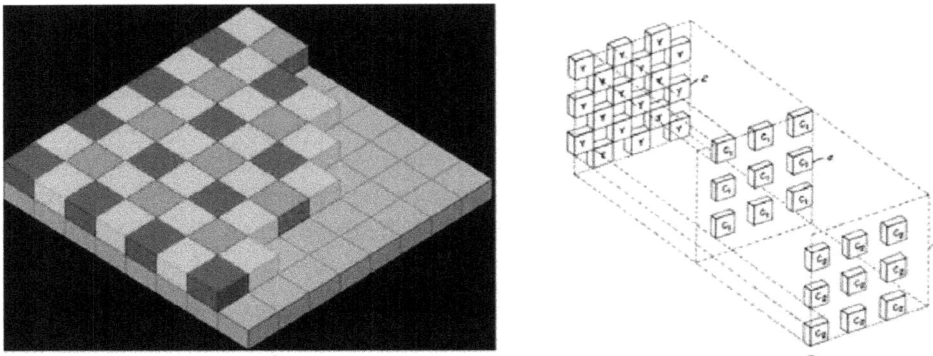

A B

Figure 2. *Depicts an arrangement of arranged individual pixels, A) Bayer mosaic, and B) Debayering Process. (http://e n.wikipedia.org/wiki/Image:US03971065_Bayer_Front.png).*

Therefore, an image obtained from a multispectral camera will be a three-dimensional array consisting of as many matrices as there are observed bands. Each pixel in the array will have the value of the intensity of the amount reflected by the photographed object, for that band of the spectrum. The same principle of operation is the basis of spectrophotometers that generally consist of a light source, a lamp, which changes typology in case it is the analysis in the visible spectrum or UV and, in some specific instruments, infrared rays. While normal photography instruments typically limit themselves to capturing the intensity of reflectance of a scene or object for a limited number of spectral bands, corresponding to those needed to produce an image that can be interpreted by the human eye, hyperspectral cameras can capture, for each pixel, the entire spectral response in a wide, almost continuous range, depending on the type of camera itself. As a result, many measurements are available for extracting useful information. Hyperspectral images, therefore, collect a considerable amount of information from the same subject or surface, but at the same time require an important amount of interpretative commitment [5]. Common cameras, which represent the most widely used image capture tool and historically early means of capture, are also optimized to capture photons of light from the visible spectrum, and from the wavebands needed to build an image that can be interpreted by the human eye, providing very limited spectral information [6]. The peaks in the reflectance spectrum, detected by specific equipment, correspond to low absorption of the incident brightness and define the so-called “spectral signature”, unique to each material [7]. Thus, the ability to exploit these differences in reflectance to characterize different materials through spectral response detection.

IMAGE ANALYSIS PROCEDURES

The steps required in the analytical procedure in the case of the study of a phenomenon through image analysis do not differ much from those of classical analytical procedures. Indeed, both methodologies involve measurements, data processing, and consequent reporting. What fundamentally changes in the preparation of the support, which in the case of classical analysis consists in the preparation of the physical sample, while in the image analysis it involves the acquisition of the image from the physical sample, properly treated [8]. The elaboration phase involves the repetition of the measurements, their statistical evaluation, and the appropriate representation of the results. Image-analysis techniques allow you to simultaneously quantify multiple visual attributes and suggest criteria for classifying certain quality performance. The images obtained from the spectrograph, therefore, represent in number the same interval between the observed and available bands. Therefore, for each observed sample, as many photos will be available as there are spectrograph reading frequencies. Samples are hardly flat, and if you want to observe in their entirety, making sure that the reading is not invasive, they must be placed in the reading compartment, without making any cuts. Therefore, for example with potatoes or apples, there is a risk of having readings that can also depend on the variable distance from the emitted source of light. For this reason, images should be observed and sampled in areas where the distance from the sensor is equal for all samples [9]. So, the first step to take is to cut out the sub-sample of photography or better than more sub-photography samples in which to read the amount read by the sensor. Processing through R software can be developed through the EBImage package (Copyright © 2003–2021, Bioconductor), importing images through the read image function, then as arrays of values where pixels are identifiable through coordinates concerning a known point of origin [10]. This allows you to prepare shapes, usually regular (squares or rectangles), to crop subplots, through coordinate extraction. So, from each sub-plot, you can read the mean values and the standard deviation of the mean value. Therefore, in R, each sample will constitute a vector that will be valued through the standard averages and deviations read from all the wavelengths used. So, the vectors of the samples will have as many values as there are wavelengths. If multiple sub-plots are used, they can be repeated for the same sample and can therefore be verified through ANOVA tests to assess the variance between the samples and that between the sub-samples. A similar image analysis path is possible even if there are few observed reflectance bands. In the case of RGB images, there are three bands (Red, Green, and Blue). In these cases, you can apply a type of analysis that relies on grayscale arrays (GLMC) [11–14]. The images are first rendered grayscale and the GLCM algorithm observes the relationship between each pixel and its neighbors **Figure 3**. In this way, parameters (homogeneity, contrast, dissimilarity, entropy) are measured that show whether the surface of the observed object is homogeneous or has irregularities.

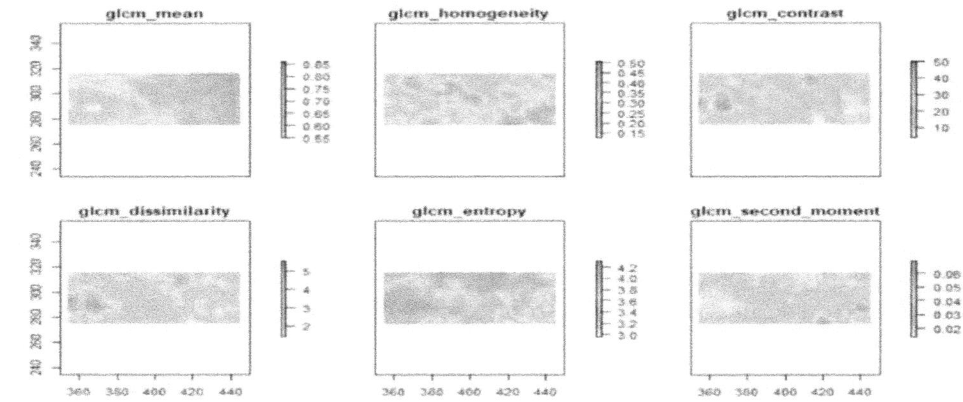

Figure 3. *GLCM on a sub-plot of a potato surface.*

SOFTWARE AVAILABLE

In recent decades, many software, or additional packages specific to generic statistical software have been developed, such as Matlab, R, or Python [15]. Most algorithms refer to multi-way analysis, such as PARAFAC, PARAFAC2, N-PLS, Tucker3, and DTLD [16]. There are also some other open-source software implementations used for the multi-way analysis of other communities. For example, the tensor toolbox [17] is powerful for analyzing a wide type of tensor, these include dense, scattered, and symmetric tensors [18] and a Matlab tensorial decomposition package called tensor box that contains various algorithms optimized for the decomposition of a tensor, such as the fast dampened CP gaussnewton algorithm [19]. Recently, some multi-way analysis software packages running in the R environment have also been developed, such as ThreeWay [20] and multiway packages [21] developed by social science statisticians. Meanwhile, there are also some Python multi-way analysis packages available, such as Tensorly [22] and TensorD [23]. Other packages were born for other applications, such as EBImage or Image. Contour detector [24–26] refer to Otsu's algorithm [27], in an intention to discriminate bodies, backgrounds, particles and are currently used for image analysis in many sectors, including agro-industrial. An example of applying the identification of the contour and then the shape is shown in **Figure 4**, [13], which shows the reading of an image of a potato tuber.

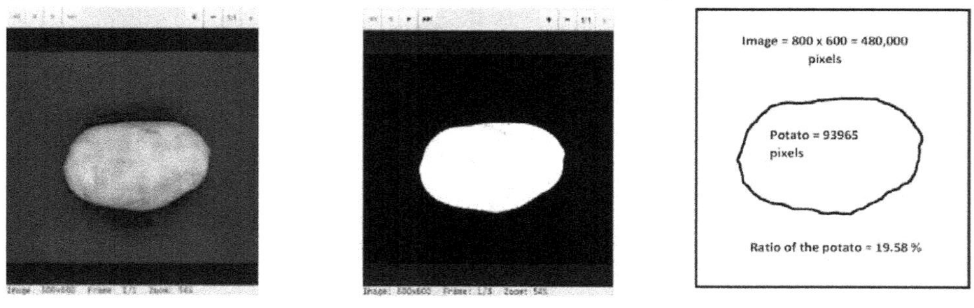

Figure 4. *Contour extraction on a potato RGB image.*

SOURCES OF DIGITAL IMAGES

Vision is the most important of the senses in human perception, so the images play the monocular most important role in enhancing this human perception. As a result of human limitations, where the human eye can identify and see objects in the visible light band in the electromagnetic spectrum. As well, in light of the evolution of consumers' desires for obtaining a high-quality and safe food product, and the inability of traditional methods to measure quality to meet the needs of the consumer, there were vigorous motives for developing imaging machines to cover almost the entire electromagnetic (EM) spectrum, ranging from gamma to radio waves. In order to be able to determine and measure the quality of food in general by identifying the appearance quality attributes, different phytochemical elements, internal structure, and detection of external and internal injuries and defects. The principal source for the images is the electromagnetic (EM) energy spectrum.

Electromagnetic radiation is an electric and magnetic disturbance that propagates through space at the speed of light (2.998 10^8 ms^{-1}). The electromagnetic spectrum is the set of all possible frequencies or wavelengths of electromagnetic waves. Depending on their frequency or wavelength, electromagnetic waves interact differently with what they encounter in their propagation. The electromagnetic spectrum is the range of all frequencies of electromagnetic radiation from the shortest to the longest wavelength that can be generated physically. This range of wavelengths can be broadly divided into regions which include gamma rays, X-rays, ultraviolet, visible light, infrared, microwaves, and radio waves as shown in **Figure 5** [28, 29]. Electromagnetic radiation from the spectrum has found multiple applications ranging from communication to manufacturing. The following is a simplified explanation of all the different types of imaging according to each wavelength on the electromagnetic spectrum.

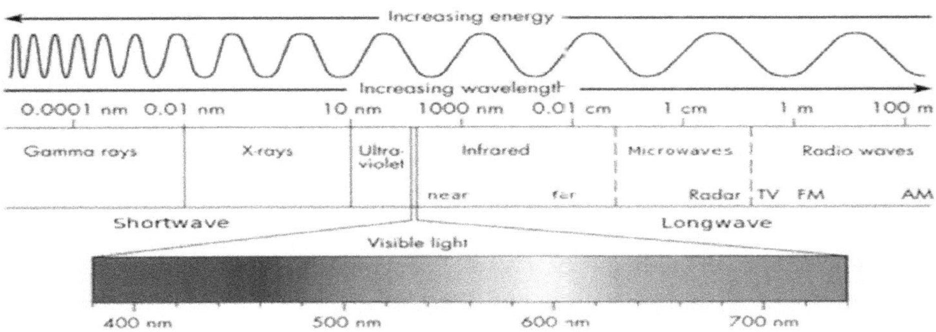

Figure 5. *Shows the different bands of electromagnetic (EM) spectrum.*

GAMMA RAYS IMAGING

Gamma rays have the smallest wavelengths (wavelength: <0.01 nm) and the most energy of any wave in the electromagnetic spectrum, that they can pass through most types of materials as shown in **Figure 6**. This high penetration property makes gamma-ray imaging technology one of the most important imaging techniques for the internal properties of extremely thick objects. Gamma waves may be generated by nuclear explosions, lightning, accelerations of charged particles by strong magnetic fields, and the less dramatic activity of radioactive decay as mentioned [30]. Gamma decay occurs when a nucleus drops to a lower energy state from a higher energy state. Unlike alpha and beta decay, the chemical element does not change and carries no charge. The resulting emission produces gamma rays. The imaging of gamma-ray photons same as any band in the electromagnetic spectrum provides the ability to determine the origin of photons in space. Predominantly, the ability of gamma-ray imaging has been used in medical applications to trace specific radioactive markers to obtain information on transport, distribution, and metabolic or more specifically, to detect cancer or to study certain dynamical behavior, such as drug additions, and recently has been applied in astrophysics applications. Gamma-ray images capture by a Gamma camera (scintillation camera) which is an instrument developed for medical diagnostics to acquiring emitted gamma radiation from internal radioisotopes to create images and this process is called scintigraphy. Gamma camera consists of a detector, collimator, photomultiplier tubes PM tubes, preamplifier, amplifier, pulse height analyzer (PHA), X–Y positioning circuit, and display or recording device [31].

The detector, PM tubes, and amplifiers are housed in a unit called the detector head. The mechanism of growth and development of agricultural and food sectors requires a major development in modern technology to monitor agricultural operations in general, in addition to food production processes. Researchers have significantly improved the performance of a gamma ray-imaging camera, which is invisible to the human eye. The new technology has potential applications in scientific research, medical treatment, and environmental monitoring. In addition to the agricultural sector, many experiments have used multiple imaging techniques with gamma rays and have reached promising results for their practical application. In this regard, [32] mentioned many studies have been carried out in the last two decades, using the gamma-ray computed tomography (CT) technique in several areas of knowledge other than medicine. As a result, used Gamma-ray computed tomography to characterize soil surface sealing and the study reached that the gamma-ray CT was able to confirm the occurrence of soil surface sealing due to the sewage sludge application and determine average densities and thickness of these layers. Through these results, concluded that the tool of gamma-ray CT allows a detailed analysis of soil bulk density profiles and the detection of very thin compacted or sealed layers. The gamma-ray computed tomography can be applied for wood density analysis, in the field for water infiltration studies and to provide information on the chemical composition of materials [33, 34]. Additionally, [30] pointed out the structure of the positron emission tomography (PET) scanners as depicted in **Figure 7** where the PET scanners detect gamma rays with a ring of gamma-ray detectors placed around the subject. Where the special tracer molecules are ingested or injected into the living tissue. The main idea is focused on preparing the tracers by especially compounds to contain one or more radioactive atoms that spontaneously emit positrons (antimatters) positively charged electrons that rapidly colloid with electrons in the neighboring atoms. Then, the collision results in the annihilation of both the positron and electron and the creation of two gamma rays with the energy of a positron or electron. Furthermore, [35] mentioned that the positron-emitting tracer imaging system is one of the powerful techniques for researching the distribution and translocation of water, photoassimilate, mineral nutrients, and environmental pollutants to plants. This system works to detects two gamma rays produced by positron-emitting nuclides with a scintillation camera and therefore enables us to study the movement of elements in intact plants in realtime.

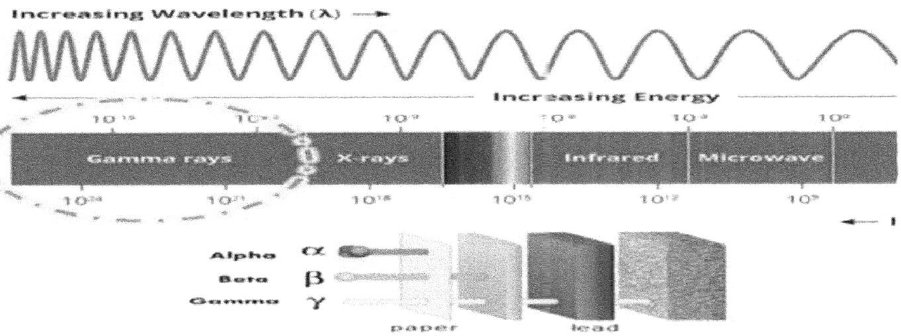

Figure 6. *Describe Gamma-ray position on the EM spectrum and the penetration property.*

Accordingly, described the PET imaging system as a more compact system and,flexible in the way to control the environment. Likewise, [36] explained that the positron-emitting tracer imaging system (PETIS) was developed to use the theory of PET in plants. It is equipped with a planar-type imaging apparatus and radioisotopes tracers such as ^{11}C, ^{13}N, ^{15}O, ^{52}Fe, ^{52}Mn, ^{64}Cu, and ^{107}Cd that are produced by a cyclotron and provides 2-D images. As well, [37] concluded that the 2-D and 3-D Gamma-ray imaging techniques have been successfully used in agriculture for the quantification and visualization of various compounds and mechanisms studies within plants such as water uptake and transportation, metal uptake, and transportation, photoassimilate translocation, etc. Modern technologies have become a necessary mechanism for growth and development in the field of agricultural production, food quality, and safety, and researchers are looking forward to achieving the best results in achieving a sustainable development strategy. From this standpoint, gamma-ray imaging has been used with success in several fields, soil analysis, [38] mentioned that the current methods for soil sampling and lab analysis for soil sensing are time-consuming and expensive. So, used hyperspectral gamma-ray energy spectra to predict various surface and subsurface soil properties. It was concluded that the developed model provided a powerful prediction of clay, course, sand, and Fe contents in the 0–15 cm soil layer and pH and course sand contents in the 15–50 cm soil layer. Also, characterized and measured the mineral uptake and translocation within plants using positron emission tomography imaging system (PETIS) such as Mn in barley [39] describe the effects of the reduced form of glutathione (GSH) and study the behavior in the roots oilseeds rape plant [40]. Also, [41] applied PETIS to describe the absorption, transportation, and accumulation of cadmium from culture to spikelet in an intact rice plant. Furthermore, [42] studied using the ^{64}Cu as a tracer in the soybean plant for the transportation from root to the leaves and concluded that the ^{64}Cu could be a useful tracer for the use in plant studies such as the distribution and translocation of copper in intact plants using the PETIS as shown in **Figure 8**. Subsequently, [43] investigated the ability of the PETIS to visualize and quantitatively analysis of the real-time Cd dynamics from roots to grains in rice cultivars that differed in grain Cd concentrations using PETIS. Moreover, the utilize of positron emission tomography imaging system (PETIS) in the field of tracking water uptake and translocation within plants was studied. Where [35] studied the effect of Aminolevulinic acid (ALA) on H ^{15}O translocation from the roots to the shoots of rice plants in real-time by PETIS technology. As well, [44] applied PETIS technology to study the effect of light on $H_2^{15}O$ flow in rice plants. Where found that the plants were exposed to low light, the H ^{15}O flow was activated more slowly. By the same token, [45] studied the visualize of ^{15}O-water flow in tomato and rice plants in light and darkness by using PETIS technology. Although the applications of Gamma-ray imaging techniques are mainly used for research and development purposes, it has extremely great potential to serve as a tool for the development of several operations in the agriculture and food sectors.

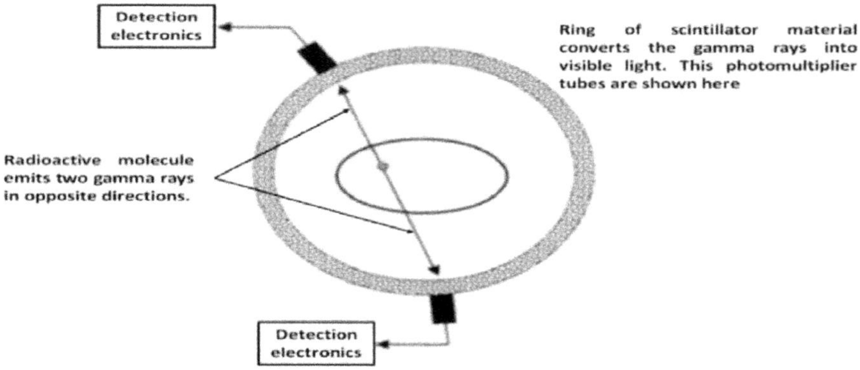

Figure 7. *Illustrative diagram of a positron emission tomography (PET) scan.*

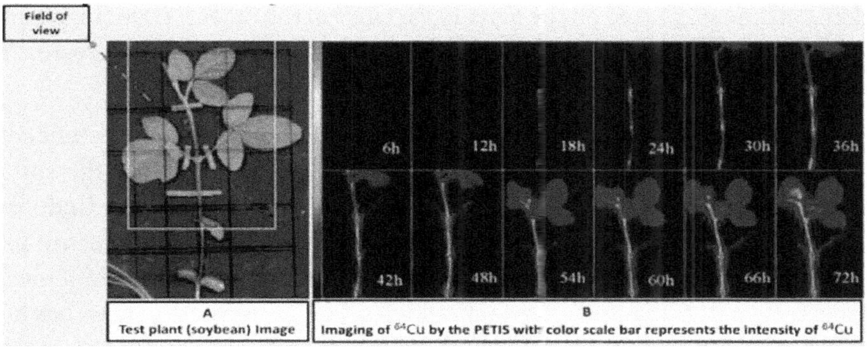

Figure 8. *Depicted the positron emission tomography imaging system setup for soybean.*

X-rays imaging

X-rays are a kind of invisible electromagnetic energy with short wavelengths ranging from 0.01 to 10 nanometers and high frequency from 3 x 10^{19} to 3 x 10^{16} Hz, and thus high energies in the range 120 electron Volt (eV) to 120 kiloelectron Volt (keV), and it falls in the range of the electromagnetic (EM) spectrum between ultraviolet radiation and gamma rays. X-rays are short electromagnetic waves that behave like particles while interacting with the matter as discrete bundles of energy and are called photons or quanta. Almost, X-rays are classified into soft X-rays and hard X-rays. Soft X-rays have relatively short wavelengths of about 10 nanometers, while hard X-rays have wavelengths of about 100 picometers [46] as shown in **Figure 9**. In general, [3] mentioned that X-rays are among the oldest sources of EM radiation used for imaging. The best-known use of X-rays in medical diagnostics, but they also are used extensively in industry and other areas, like astronomy. Besides medical diagnostics imaging and astronomy, there are other applications of X-rays such as checking luggage at the airport, inspecting industrial ingredients, and security.

As a result of the powerful penetrating X-ray, it has become one of the most important modern applications used in the inspection of agricultural products and food in general. X-ray imaging techniques are the least used in non-destructive methods for internal quality evaluation which are gaining popularity nowadays in various fields of agriculture and food quality evaluation. Although, X-ray techniques, so far predominantly used in medical applications but also have been explored for internal quality inspection of several agricultural products nondestructively when quality attributes are invisible on the surface of the products. Given considerations of product safety, consumer health, and meeting market needs, the non-destructive nature of these techniques has great potential for wide applications on agricultural and food products. In short, the action idea of an X-ray imaging system is based on the principle of transmission imaging technique, that the X-ray beam penetrates the object and attenuates based on the density variance of the object. Then this attenuated energy that passed through the object is identified through a photodetector, a film, or an ionization chamber on the other side.

Thus, the attenuation coefficients of the object components lead to different contrast between these components [46–50]. Accordingly, [51] reported that the soft X-ray method was rapid and took only 3–5 s to produce an X-ray image. Undoubtedly, X-ray inspection systems are becoming one of the best solutions to ensure product quality, safety and prevent risks in the food sector. Based on the combination of multispectral and X-ray imaging technologies [52] presented a new method for automatic characterization of seed quality. This new method included the application of a normalized canonical discriminant analyses (nCDA) algorithm to obtain spatial and spectral patterns on different seed lots. Reflectance data and Xray classes based on linear discriminant analysis (LDA) were used to develop the classification models. Concluded that multispectral and X-ray imaging has a strong relationship with seed physiological performance. Reflectance at 940 nm and X-ray data showed high accuracy (>0.96) to predict quality traits such as normal seedlings, abnormal seedlings, and dead seeds as shown in **Figure 10**. These techniques can be alternative methods for rapid, efficient, sustainable, and nondestructive characterization of seed quality in the future, overcoming the intrinsic subjectivity of the conventional seed quality analysis. In a serious study for nondestructive inspection and detection of foreign materials in food products, [53] demonstrate a method for novelty detection of foreign objects **Figure 11** such as wood chips, insects, and soft plastics in food products using grating-based multimodal X-ray imaging. Through using X-ray imaging technique with three modalities absorption, phase contrast, and dark field to pixel correspondence and enhancing organic materials such as wood chips, insects, and soft plastics not detectable by conventional X-ray absorption radiography.

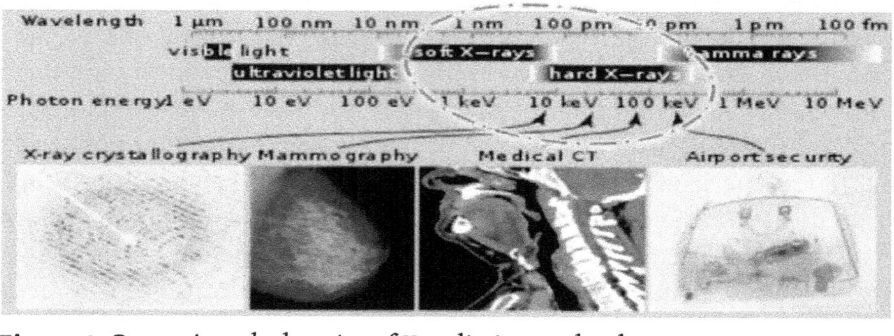

Figure 9. *Determines the location of X-radiation on the electromagnetic spectrum.*

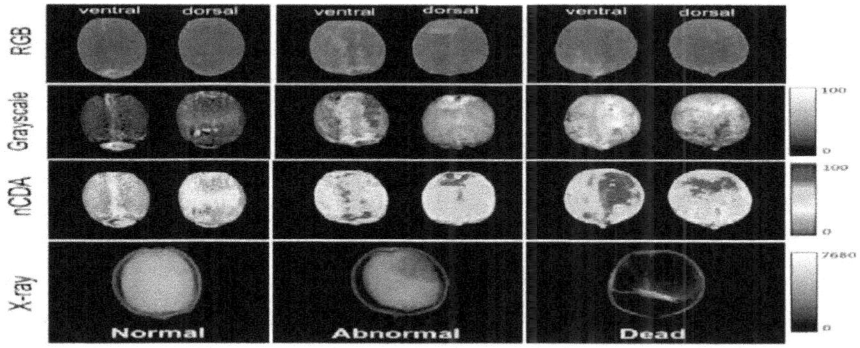

Figure 10. *Illustrates raw RGB images, reflectance images captured at 940 nm (grayscale and transformed images using nCDA algorithm), and X-ray images of ventral and dorsal surfaces of Jatropha curcas seeds.*

Figure 11. *Describes eight different foreign objects in three size groups.*

An example of X-ray images obtained of all food products with foreign objects from size group 2 at absorption, contrast, and dark field, from top to bottom, respectively as shown in **Figure 12.** It is clearly visible that there is a different contrast between the three imaging modalities. Concluded that the results give a clear indication of superior detection results from the grating-based method, and especially show promising detection results of organic materials. At the same time,

[54] used the X-ray Imaging technique in a study conducted to detect the Infestation by Saw-Toothed Beetles of stored dates fruits, where its main goal was to investigate the capability of X-ray imaging in detecting internal infestations caused by the saw-toothed beetle in stored date fruits.

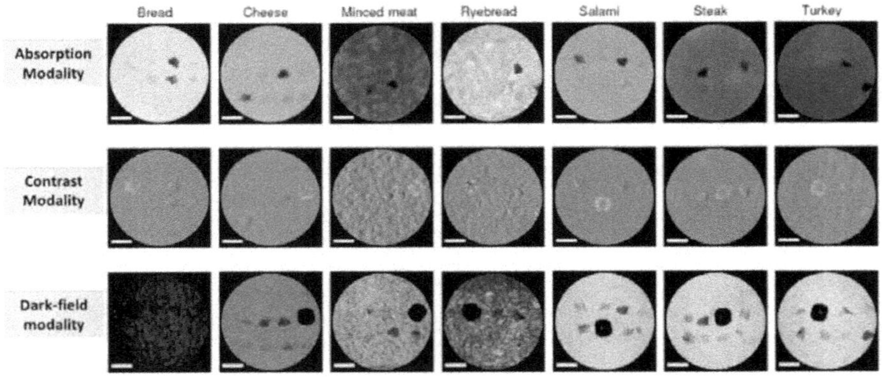

Figure 12. *Shows the X-ray images obtained for seven food products with foreign materials from size group 2, and the white bar represents 1 cm.*

X-ray images of the dates were acquired at 40 kV potential and 1.6 mAs with a resolution of 512 x 512 by using an X-ray machine as shown in **Figure 13.** In the final analysis, the X-ray imaging system yielded around 97% accuracy in detecting internal infestation of dates with an adult beetle while using a pairwise classification method. Similarly, [55] presented an approach for visual detection of organic foreign objects such as paper and insects in food products using X-ray dark-field imaging. The results proved that the dark-field modality gave larger contrast-tonoise ratios than absorption radiography for organic foreign objects. Additionally, [56] developed an adaptive X-ray image segmentation algorithm based on the local pixels intensities and an unsupervised thresholding algorithm for the determination of infestation sites of several types of fruit such as citrus, peach, guava, etc.

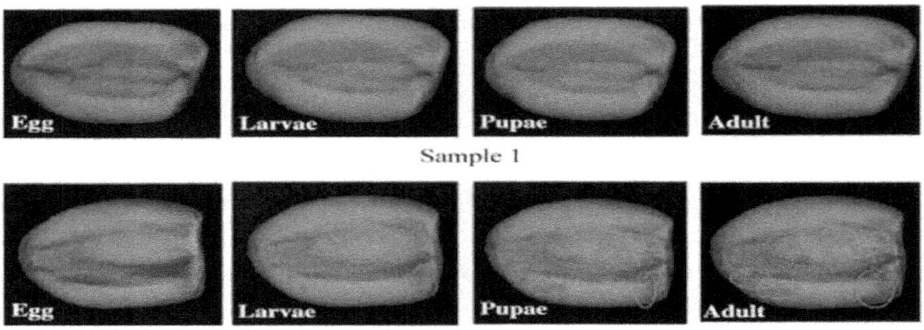

Figure 13. *Shows X-ray images of two dates infested.*

The X-ray images were acquired through an X-ray imaging system which consists of a microfocus X-ray source, a line-scan sensor camera both of which are controlled by a desktop computer, and a frame grabber board to acquire and transfer the signal from the line-scan sensor to the host computer as shown in **Figure 14.** The developed algorithm proved fast in computation time and was implemented in the X-ray scanner for real-time quarantine inspection at a scanning rate of 1.2 m/min.

Thus, suspected sites of infestation inside the fruit can be accurately marked on the acquired X-ray image to aid the quarantine officer during the inspection **Figure 15** for guava and peach. In conclusion, the detection accuracies of the infestation detection experiments for guava and peach fruits were apparently affected by the selection of sub-image and it decreases as the sub-image size increases. Where, the detection accuracy slightly increased to 95% (guava) and 98% (peach), by reducing the sub-image size to 12 x 12 in the adaptive segmentation procedure. Furthermore, there are many studies on the applications of X-ray imaging techniques for inspection and evaluation quality that have been reported in the field of food and agriculture.

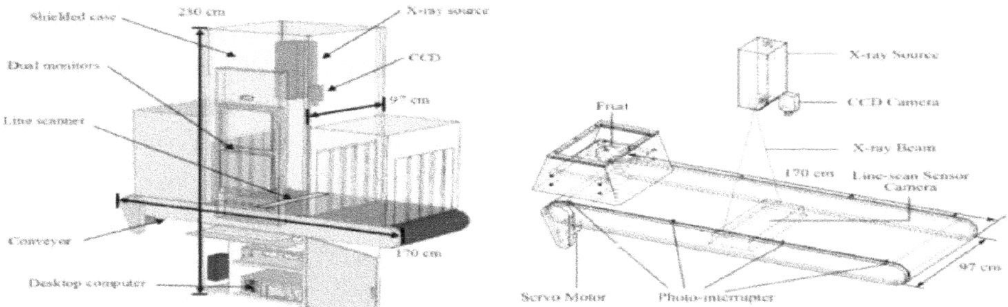

Figure 14. *Schematic drawing of the X-ray imaging system.*

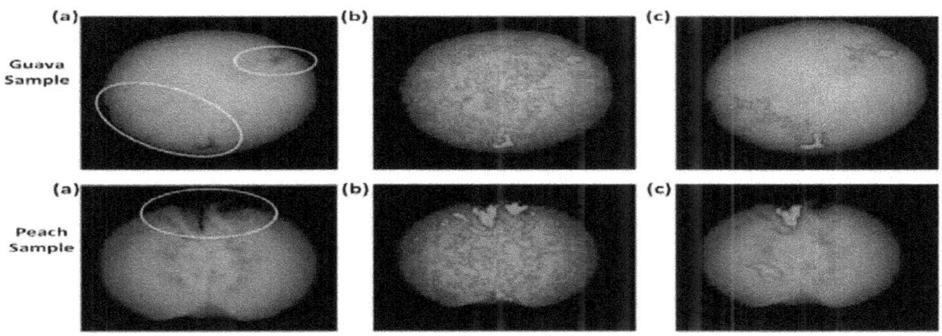

Figure 15. *Illustrates effect of morphological filtering: (a) X-ray image of a sample, (b) segmented spots after adaptive thresholding of (a), and (c) morphological filtering of (b) with three iterations.*

For example, [57] used the X-ray image to create a method of measuring the mass of wheat grains via calculating the total grey value. Detecting internal defects in grains or seeds by applying X-ray imaging has also shown promising results where, [58,

59] proved that the X-ray imaging technique can identified wheat grains infested by weevils. Also, the apple bruises were detectable using X-ray imaging and the extracted image features can be used to sort defective apples [60]. As well, [61] concluded that digital X-ray images can detect the internal disorder that leads to tissue breakdown such as the watercore in apples.

Ultraviolet (UV) imaging

Ultraviolet (UV) radiation has a shorter wavelength and higher energy than visible light band covers the wavelength range 100–400 nm. Moreover, UV radiation was divided into three bands UVA (320–400 nm), UVB (290–320 nm), and UVC (100–290 nm) is the most damaging type of UV radiation. However, it is completely filtered by the atmosphere and does not reach the earth's surface. Indeed, [30] mentioned that the imaging cannot be used in the region below 290 nm while UVB scattered more than the UVA and visible light. In reflected-UV imaging, UV illumination reflects off a scene then recorded by a UV sensitive camera while in UV fluorescence imaging. Also, UV illumination stimulates fluorescence at a longer wavelength UVA than the UV excitation source. The resulting fluorescence is typically in the visible light band. Applications of the ultraviolet band are varied. They include lithography, industrial inspection, microscopy, lasers, biological imaging, and astronomical observations [3]. In addition, [30] mentioned that UV light tends to be absorbed strongly by many organic materials and makes it possible to visualize the surface topology of an object without the light penetrating the interior parts. In the UV imaging field, little research on the UV camera as the main part of computer vision system based on image processing system has also been carried by the researchers. For example, [62] pointed to that running research focused on the study of reflected ultraviolet imaging (UV) technique, its potential of detecting defects in mangoes, and to develop a computer vision system that could find the reflected area on injured or defected mango's surface. So, they studied the possibility of a reflected UV imaging technique for the detection of defects on the surface area of mango. concluded that the distinction between RGB color and reflected UV imaging is very clear as shown in **Figure 16**. The band-pass filter of 400 nm wavelengths was found more suitable to detect the defected or ruptured tissues of mangoes. It might be due to the high photographic value of the UV-A band and since the reflected UV photography well performed over 360 nm as mentioned by [63]. Accordingly, an algorithm for defect segmentation can be developed and CVS could combine with a UV camera and a software algorithm to detect injuries. In this context, [64] developed and tested a prototype UV-based imaging system for real-time detection and separation of dried figs contaminated with aflatoxins as shown in **Figure 17**. The prototype system was tested by using 400 dried figs.

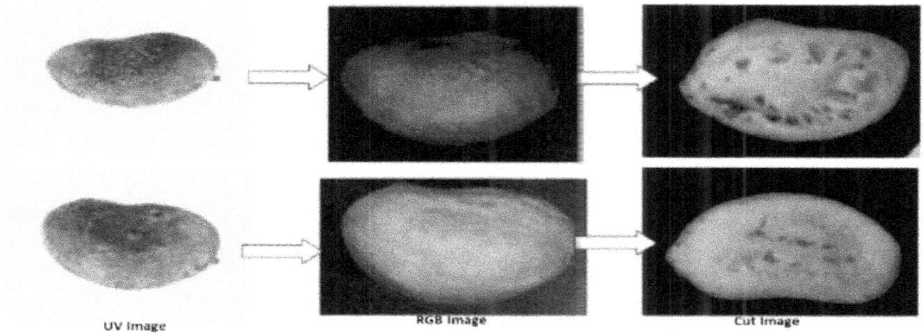

Figure 16. *Shows different images of shriveled mangoes.*

In the final analysis, the prototype system achieved a 98% success rate in the detection and separation of the dried figs contaminated with aflatoxins. Also, [65] have built a simple computer vision system to detection of anthracnose infection and latex stain by using a low-cost webcam under UV-A illumination. The UVfluorescence imaging technique has been selected for detecting areas on dried figs that are contaminated with aflatoxin [66]. Similarly, freeze-damaged oranges were also detected using the ultraviolet (UV) fluorescence method by [67] at 365 nm. Furthermore, [68] found that a UV-based computer vision system was effective in identifying stem end injuries in citrus fruits, which was used for fruit sorting. Likewise, [69] introduced a modern method based on UV imaging technique and processing images under a 365 nm UV light for separating pistachio nuts contaminated with aflatoxins. Accordingly, [70] indicated that various hidden defects inside fruits and vegetables can't be recognized by conventional systems, in contrast, can be identified by the reflected UV imaging technique. In this regard [71] mentioned that there is an important band (i.e., 365 nm) was identified during UV band selection in the application of UV-fluorescence imaging technique for inspecting aflatoxin contamination. Also, for more than 30 years [72] used UV photographs for aflatoxin-producing molds that were identified as gray or black colonies, whereas molds not producing aflatoxins appeared as white colonies.

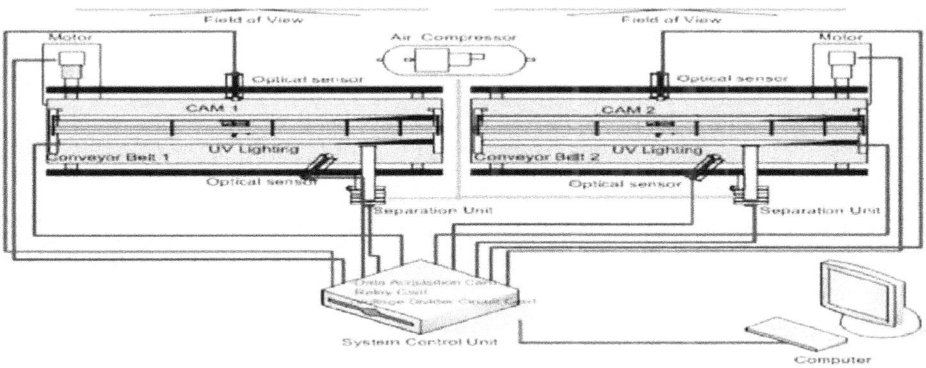

Figure 17. *Schematic of UV-based imaging system for aflatoxin contamination detection.*

Visible light imaging

The visible light spectrum is defined as the segment of the electromagnetic spectrum that the human eye can view. This visible light band is located in between ultraviolet (UV) and infrared (IR) regions, whose wavelength ranges from 400 to 700 nm as shown in **Figure 18.** Visible light is partly absorbed or diffused and partly reflected from the surfaces of objects, giving them the color, we perceive. This visible light region consists of red, orange, yellow, green, blue, and violet waves.

Obviously, each color wave is defined with a specific wavelength where violet-blue is in the area of from 400 to 475 nm, the yellow-green color of about average 550 nm, and red is located in the area of 700 nm [28, 73]. Additionally, [74] mentioned that when the light falls on an object, it is usually reflected, absorbed, or transmitted. The intensity of these phenomena depends on the nature of the material and that specific wavelength region of the electromagnetic spectrum that is being used.

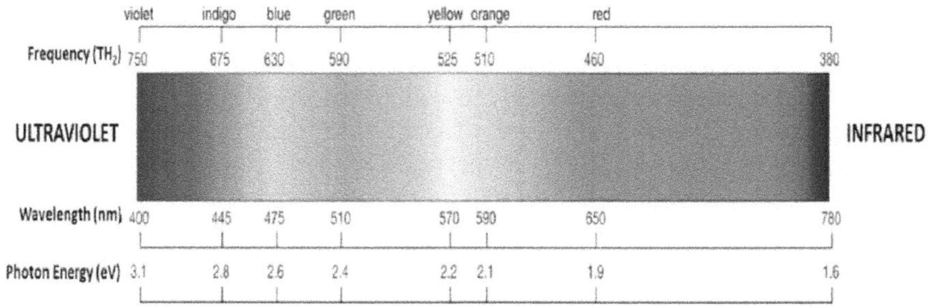

Figure 18. *Shows the characteristics of visible light on the electromagnetic spectrum.*

The visual band of the electromagnetic spectrum is the most familiar in all our life activities, it is not surprising that imaging systems based on this visible light band outweigh by far all the others in terms of scope of application [3]. More simply, the emitted, transmitted or reflected visible light from an object carries information about that object which facilitates the quality inspectors to get information concerning the quality. So, visible-light imaging systems, play a significant role to see clearer, farther, and deeper and gaining detailed information about different objects [75]. Imaging machines base on the visible light spectrum or as called color imaging systems has become an extremely significant technique for nondestructively inspecting and assessing the quality of agricultural and food products. So, the color imaging machines are considering a promising technique currently applied for quality measurement of fresh and processed food. Visible light imaging machines operation is summarized, by acquiring images under illumination standard conditions, pixels processing, and analyzing the whole image which can classify and quantify objects. Also, visible light machine vision systems scan and sort millions of items per minute and provide fast, objective, robust measurement, and detailed characterization of color uniformity at a pixel-based level [76–78]. The simplest machine vision system is mainly composed of a lighting system attached with a camera, and a computer equipped with an image acquisition board as shown in **Figure 19**, [79, 80]. The accuracy, speed, and consistency of these technological developments represented in visible light imaging have greatly increased their applications in multiple fields in agriculture and food such as applications of preand post-harvest, food industry, baking industry, cereals, meat, fish, poultry, fruits and vegetable industry, and liquids. For instance, some agricultural research has been focusing on using machine vision systems based on color imaging and developing algorithms to count agricultural elements, mainly vegetables, and fruits to determine the full maturity, production, and harvest dates [81, 82]. In this regard, [83] presented a method for identifying and counting fruits from images acquired in cluttered greenhouses. The results showed a strong correlation, 94.6%, between the automatic and manual counting data. As well, [84] estimated the mango crop yield using image analysis to count mango from orchard images, and that is through segment the pixels of the images into two groups, fruit, and background, utilizing color and texture information. Then, the mangos were identified to count the number of fruits in the image. The automatic results achieving a strong correlation of 0.91.

The number of green apples was determined by using RGB color images under natural illumination [85]. Similarly, image analysis was used before harvesting to counting the number of ripe and unripe fruits [86]. Also, [87] proposed a machine vision-based visible and NIR hyperspectral imaging method for automating yield estimation of golden delicious apples on trees at different growth stages. Subsequently, many successful attempts were recorded to use automatic vision systems based on image processing for quality control in post-harvest stages. In this context, [13] inspected potato tubers according to some sensitive quality features such as color, size, mass, firmness, and the texture homogeneity of potato surface.

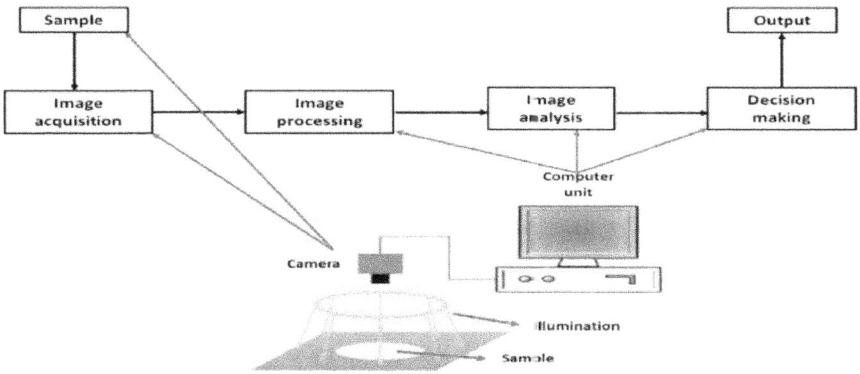

Figure 19. *Illustrates the basic components of the machine vision system.*

Figure 20 through a developed automated vision system. Concluded that the vision system can be applied as a non-destructive, precise, and symmetric technique inline inspection. Additionally, [88] applied a computer vision system and machine learning algorithms to obtain a prediction model for cherry tomato volume and mass estimation and the results achieved an accuracy of 0.97. Also, the carrot was graded using a machine vision system and the results showed that the constructed image acquisition system success to extract the feature parameters of the carrot accurately [89]. As well, [6] sorted irregular potatoes using the RGB color imaging technique. Furthermore, ripeness determination of grape berries and seeds was performed using image analysis [90]. In a similar trend, the visual quality of agricultural grain is one of the extremely important issues in grain commercialization, which is assessed based on color, shape, and size, which generally impact the product's market price.

The main problems associated with the process of grain quality inspection are the high probability of error occurrence and the difficulty of standardizing the results. So, many proposals have been presented in the field of computer vision systems to assist visual inspection quality of several agricultural grains such as rice [91–94]. Also, about beans grains, many studies show the need and the importance of computer vision systems based on image processing for bean inspection [95–99]. In this context, [100] presents a machine vision system (MVS) for visual quality inspection of beans composed of a set of hardware consists of a board that includes an image acquisition chamber, a conveyor belt controlled by a servo motor, and a feeding mechanism and software for segmentation, classification, and defect detection as shown in **Figure 21**.

The results of offline experiments for segmentation, classification, and defect detection achieved, respectively, the average success rates of 99.6%, 99.6%, and 90.0%. While the results obtained in the online mode demonstrated the robustness and viability of this machine vision system, with average success rates of 98.5%, 97.8%, and 85.0%, respectively, to segment, classify, and detect defects in the grains contained in each analyzed image. In the field of inspection quality of meat, imaging methods have been recently applied to visually assess meat and foodstuff quality on the processing line based on color, shape, size, surface texture features [101, 102]. A machine vision system with a support vector machine was utilized to grade the beef fat color. The highest performance percentage of the SVM classifier obtained was 97.4% [103]. Moreover, [104] mentioned that RGB color imaging has been a promising technique for predicting the color of meat. The moisture content of cooked beef joints was correlated with its color, using an RGB color imaging system [105]. The combination of machine vision, linear and nonlinear classifiers was employed for the automatic sorting of chicken pieces like breast, leg, fillet, wing, and drumstick.

The results revealed that the total accuracy of online sorting (highest speed about 0.2 $m.s^{-1}$) was 93% [106]. As well, in the case of fish, visible-light imaging technology has been able to successfully predict the breed, species, quality, and gender of the fish [102]. Also, a machine vision method was used to evaluate the freshness of some fish. The best classification performance was achieved by the support vector machine classifiers with an 86.3% accuracy rate in the assessment of the carp fish based on its freshness [107].

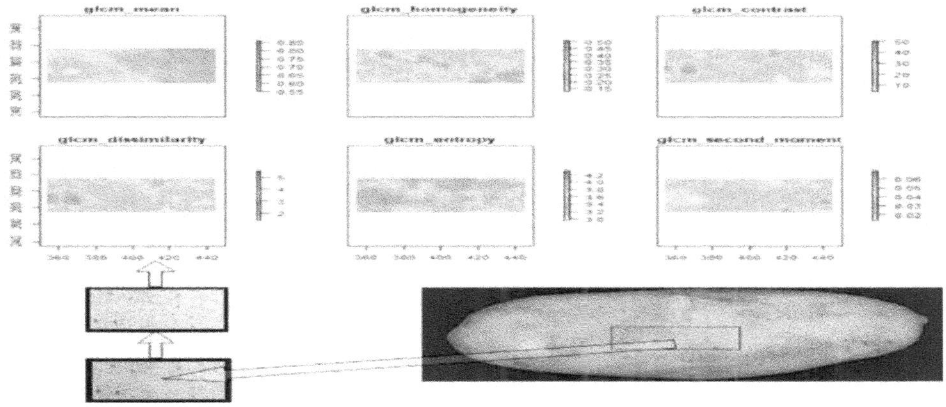

Figure 20. *Describe the extracting distinctive texture features of potato.*

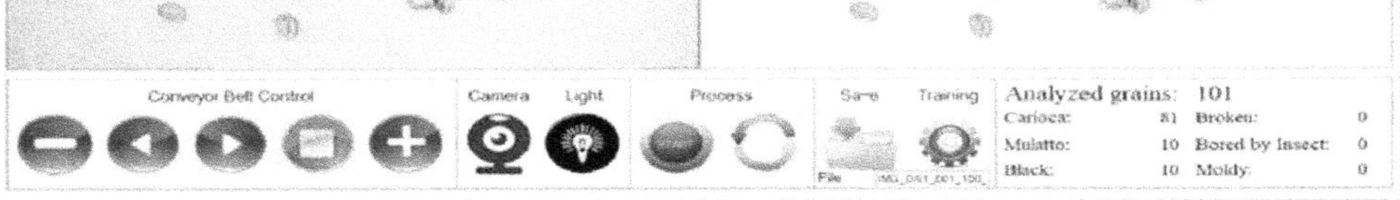

Figure 21. *Interface of the software of machine vision system for beans grain.*

Infrared (IR) imaging

Infrared (IR) radiation is a type of electromagnetic spectrum, a continuum of frequencies produced when atoms absorb and then release energy that's invisible to human eyes but that we can feel as heat. IR radiation is emitted by any object with a temperature above absolute zero and the most common sources of infrared radiation are the sun and fire. IR radiation exists in the electromagnetic spectrum at frequencies above those of microwaves and exactly below those of the red visible light band, hence it was called "infrared" as shown in **Figure 22.** IR frequencies range from about 300 (GHz) up to about 400 (THz). Waves of infrared radiation are longer than those of visible light, ranging from 0.75 to 1000 µm, and are divided into near (NIR, 0.78–3 µm), Mid-Infrared (MIR, 3–50 µm), and Far-Infrared (FIR, 50–1000 µm) as defined by the International Organization for Standardization (ISO 20473, 2007) optics and photonics-spectral bands, [108, 109]. The infrared spectrum (IR) is invisible to the human eye but has a wide range of uses in modern technology.

Different wavelengths (NIR, MIR, and FIR) of IR radiation have many different applications. The sources of IR introducing great technological advancements in imaging, thermal imaging, motion detection, gas analyzing, monitoring, and environmental health analysis, etc. IR imaging is widely used in the military, medical, scientific, and industrial fields, since it is able to create a visual with an otherwise non-visible wavelength band to the human eye [110, 111]. Recently, multispectral, and hyperspectral imaging systems based on the IR spectrum, have been used for developing and evaluating most agricultural and food processing operations, such as tracking and estimating the quality of agricultural and food products.

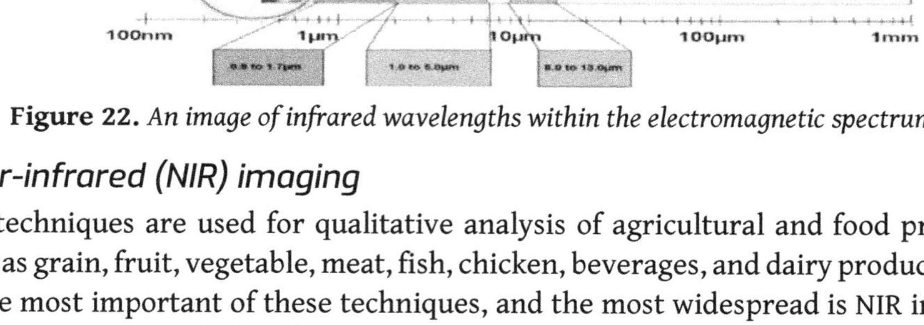

Figure 22. *An image of infrared wavelengths within the electromagnetic spectrum.*

Near-infrared (NIR) imaging

NIR techniques are used for qualitative analysis of agricultural and food products such as grain, fruit, vegetable, meat, fish, chicken, beverages, and dairy products. One of the most important of these techniques, and the most widespread is NIR imaging and spectroscopy, which offers a rapid, non-destructive, and cost-effective method. Development in instrumentation and data analysis techniques of NIR imaging and spectroscopy, expanded the application range to chemical analysis, agricultural and food product analysis, and more. So, NIR imaging is one of the preferred quality monitoring methods in the food industry [112, 113]. Conventional methods of agricultural and food product monitoring are time-consuming, expensive, and require sample destruction. So, the trend was towards fast, accurate, and non-destructive methods. NIR spectroscopy was established as a non-destructive method for quality analysis of food materials as mentioned by [114]. Ordinarily, when the IR radiation interacts with matter, the energy can be absorbed and result in molecular vibrations for example stretching, bending, rocking, wagging, and twisting. Hence, a change occurs in the electric dipole moment (change in the positive–negative charge separation) of the molecule, and the molecules transition to different vibrational levels as shown in **Figure 23**. These, transitioning from 1st to the 2nd, 3rd, or 4th excited state are known as overtones, and NIR spectroscopy measures these overtones [115]. So, NIR spectroscopy can therefore be used to study organic samples, which contain chemical bonds such as (C-H, O-H, N-H) because these functional groups absorb the energy from radiation in this region [116].

Therefore, [75, 117] mentioned that instead of individual compounds, major functional groups were assigned to specific NIR regions, where at a given wavelength range, a chemical bond will absorb the energy at a specific frequency when the energy matches the energy required to induce a vibrational response **Figure 24**.

Hyperspectral imaging based on the NIR band is the most widely used in the quality determination of agricultural and food products. However, NIR spectroscopy assessments do not contain spatial information, which is important to many food inspection applications. Furthermore, the inability of NIR spectrometers to capture internal constituent gradients within food products may lead to discrepancies between predicted and measured composition. Also, conventional Vis/NIR imaging provides only spatial information and does not supply any spectral information, which may lead to deficiencies in monitoring and evaluating the quality of products [118–120]. To overcome this, multispectral and hyperspectral imaging systems have been developed to combine images that contain spatial and spectral information, acquired at narrow wavebands, sensitive to features of interest on the object.

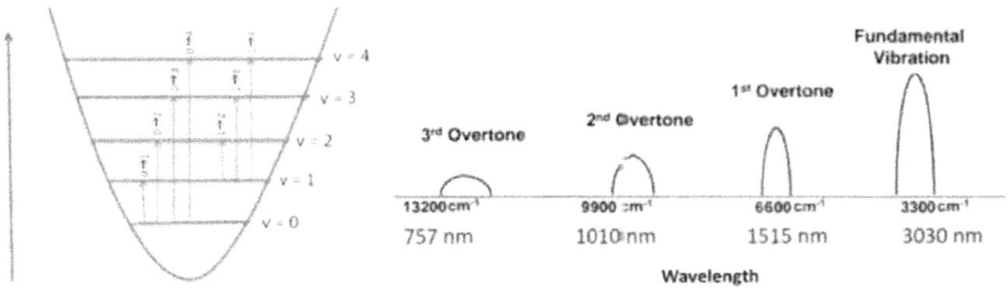

Figure 23. *Shows different vibrational levels for molecules and overtones transitions.*

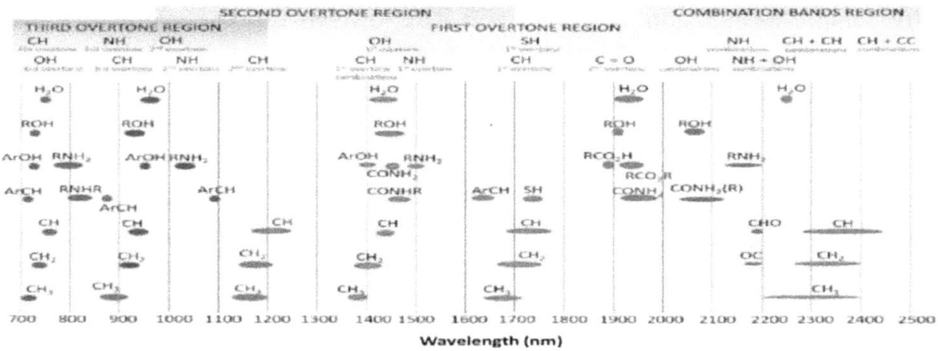

Figure 24. *Shows major analytical bands and relative peak positions for major NIR absorptions.*

Hyperspectral imaging (HSI)

Hyperspectral imaging (HSI) or spectroscopic imaging is one of the most promising emerging technologies that integrates conventional imaging and spectroscopy to acquire both spatial and spectral information from an object. Although HSI was originally developed for remote sensing, it has recently emerged as a powerful process analytical tool for automatic non-destructive analysis of agricultural and food products [6, 121–125]. Where, the non-destructive, and flexible nature of HSI makes it an attractive process analytical technology for the identification of critical control parameters that impact finished product quality. As a result, expected [126, 127] that HSI will be increasingly adopted as a process analytical technology for quality monitoring of agricultural products and the food industry, as has already been the case in the pharmaceutical industry. There is an equally significant aspect, where the importance of the HSI system is that it consists of hundreds of neighboring wavebands for each spatial position (pixel) within the image. Hence, the spectrum considers like a fingerprint that can be used to characterize the composition of that pixel. HSI images are three-dimensional blocks of data, including two dimensions as spatial position and one spectral dimension, so this HSI is known as hypercubes, as clarified in **Figure 25**. Each hypercube consists of 50–300 images acquired at different wavelengths with a spectral resolution of 1–10 nm. Another significant factor is that the hypercubes (HSI) permit the visualization of biochemical constituents of a sample, as separated forms into areas of the hyper image [6, 122, 128–132]. In brief, the main idea of the HSI imaging system running is that when the electromagnetic spectrum beam incident on the sample during sample analysis, the radiation turns into forms of reflection, scattering, absorption, and emit electromagnetic energy obtaining different patterns in specific wavelengths, due to the difference in chemical composition and physical structure of the sample. As a consequence, each element has a spectral fingerprint declaring its chemical composition. So, differences in the chemical concentration of the constituents of the sample lead to different reflectance or absorbance values in some main wavelengths [130, 131, 133].

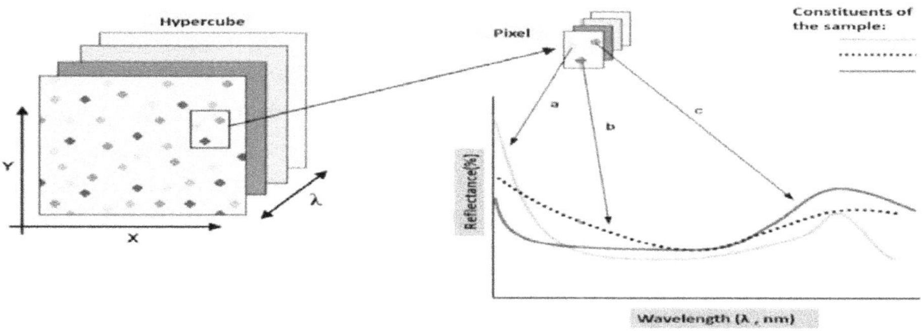

Figure 25. *Schematic of HSI hypercube, the spectral and spatial dimensions relationship.*

Generally, the structure of the HSI system consists of some major components: lens, spectrograph, camera, translation stage, illumination unit, and computer system **Figure 26.** Then, when the sample is highlighted by diffuse illumination such as tungsten-halogen or LED source. then, the sample reflects the light to the lens and is separated into its component wavelengths by diffraction optics contained in the spectrograph, then a two-dimensional image (spatial and spectral dimensions) is formed on the camera and saved on the computer system [122, 134].

Consequently, these technological developments in HSI techniques based on NIR as a measuring non-destructive method, accurate, reliable, and fast for quality and safety analysis, have greatly increased the applications in a wide range of agricultural and food products. In this paragraph, some applications will be listed that demonstrate the capability of HSI in the field of food to perform classification, defect and disease detection, and assessment of some chemical characteristics.

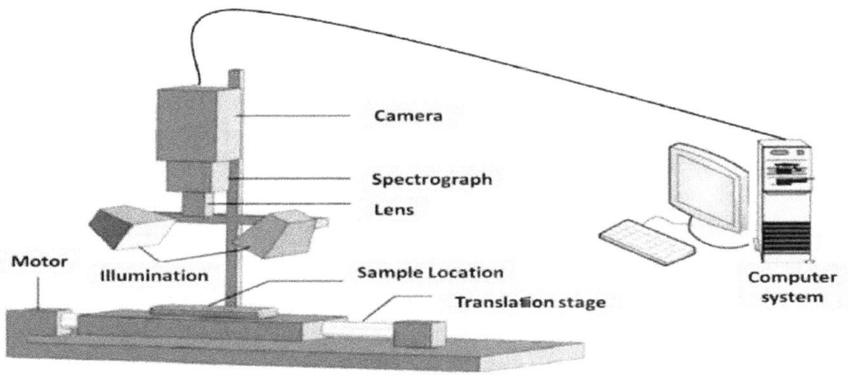

Figure 26. *Diagram of the hyperspectral imaging system components.*

Furthermore, [135] clarified that the HSI systems can be used for the discrimination of different types of grains, including maize, wheat, barley, oat, soybean, and rice seed, etc. For instance, [136] developed indices for Norway spruce (Picea abies) seeds screening through applying HSI at different wavelengths 1310, 1710, and 1985 nm and the results showed a good classification, recommending the possibility to build inexpensive devices. As well, [137] used HSI based on the NIR band to explore the influence of grain shape and texture on the spectral variation represented in three kinds of cereal barley, wheat, and sorghum using PCA and gradients classification. Concluded that the results of classification gradient images and PC score plots were 91.18, 89.43, and 84.39% respectively, and all were influenced by kernel topography. An equally significant aspect is determining the viability of seeds by applying HSI at different spectral ranges (400–1000, and 1000–2500 nm). Visualization of treated and non-treated corn seeds was also achieved with HSI. The results demonstrated that the spectral range in the 1000– 2500 nm performed better in exploring the seed viability [138]. Also, [139] classified viable and non-viable kernels of different cultivars of barley, wheat, and sorghum by using the NIR-HSI system. The results showed that NIR hyperspectral imaging is capable to identify viable and non-viable kernels of different cultivars. In a study for industrial baking of sponge cakes [140], the production process required various quality indicators to be measured continuously such as moisture content and sponge hardness. The existing techniques for performing these measures, randomly selected sponges are removed from the production line, and then samples are manually cut from each sponge by a destructive method to test as shown in **Figure 27A**. In contrast, the authors used the NIR-HSI system with a spectral range of 900–1700 nm as a non-destructive method to predict both moisture and hardness of cake **Figure 27B**. The results showed that the moisture and hardness prediction models when using a PLS-R model were 0.99 and 0.98. Accordingly, concluded that HSI is a valid method for predicting sponge cakes' moisture content and hardness. This study established a proof of concept for a new stand-off cake moisture and hardness monitoring system. Additionally, this HSI system would provide the added advantage to record every product in an HS image, which leads to detect variations in the production process. Also, HSI systems were applied for the ripeness monitoring of a large number of different fruit varieties [141–147]. Also, defects or blemishes detection such as bruising in fruit [147–155]. Recent studies on the safety inspection of agricultural products and livestock use multispectral imaging and HSI technologies. HSI methods have been used to determine the contamination of internal secretions on the surface of chickens, surface contamination in food processing, and fecal or foreign contamination of matter for apples and lettuce [118, 156, 157]. While studying the potential application of HSI for defect identification, apple and cucumber are two of the most popular food products that are being studied for bruises and frost injury defects, respectively.

Figure 27. *Illustrated traditional measuring technique (A), and (B) NIR-HSI, (1) single band at 1450 nm, (2) binary image obtained indicating the location of the cavities (in black), (3) Binary mask selecting the center of the cake (white), air bubbles (black), and (4) cake image ready for spectral data extraction.*

Moreover, [158] took three varieties of apples to study the damage in apples and noted that the NIR region (700 and 900 nm) was more efficient at determining it. As a result, the NIR-HSI system from 900–1700 nm, to examine its application in the identification of bruises during various periods of storage after bruising was subsequently implemented by [159]. The spectrally reflective image analysis system has also been developed to assess defects on lettuce cut in the processing line. In particular, [160] algorithms have been identified to detect snails and worms.

Another significant factor, where HSI systems proved not only to detect nonobvious bruises of fruits but also capable of assessing internal quality parameters such as soluble solid content, firmness, pH value, antioxidant, etc. [161–171].

Mid and far infrared imaging

Mid and far-infrared bands of EM radiation are an extremely useful part of the spectrum. Where, it can provide imaging in the dark, trace heat signatures, and provide sensitive detection of many biomolecular and chemical signals. However, the mid-infrared (MIR) band of the electromagnetic spectrum seems to contain valuable new information about some of the features needed to differentiate the samples, for example, the samples with diseases or some contamination or for quality inspection. Also, the recent development of light sources and imaging systems in MIR allows the use of multi/hyperspectral MIR imaging in many new applications as mentioned by [172, 173]. Also, signals of all IR radiation are known to be sensitive to leaf compounds such as water, lignin, and cellulose, which are essential to the functioning and structure of the leaf [173–175]. The thermal imaging technique is defined as a non-destructive, contactless, and rapid method for capturing the IR radiation from the object's surface. Where the surfaces of the hot objects emit electromagnetic waves in the IR region. Thermal imaging systems commonly capture radiation data from

7.5 μm up to 14 μm [176, 177]. This IR range is defined as the transmission window of the atmosphere characterized by the minimum attenuation of radiation [178]. Where the idea of a thermal imaging system based on captures temperature and spatial information simultaneously.

Then, delivers the MIR data to be processed through a computer unit and provided in matrices forming called thermograms. From this point, there have been many successful attempts to apply thermal imaging systems as non-destructive and contactless methods to monitor the quality of many agricultural and food products. For example, but not limited, [37] developed an infrared thermal imaging system to detect infestation by Cryptolestes ferrugineus under the seed coat on the germ of the wheat kernels. Found that the overall classification accuracy for a quadratic function was 83.5% and 77.7% for infested and sound kernels, respectively, and for a linear function, it was 77.6% and 83.0% for infested and sound kernels, respectively, in pairwise discriminations. As well, [179] studied the feasibility of applying an IR thermal imaging system to classified fungal infections of stored wheat, the results prove that a thermal imaging system could be a useful tool to find if the wheat grain is infected by fungi or not, where the classification models gave a maximum accuracy of 100% for healthy samples and more than 97% and 96% for infected samples, respectively. Additionally, [180] developed a method to early detect apple bruising based on pulsed-phase thermography. The results indicated the high possibilities of the active thermography method for detecting defects up to several millimeters. Also, [181] conducted a follow-up study in which hyperspectral cameras were used equipped with sensors working in the visible and NIR (400– 1000 nm), short-wavelength (1000–2500 nm), and thermal imaging camera in the MIR range (3500–5000 nm) to producing visualizations of bruises and providing information about bruise depth. the results obtained confirmed that the broadspectrum range (400–5000 nm) of fruit surface imaging can improve the detection of early bruises with varying depths. Likewise, [157] adjusted an infrared lock-in thermography technique for the detection of early bruises on pears, the thermal emission signals from pears were measured using a highly sensitive MIR thermal camera. Found that the phase information of thermal emission from pears provides good metrics to identify quantitative information about both the size and the depth of damage for pears. In the same context, [182] developed a pulsed thermographic imaging system and explore its feasibility in non-destructively detecting bruised blueberries. The results demonstrated the feasibility of pulsed thermography to discriminate between bruised and healthy blueberries. Most recently, in the food processing sector, a study conducted by [183] indicated the possibility of monitoring and evaluating ovens systems through MIR imaging. Where this study aims to demonstrate the applicability of thermal imaging with image processing for the real-time evaluation of oven systems. A thermal camera was adapted to two different oven systems: a standard electric deck oven and a novel gas-fired baking oven with integrated volumetric ceramic burners as shown in **Figure 28.**

Figure 28. *Shows MIR system for monitoring the baking process: A) MIR camera, B) electric deck oven, and C) a thermogram captured during baking.*

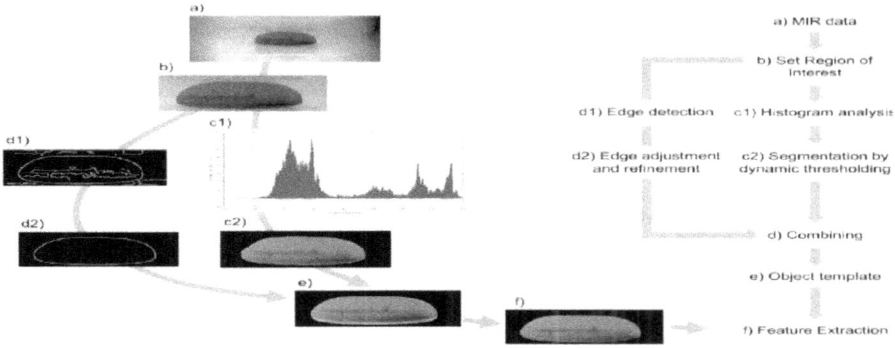

Figure 29. *Image processing steps with major operations.*

MIR data with image processing are used to accomplish a time-resolved and automated monitoring of the baking process for oven system evaluation. Therefore, items to be baked were captured by the thermal camera, detected and feature extraction was performed to calculate the quality feature relevant such as texture homogeneity, temperature distribution, spatial dimensions (width and height), and the corresponding growth kinetic as shown in **Figure 29**. The results of the proposed study proved its fundamental qualification for comparing, monitoring, and evaluating different oven systems. In the final analysis, concluded that thermal imaging is an emerging and promising technique for the food industry and offers promising possibilities for inline process sensing and monitoring in the food sector.

Microwaves imaging (MWI)

Microwave radiation appears on electromagnetic radiation, between IR and radio waves. Where, microwaves refer to alternating current signals in the frequency range from 300 MHz to 300 GHz and (3 x 108 m/sec)/frequency, which gives you a wavelength range from 1 mm to 1 meter. These dimensions allow penetrating deep inside many optically not transparent mediums such as biological tissues, concrete, soil, wood, etc. In this regard, [3] indicated that radar is the dominant application of microwave imaging techniques. Because the imaging radar technique in the microwave band can collect data over any region at any time, regardless of the weather or ambient lighting conditions. Some radar waves can penetrate clouds, and can also see-through vegetation, ice, and extremely dry sand under non-standard conditions. The imaging radar works like a flash camera that provides microwave pulses to illuminate the target area and take a snapshot image. Where, imaging radar uses an antenna instead of a camera lens, attached with digital computer processing to record its images. In a radar image, one can see only the microwave energy that was reflected toward the radar antenna. There are many similarities between optical imaging, using a digital camera, and microwave imaging, using an antenna array as highlighted in **Figure 30**. In this type of imaging known as microwave holography, one or more antennas in the array illuminate the scene with a radiofrequency (RF) signal. Part of this signal is reflected in the other antennas, which record both the amplitude and phase of the reflected signal. These reflected RF signals are then processed to form an image of the scene [184–187].

Microwave imaging techniques have shown excellent capabilities in various fields such as civil engineering, biomedical diagnostics, safety, industrial applications, and have in the latest decades experienced strong growth as a research topic in the agricultural and food fields.

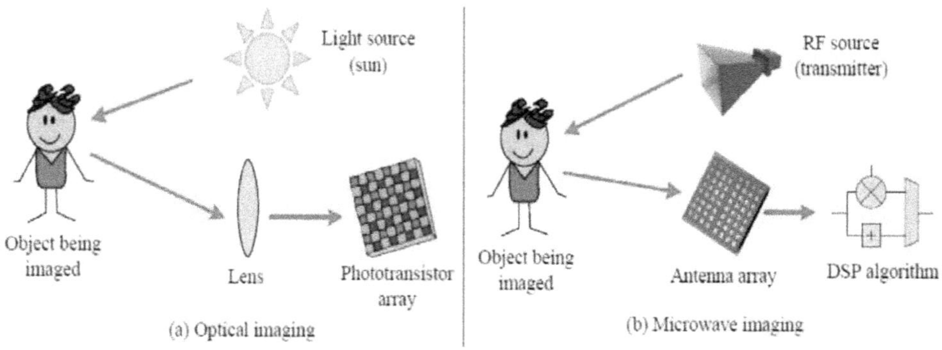

Figure 30. *Highlights similarities between visible light imaging, and microwave imaging.*

Microwave imaging technology means the initial rapid screening of the hidden objects in an object's internal structure employing electromagnetic fields at microwave frequencies (300 MHz-30 GHz). Microwave images are maps of the electrical property distributions in dielectric samples [188, 189]. Therefore, microwave imaging for agricultural and food applications is nowadays of great interest, having the potential of providing information about the internal quality of agricultural and food products. There are three main reasons for the growing interest and rapid development of microwave-based methodologies, starting with the idea that the microwave band can penetrate all materials (unless ideal conductors), and the related scattered fields are representative of the overall volume of the object under test and not only of its surface; the second main interest reason that the microwave imaging modalities are very sensitive to the water content of the specimen, which makes them extremely suitable by particularly for food processing techniques; and thirdly it contactless concerning the specimen. Microwave imaging (radar tomography) has been used to evaluate the physical properties of food. In particular, the microwave imaging technique, able to identify the composition and the shape of biological materials, for the quality control of packed foods, and identify the degree of ripeness of fruits [190–192]. It could also be said through several investigations focused on the use of microwave technologies that microwave imaging techniques are used to probe inaccessible domains and to reveal the dielectric properties of the media that they penetrate. This technique aims to fully characterize the area in terms of positions, shapes, and complex permittivity profiles of the dielectric discontinuities (i.e., the scatterers). This aim is achieved by using inverse scattering algorithms to be analyzing the scattered field reflected by the material under consideration. Therefore, inverse scattering methods have been applied in many applications such as medical diagnosis, subsurface monitoring or geophysical inspection, and nondestructive evaluation and testing in various fields [193–198]. Accordingly,,[199] focused on the application of microwave imaging technology for food contamination monitoring where the mechanism of this technology is based on transmission across the food sample to exploit the local dielectric proprieties variation that means a foreign object detection. Ordinarily, the microwave imaging system is composed of two main components hardware and software **Figure 31**, where the hardware part collects data, and the software process them to generate the output. Where the transmitter antenna generates EM waves toward the sample, that in food ambient it can be reasonably considered a homogeneous material, and a receiver antenna collects them. After their acquisition, dedicated software processes the data to generate the outputs where the detected intrusion is reported.

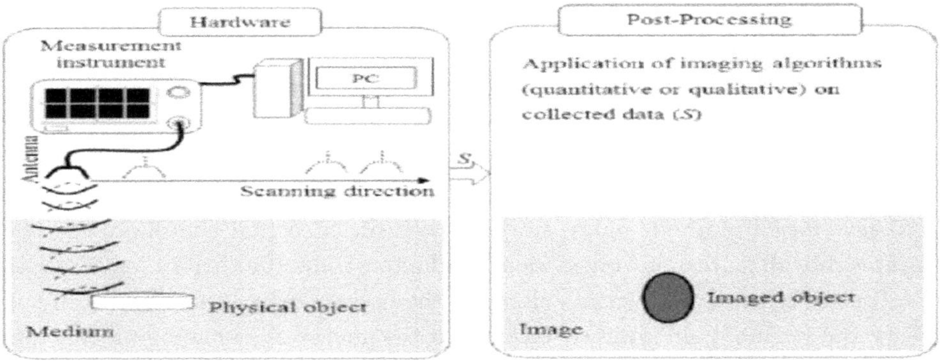

Figure 31. *Illustrates components of antenna microwave imaging system.*

By the same token, [200] investigated the dielectric properties of fresh eggs during storage through frequency range 20–1800 MHz using an open-ended coaxial probe on thick albumen and yolk of eggs after 1–15 days of storage at room temperature. Also, [201] concluded that the dielectric properties of egg albumen and yolk were distinguished over the frequency range of 10–1800 MHz. Also, [202] presented a form of food security sensing using a waveguide antenna microwave imaging system to identify the health status of eggs. Therefore, proposed a waveguide antenna system with a frequency range of 7–13 GHz and a maximum gain of 17.37 dBi, with a scanning area of 30x30 cm^2. The results found that the proposed waveguide antenna microwave imaging sensing system could effectively identify the health status of many eggs very quickly. As a consequence, concluded that the waveguide antenna microwave imaging sensing system provides a simple, nondestructive, effective, and rapid method for food security applications. Images are undoubtedly the optimum technique in representing concepts to the human brain. Regardless of whether the product is fresh fruits or prepared food, color and moisture content are important attributes that food and agricultural engineers regularly look for. Therefore, [203] suggested an investigation focused on image acquisition technologies that can reveal the information of interest in 2-D using the visible, and non-visible (radar tomography) bands of radiation. The visible band was applied for color grading of oil palms and the computerized radar tomography was used to map the moisture content in grain. The results of this study found that the vision system correctly classified 92% of oil palms by four-color categories, and the radar tomography at 1 GHz frequency accurately mapped the homogeneity and heterogeneity in moisture content of grain over the moisture range 12–39%. At the same time, the microwave imaging technique is particularly useful for monitoring foods also after the packaging, without the necessity of opening the package. This is due to the microwave's ability to easily penetrate any type of non-metallic packages. Furthermore, it has the ability to identify unwanted or extraneous objects (such as glass or plastics pieces) embedded in food that cannot be detected with standard metal detectors. Microwave imaging technique in the frequency band from 8–12 GHz has been used to assess the contents of a package of cookies. The main purpose of applying this technique was to assess and ensure whether all the cookies within the package and ensure if their shape is preserved after the distribution or not. Through **Figure 32a,** the results concluded that the imaging technique in the microwave radiation range is capable of reconstruction of a package of cookies, and it can be noticed from the microwave image that one cookie is missed, and another is broken. Also, in a study to show the potentialities and abilities of microwave imaging techniques for food processing applications. A sample of cheese was corrupted by placing a small piece of plastic material inside it to verify the ability of microwave imaging technology to detect this sample in a non-destructive manner. In this inspection technique, **Figure 32b** clearly shows the reconstruction of the dielectric distribution of the cheese piece through a microwave image, which clearly showed the presence of a small piece of plastic material and identified it as a yellow area [191].

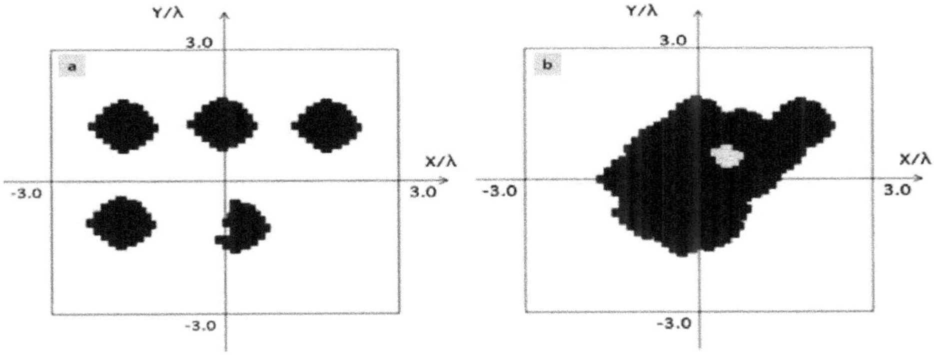

Figure 32. *Highlights microwave images, a) reconstruction of a package of cookies and b) identification of an abnormal object inside a piece of cheese.*

4.7 Radio waves imaging

Radio waves are a type of electromagnetic radiation best-known for their use in communication technologies, such as television, mobile phones, and radios.

According to NASA, radio waves have the longest wavelengths (1 mm to more than 100 km), also have the lowest frequencies, from about 3 kHz, up to about 300 GHz in the EM spectrum. The National Telecommunications and Information Administration generally divides the radio spectrum into nine bands **Table 1.** Low to medium frequencies, the lowest of all radio frequencies, have a long-range and are useful in penetrating water and rock. While, the high, very high, and ultra-high bands of radio frequencies include FM radio, broadcast television sound, public service radio, cellphones, and global positioning system (GPS). Moreover, super, and extremely high frequencies perform the highest frequencies in the radio band and are sometimes considered to be part of the microwave band. Imaging in the radio band, as in the case of imaging at the other end of the electromagnetic wave (gamma rays). Medicine and astronomy are the major applications of imaging in the radio band. Magnetic resonance imaging (MRI), or nuclear magnetic resonance scanner (NMR), is mostly known as a magnetic resonance imaging device. Because of its strong magnetism, the efficient polarization and further excites the focused proton singly included in water molecules present in the tissue. The technique of magnetic resonance imaging (MRI) is based on the magnetic field and pulses of radio radiation energy to evaluate the properties of objects, mostly applied for the diagnosis of various ailments internal to human and animal bodies. The main idea of the MRI technique is based on the magnetization of the atomic nuclei of the object using strong magnets and the nuclei rotate the magnetic field at variable speeds, which can be detected by the scanner and converted into usable data through Fourier transform. In the MRI technique, the hydrogen atom is used as a base atom because water is plentiful in all biological systems [3, 191, 204, 205].

Table 1. *Illustrates the nine bands classified of the radio spectrum.*

Numbers	Radio bands	Frequency range	Wavelength range
1	Extremely Low Frequency (ELF)	<3 kHz	>100 km
2	Very Low Frequency (VLF)	3 to 30 kHz	10 to 100 km
3	Low Frequency (LF)	30 to 300 kHz	1 m to 10 km
4	Medium Frequency (MF)	300 kHz to 3 MHz	100 m to 1 km
5	High Frequency (HF)	3 to 30 MHz	10 to 100 m
6	Very High Frequency (VHF)	30 to 300 MHz	1 to 10 m
7	Ultra-High Frequency (UHF)	300 MHz to 3 GHz	10 cm to 1 m
8	Super High Frequency (SHF)	3 to 30 GHz	1 to 1 cm
9	Extremely High Frequency (EHF)	30 to 300 GHz	1 mm to 1 cm

Noting the magnetic nature of this MRI [204, 206] has mentioned that the low magnetic nature of the hydrogen protons which have different behaviors depending on the type of the tissues (e.g., lipids and water). The inspected object is placed within the magnet usually having 0.2–3.0 Tesla magnetic field power (T). This constant magnetic field is produced by radio-frequency pulses on the appropriate resonant frequency known as the Larmor frequency. It causes an excited state for the protons in the sample due to energy absorption. These protons generate radio waves, the emission can be detected by the receiver coil, producing an NMR signal. The basis for MR imaging is measuring the intensity signal of MR, accurate spatial placement of signal intensities, and cross-sectional representation of the signal intensities with the greyscale. Having high moisture content, agricultural products yield strong signals when applying an MRI technique. Therefore, [207] monitor the ripening of mangoes by using MRI technique and found that signal magnetic resonance intensity of the pericarp in MR images varied with the ripening stage. Also, studied the prediction of sensory texture quality attributes of raw and cooked potatoes by NMR-imaging. MRI analysis on the obtained data and subsequent sensory analysis of the cooked potatoes displayed the high potential of employing advanced image analysis on MR-imaging data from raw potatoes to predict sensory attributes related to the texture of cooked potatoes. In short, concluded that MRimaging besides giving well-known information about water distribution also gives information about anatomic structures within raw potatoes, which are considered important for the perceived textural properties of the cooked potatoes. Moreover, designed an MRI apparatus characterized by its small, lightweight, and usable in an ordinary research room was devised for developmental research and quality estimation of foods and agricultural products. The proton-specified MRI was easy to operate and provided well-depicted images of internal structures, the distribution and mobility of water and oils, and susceptibility differences inside materials, demonstrating that the devised machine is useful for food and agricultural research. As well, [210] examined the changes in kiwi fruit tissue structure to evaluate the effect of storage conditions and found water migration in the direction of the outer region in the pericarp during storage. Also, [211, 212] applied MRI techniques to inspect the physicochemical changes of cherry tomatoes and found the potential of MRI techniques for tomato classification according to maturity. MRI is a technique that permits watercore detection without destroying the sample. From this point, [213] investigated the watercore distribution inside apple fruit (block or radial), and its incidence (% of tissue) by the non-invasive and non-destructive technique of MRI to obtain 20 inner tomography slices from each fruit and analyze the damaged areas using an interactive 3D segmentation method as shown in **Figure 33**. Apples with block watercore were grouped in Euler numbers between -400 and 400 with a small evolution. For apples with radial development, the Euler number was highly negative, up to -1439. Significant differences were also found regarding sugar composition, with higher fructose and total sugar contents in apples from the upper canopy, compared to those in the lower canopy location. Also, noted significantly higher sorbitol and lower sucrose and fructose contents were found in watercoreaffected tissue compared to the healthy tissue of affected apples and compared to healthy apples. Additionally, [214] mentioned that by using MRI, the results of additional tests such as chemical analysis, oil and moisture distribution, sugar level, pH and physical analysis of structure, voids, the thickness of filling and coating, are immediately tested within seconds on the production line. Thus, the idea about the value of the MRI technique and its application in the food industry is going to improve and maintain the quality in processing, testing, and optimizing the parameters.

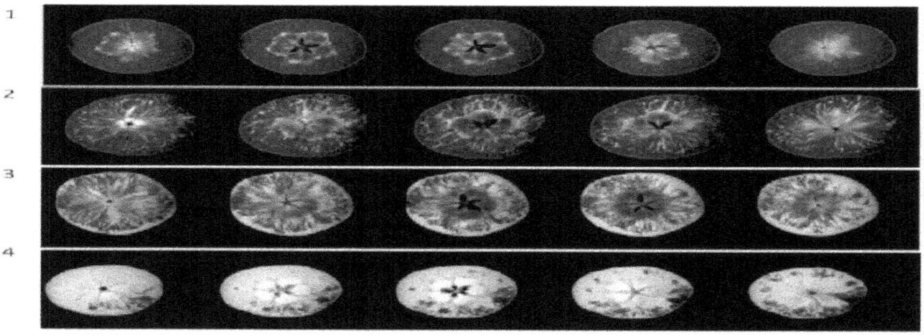

Figure 33. *Shows the MR images of central apple slices belonging to the four watercore levels, classified by three experts. (1) Sound apple; (2) light watercore; (3) medium watercore; (4) strong water core.*

However, the high cost of imaging facilities is another barrier to the exploitation of MRI in the food industry. As well, [215] presented a detailed discussion on the fundamentals of MRI in the study of food materials. Also, [216] pointed to that the MRI is done with the use of an NMR instrument equipped with magnetic gradient coils. Where these coils have the capability to collect data spatially and create twodimensional and three-dimensional images displaying diverse physicochemical characteristics. Likewise, [217] adopted the idea that fast and non-destructive solutions for sensing watercore would be readily accepted in the postharvest industry. Therefore, conducted a comparative study between X-ray CT and MRI as potential imaging technologies for detecting watercore disorder of different apple cultivars. After the acquisition of X-ray and MR images the 3D datasets of X-ray CT and MRI were matched, the images obtained on quantitatively identical fruit were compared. The results indicated that both MRI and CT were able to detect watercore disorder of different apple cultivars, however, the contrast in MRI images was superior as shown in **Figure 34.** Finally, concluded that the mean and variance of the frequency distribution of MRI and X-ray CT intensity appeared to be a parameter that allows the identification of healthy apples from affected fruit. A study by [218] provided a potential and detailed description of all components of the MRI system in agricultural fruits and vegetables for the assessment of maturity and quality parameters. As well, [219] used the MRI technique for non-invasive imaging of plant roots in different soils. Where used barley as a model plant to investigate the achievable image quality and the suitability for root phenotyping of six natural soil substrates of commonly occurring soil textures.

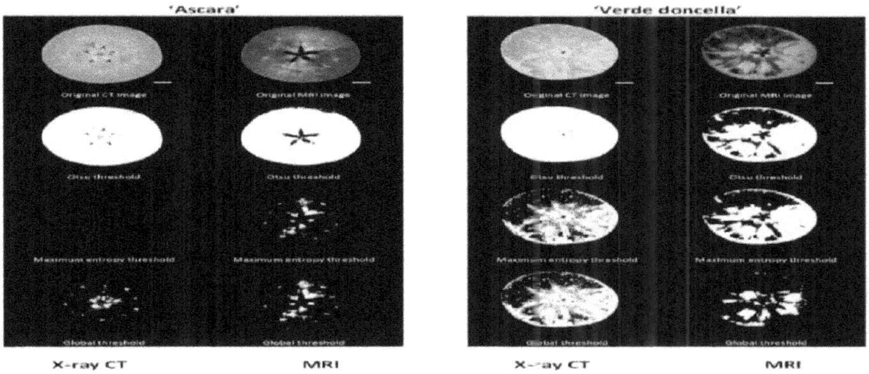

Figure 34. *Shows the X-ray CT (left) and MR images (right) cross-sections of sound Ascara and watercore Verde doncella fruit and their segmentation results.*

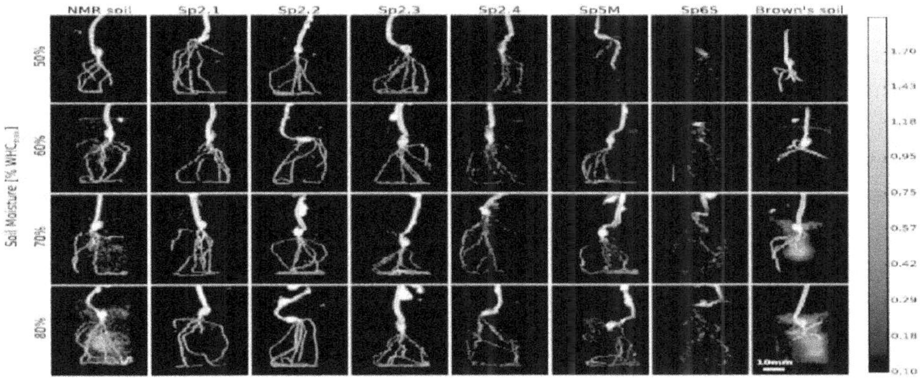

Figure 35. *MR Image for barley seedlings 3 days after sowing in eight different substrates at four different soil moisture levels.*

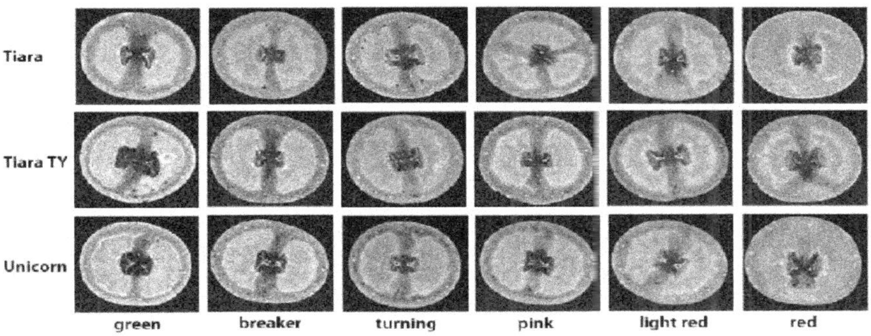

Figure 36. *Demonstrates the MR images of three cherry tomato cultivars Tiara, Tiara TY, and Unicornat, at different maturity stages.*

The results are compared with two artificially composed substrates previously documented for MRI root imaging as shown in **Figure 35**. The results demonstrated that only one soil did not allow imaging of the roots with MRI. In the artificially composed substrates, soil moisture above 70% of the maximal water holding capacity (WHCmax) impeded root imaging. For the natural soil substrates, soil moisture did not affect MRI root image quality in the investigated range of 50–80% WHCmax. Concluded that with the characterization of different soils, investigations such as trait stability across substrates are now possible using non-invasive MRI. Subsequently, [220] presented a review conducted on the use of NMR/MRI techniques, for inspection of some agricultural fruits and vegetables, and explained the benefits of their implementation in the assessment of internal quality attributes such as internal defects, water content, nutrition content, maturity, fruit firmness, seed detection, physicochemical and microbiological quality in both commercial and industrial applications. Accordingly, concluded that the low-field nuclear magnetic resonance (LF-NMR) and MRI are viable technologies in assessing water status, which can significantly impact the quality of fruits and vegetables' texture, tenderness, and microstructure. Despite considerable developments in the quality measurement of fruits and vegetables and their products, the implementation of these techniques at an industrial level has been unsatisfactory. As well, [221] used MRI to study the changes in the internal structure of tomato fruit during development as a function of maturity. The internal structure of intact cherry tomato fruit at six different maturity stages (green, breaker, turning, pink, light red, and red) was measured using a series of two-dimensional (2D) MR images as shown in **Figure 36**. water content appears evenly distributed in the pericarp region from breaker to light red maturity stages.

MR signal intensity changes when different maturity stages are observed. Especially, signal intensity variation between the pericarp and locule regions is observed. Quantifying variations of signal intensity using a ratio of signal between pericarp and locule different regions enables the assignment of the maturity of cherry tomato. Additionally, concluded that since MRI provides detailed internal structure information, characterization of internal defects (e.g., bruises, voids, impact damage) and other quality factors is possible.

CONCLUSIONS

Inspecting and measuring the external and internal quality of agricultural and food products and assuring their safety from diseases and contamination, is one of the most important issues facing the food sector at present. This is a result of multiple and repeated complaints against agricultural producers and food manufacturers for the inability to meet quality requirements that meet the consumer's desires.

When agricultural and food products do not meet quality standards and safety criteria, consumers lose faith in producers leading to the loss of these products' competitiveness in the market, and thus significant economic loss. With consumers rapidly growing demand for safer and better-quality food. So, agricultural producers and food manufacturers are working hard to eliminate sources of food contamination and achieve better quality. Although some systems are proposed to achieve food safety and quality by achieving a set of conditions that fall under the so-called good manufacturing practices (GMP) and hazard analysis and critical control point (HACCP) which represents the best way to achieve food security through all production steps. Unfortunately, with all these requirements for GMP and HACCP systems and others, they may not completely ensure the production of safe food free of contaminants and defects. So, it has become necessary to introduce modern technologies to quality inspect and detect blemishes and contamination and then reject these products that are not fit for human consumption. Therefore, the focus was on the development of non-destructive, modern, fast, reliable, and applicable methods that meet the needs of both food manufacturers and producers, as well as the desires of the consumer. So, the majority of all quality detection systems use electromagnetic wave measurements across all regions of the electromagnetic spectrum through imaging technologies. Gamma-ray and X-ray imaging technologies have high frequency and energy and are often used for irradiation, plant breeding applications, determining food quality, and food safety. Although applications of this technique are mainly used for research and development work, it has great potential to serve as a tool for the development of various plant varieties, assessment and quality assurance, and management practices for a wide range of agro-food practices. As well, UV and visible light imaging has proved itself to be very reliable and efficient for performing several tasks such as evaluating color, shape, size, and detect external defects. Additionally, these imaging systems can empower the agricultural and food industry with a new tool to detect defects and contaminations to ensure food safety and quality. Undoubtedly, with the evolution of a new generation of detectors and cameras, imaging within UV and visible light regions will have great potential in food defense and safety. Furthermore, IR and HSI systems can combine spectral and spatial data of a sample. For this reason, the HSI system became a standalone unit for non-destructive analysis of the physical,

textural, and chemical parameters of the sample. So, the IR-HIS system has gained fame and a good reputation and is elaborately tested to predict chemical composition, detect defects, and adulterate agricultural and food products. Also, microwave and radio-wave imaging can improve the efficiency of real-time monitoring of food production, storing, and control quality chain. There appears to be an acceleration in the growth of hardware and software of imaging systems to overcome the limitations of this technology will help the agricultural and food industry in implementing the different imaging systems for rapid and in-line quality monitoring applications such as foreign material detection, discrimination external and internal quality attributes of agricultural and food products and detecting various defects and diseases. In conclusion, in this chapter, it has been shown that the different modern imaging technologies could provide unquestionably advantages for monitoring the quality and safety of agri-food production and processing.

ABSTRACT

In recent decades, the quality and safety of fruits, vegetables, cereals, meats, milk, and their derivatives from processed foods have become a serious issue for consumers in developed as well as developing countries. Undoubtedly, the traditional methods of inspecting and ensuring quality that depends on the human factor, some mechanical and chemical methods, have proven beyond any doubt their inability to achieve food quality and safety, and thus a failure to achieve food security. With growing attention on human health, the standards of food safety and quality are continuously being improved through advanced technology applications that depend on artificial intelligence tools to monitor the quality and safety of food. One of the most important of these applications is imaging technology. A brief discussion in this chapter on the utilize of multiple imaging systems based on all different bands of the electromagnetic spectrum as a principal source of various imaging systems. As well as methods of analyzing and reading images to build intelligence and non-destructive systems for monitoring and measuring the quality of foods.

REFERENCES

1. Kingdom FAA. Mach bands explained by response normalization. Frontiers in Human Neuroscience. 2014; 8, 843. DOI:10.3389/fnhum.2014.00843
2. Land EH, McCann JJ. Lightness and Retinex Theory. Journal of the Optical Society of America. 1971; 61: 1-11. DOI.org/10.1364/JOSA.61.000001
3. Gonzalez RC, Woods, RE. Digital Image Processing. 2nd ed. Prentice Hall, Upper Saddle River; 2002. ISBN 10: 0201180758 / ISBN 13: 9780201180756
4. Cheremkhin PA, Lesnichii V V, Petrov NV. Use of spectral characteristics of DSLR cameras with Bayer filter sensors. Journal of Physics: Conference Series. 2014; 536: 012021. DOI:10.1088/1742-6596/536/1/012021
5. Bellman R. Adaptive Control Processes: A Guided Tour, Princeton University Press. 1961; DOI.org/10.1002/zamm.19620420718
6. ElMasry G, Cubero S, Molto E, Blasco J. In-line sorting of irregular potatoes by using automated computer-based machine vision system. Journal of Food Engineering. 2012; 112: 60-68. DOI: 10.1016/j.jfoodeng.2012.03.027

7. Ceamanos X, Valero S. Processing hyperspectral images. In: Nicolas B, Mehrez Z, editors. Handbook of Optical Remote Sensing of Land Surface. Elsevier; 2016. p. 163-200. DOI:10.1016/C2015-0-01220-5

8. Yao-Ze F, Sun D-W. Application of Hyperspectral Imaging in Food Safety Inspection and Control: A Review. Critical Reviews in Food Science and Nutrition. 2012; 52: 1039-1058. DOI: 10.1080/10408398.2011.651542

9. Kumar S, Kapil P, Bhardwaj Y, Acharya US, Gupta C. Anomaly Detection in Fruits using Hyper Spectral Images. International Journal of Trend in Scientific Research and Development. 2019; 3(4): 394-397. DOI: 10.31142/ijtsrd23753

10. Pau G, Fuchs F, Sklyar O, Boutros M, Huber W. EBImage—an R package for image processing with applications to cellular phenotypes. Bioinformatics. 2010; 26(7): 979-981. DOI: 10.1093/bioinformatics/btq046

11. Haralick RM, Shanmugam K, Dinstein I. Textural Features for Image Classification. IEEE Transaction on Systems, Man and Cybernetics. 1973; 3(6): 610-621. DOI: 10.1109/TSMC.1973.4309314

12. Park Y, Guldmann JM. Measuring continuous landscape patterns with Gray-Level Co-Occurrence Matrix (GLCM) indices: An alternative to patch metrics? Ecological Indicators. 2020; 109: 1-18. DOI: 10.1016/j.ecolind.2019.105802

13. Ibrahima A, El-Bialeea N, Saad M, Romano E. Non-Destructive Quality Inspection of Potato Tubers Using Automated Vision System. International Journal on Advanced Science Engineering Information Technology. 2020; 10(6): 2419-2428. DOI: 10.18517/ijaseit.10.6.13079

14. Romano E, Bergonzoli S, Pecorella I, Bisaglia C, De Vita P. Methodology for the definition of durum wheat yield homogeneous zones by using satellite spectral indices. Remote Sens. 2021;13: 2036. DOI.org/10.3390/rs13112036

15. Yu H, Guo L, Kharbach M, Han W. Multi-Way Analysis Coupled with Near-Infrared Spectroscopy in Food Industry: Models and Applications. Foods. 2021; 10: 802. DOI.org/10.3390/foods10040802

16. Gourvénec S, Tomasi G, Durville C, Di Crescenzo E, Saby C, Massart D, Bro R, Oppenheim G. CuBatch, a MATLAB® interface for n-mode data analysis. Chemometrics and Intelligent Laboratory Systems. 2005; 77 (1-2): 122-130. DOI: 10.1016/j.chemolab.2004.01.027

17. Kolda TG, Bader BW. Algorithm 862: MATLAB tensor classes for fast algorithm prototyping. ACM Transactions on Mathematical Software. 2006; 32(4): 635-653. DOI: 10.1145/1186785.1186794

18. Phan AH, Tichavský P, Cichocki A. Tensorbox: A Matlab Package for Tensor Decomposition. LABSP RIKEN: Saitama, Japan. 2012. http://www.bsp.brain.riken.jp/~phan/tensorbox.php

19. Phan AH, Tichavsky P, Cichocki A. Low complexity damped Gauss–Newton algorithms for CANDECOMP/PARAFAC. SIAM Journal on Matrix Analysis and Applications. 2013; 34: 126-147. DOI.org/10.1137/100808034

20. Giordani P, Kiers HA, Del Ferraro MA. Three-way component analysis using the R package ThreeWay. Journal of Statistical Software. 2014; 57: 1-23. DOI: 10.18637/jss.v057.i07

21. Helwig NE. Multiway: Component Models for Multi-Way Data. R Package Version 1.0-5. 2021; https://CRAN.Rproject. org/package=multiway. http://journal.ashspublications.org/content/123/5/898

22. Kossaifi J, Panagakis Y, Anandkumar A, Pantic M. Tensorly: Tensor learning in python. Journal of Machine Learning Research. 2019; 20: 1-6. https://creativecommons.org/licenses/by/4.0/

23. Hao L, Liang S, Ye J, Xu Z. TensorD: A tensor decomposition library in TensorFlow. Neurocomputing. 2018; 318: 196-200. DOI: 10.1016/j.neucom.2018.08.055

24. Gregoire P, Florian F, Oleg S, Michael B, Wolfgang H. EBImage - an R package for image processing with applications to cellular phenotypes. Bioinformatics. 2010; 26(7): 979-981. DOI: 10.1093/bioinformatics/btq046

25. Grompone von Gioi R, Randall G. Unsupervised smooth contour detection. Image processing on line. 2016; 6: 233-267. DOI: 10.5201/ipol.2016.175

26. Grompone von Gioi R. A Contrario Line Segment Detection, Springer. 2014; DOI: 10.1007/978-1-4939-0575-1

27. Otsu N. A threshold selection method from gray-level histograms. IEEE Transactions on Systems, Man, and Cybernetics. 1979; 9 (1): 62-66. DOI: 10.1109/TSMC.1979.4310076

28. Starr C. Biology: Concepts and Applications. Thomson Brooks Cole; 6th edition. 2005; ISBN 0-534-46226-X

29. Williamson SJ, Cummins H. Light and Color in Nature and Art. American Journal of Physics. 1983; 1(2). DOI: 10.1119/1.13928

30. Richards A. X rays and gamma rays: crookes tubes and nuclear light. In Alien Vision: exploring the electromagnetic spectrum with imaging technology. SPIE Press, Washington. 2001. ISBN-10 : 0819441422

31. Saha GB. Physics and radiobiology of nuclear medicine. Springer, New York. 2006. DOI: 10.1007/978-0-387-36281-6

32. Pires LF, de Macedo JR, de Souza MD, Bacchi OS, Reichardt K. Gamma-ray computed tomography to characterize soil surface sealing. Applied Radiation and Isotopes. 2002; 57: 375-380. DOI: 10.1016/S0969-8043(02)00098-2

33. Naime JM, Cruvinel PE, Silva AM, Crestana S, Vaz CMP. Applications of X and g-rays dedicated computerized tomography scanner in agriculture. In: Paulo, E.C., Luiz, A.C. (Eds.), Agricultural Tomography. Embrapa - CNPDIA, Sao Carlos. 2000; 96-104

34. Macedo A, Vaz, CMP, Naime JM, Jorge LAC, Crestana S, Cruvinel PE, Pereira JCD, Guimaraes MF, Ralisch R. Soil management impact and wood science recent contributions of Embrapa Agricultural Instrumentation Center using CT imaging. In: Paulo, E.C., Luiz, A.C. (Eds.), Agricultural Tomography. Embrapa-CNPDIA, Sao Carlos. 2000; 44-54

35. Tsukamoto T, Uchida H, Nakanishi H, Nishiyama S, Tsukada H, Matsuhashi S, Nishizawa NK, Mori S. H215O translocation in rice was enhanced by 10 µm 5-aminolevulinic acid as monitored by positron-emitting tracer imaging system (PETIS). Soil Science and Plant Nutrition. 2004; 50(7):1085-1088. DOI: 10.1080/00380768.2004.10408578

36. Kawachi N, Kikuchi K, Suzui N, Ishii S, Fujimaki S, Ishioka NS, Watabe H. Imaging of carbon translocation to fruit using carbon-11-labeled carbon dioxide and positron emission tomography. IEEE Transactions on Nuclear Science. 2011; 58(2): 395-399. DOI: 10.1109/TNS.2011.2113192

37. Manickavasagan A, Jayas DS, White, NDG. Thermal imaging to detect infestation by Cryptolestes ferrugineus inside wheat kernels. Journal of Stored Products Research. 2008; 44(2): 186-192. DOI: 10.1016/j.jspr.2007.10.006

38. Rossel RAV, Taylor JJ, Mcbratney AB. Multivariate calibration of hyperspectral g-ray energy spectra for proximal soil sensing. European Journal of Soil Science. 2007; 58: 343-353. DOI: 10.1111/j.1365-2389.2006.00859.x

39. Tsukamoto T, Nakanishi H, Kiyomiya S, Watanbe S, Matsuhashi S, Nishizawa NK, Mori S. 52Mn translocation in barley monitored using a positron-emitting tracer imaging system. Soil Science and Plant Nutrition. 2006; 52(6): 717-725. DOI.org/10.1111/j.1747-0765.2006.00096.x

40. Nakamura S, Suzui N, Nagasaka T, Komatsu F, Ishioka NS, Tanabata SI, Kawachi N, Rai H, Hattori H, Chino M, Fujimaki S. Application of glutathione to roots selectively inhibits cadmium transport from roots to shoots in oilseed rape. Journal of Experimental Botany. 2013; 64(4): 1073-1081. DOI: 10.1093/jxb/ers388

41. Fujimaki S, Suzui N, Ishioka NS, Kawachi N, Ito S, Chino M, Nakamura S. Tracing cadmium from culture to spikelet: noninvasive imaging and quantitative characterization of absorption, transport, and accumulation of cadmium in an intact rice plant. Plant Physiology. 2010; 152(4): 1796-1806. DOI: 10.1104/pp.109.151035

42. Watanabe S, Iida Y, Suzui N, Katabuchi T, Ishii S, Kawachi N, Hanaoka H, Watanabe S, Matsuhashi S, Endo K, Ishioka NS. Production of no-carrier-added 64Cu and applications to molecular imaging by PET and PETIS as a biomedical tracer. Journal of Radioanalytical and Nuclear Chemistry. 2009; 280: 199-205. DOI:10.1007/S10967-008-7443-9

43. Ishikawa S, Suzui N, Ito-Tanabata S, Ishii S, Igura M, Abe T, Kuramata M, Kawachi N, Fujimaki S. Real-time imaging and analysis of differences in cadmium dynamics in rice cultivars (Oryza sativa) using positron-emitting107Cd tracer. BMC Plant Biology. 2011; 11(1): 1-12. DOI: 10.1186/1471-2229-11-172

44. Kiyomiya S, Nakanishi H, Uchida H, Nishiyama S, Tsukada H, Ishioka NS, Watanabe S, Osa A, Mizuniwa C, Ito T, Matsuhashi S, Hashimoto S, Sekine T, Tsuji A, Mori S. Light activates H215O flow in rice: detailed monitoring using a positron-emitting tracer imaging system (PETIS). Physiol Plant. 2001; 113(3): 359-367. DOI: 10.1034/j.1399-3054.2001.1130309.x

45. Mori S, Kiyomiya S, Nakanishi H, Ishioka NS, Watanabe S, Osa A, Matsuhashi S, Hashimoto S, Sekine T, Uchida H, Nishiyama S, Tsukada H, Tsuji A. Visualization of 15O-water flow in tomato and rice in the light and dark using a positron-emitting tracer imaging system (PETIS). Soil Science and Plant Nutrition. 2000; 46(4): 975-979. DOI: 10.1080/00380768.2000.10409163

46. Kotwaliwale N, Singh K, Kalne A, Jha SN, Seth N, Kar A. X-ray imaging methods for internal quality evaluation of agricultural produce. Journal of Food Science and Technology. 2014; 51(1): 1-15. DOI: 10.1007/s13197-011-0485-y

47. Adedeji AA, Ngadi MO. Microstructural characterization of deep-fat fried breaded chicken nuggets using X-ray micro-computed tomography. Journal of Food Process Engineering. 2010; 34(6): 2205-2219. DOI: 10.1111/j.1745-4530.2009.00565.x

48. Vavrik D, Jakubek J. Radiogram enhancement and linearization using the beam hardening correction method. Nuclear Instruments and Methods in Physics Research. 2009; 607(1): 212-214. DOI: 10.1016/j.nima.2009.03.156

49. Ammari H. An Introduction to Mathematics of Emerging Biomedical Imaging. 1st ed; Springer-Verlag Berlin Heidelberg, Germany. 2008; X, 198p. DOI: 10.1007/978-3-540-79553-7

50. Patton JA, Turkington TG. SPECT/CT physical principles and attenuation correction. Journal of Nuclear Medicine Technology. 2008; 36(1): 1-10. DOI: https://doi.org/10.2967/jnmt.107.046839

51. Neethirajan S, Jayas DS, White NDG. Detection of sprouted wheat kernel using soft X-ray image analysis. Journal of Food Engineering. 2007; 81(3): 509-513. DOI: 10.1016/j.jfoodeng.2006.11.020

52. Bianchini VJM, Mascarin GM, Silva LCAS, Arthur V, Carstensen JM, Boelt B, da Silva CB. Multispectral and X-ray images for characterization of Jatropha curcas L. seed quality. Plant Methods. 2021; 17(9): 1-13. DOI: 10.21203/rs.3.rs-28449/v1

53. Einarsdottir H, Emerson MJ, Clemmensen LH, Scherer K, Willer K, Bech M, Larsen R, Ersbøll BK, Pfeiffer F. Novelty detection of foreign objects in food using multi-modal X-ray imaging. Food Control. 2016; 67: 39-47. DOI: 10.1016/j.foodcont.2016.02.023

54. Al-Mezeini N, Manickavasagan A, Al-Yahyai R, Al-Wahaibi A, Al-Raeesi A, Khriji L. X-ray Imaging of Stored Dates to Detect Infestation by Saw-Toothed Beetles. International journal of fruit science. 2016; 16(1): 42-56. DOI: 10.1080/15538362.2015.1044692

55. Nielsen MS, Lauridsen T, Christensen LB, Feidenhans'l R. X-ray darkfield imaging for detection of foreign bodies in food. Food Control. 2013; 30(2): 531-535. DOI: 10.1016/j.foodcont.2012.08.007

56. Jiang JA, Chang HY, Wu KH, Ouyang CS, Yang MM, Yang EC, Chen TW, Lin TT. An adaptive image segmentation algorithm for X-ray quarantine inspection of selected fruits. Computers and electronics in agriculture. 2008; 60(2): 190-200. DOI: 10.1016/j.compag.2007.08.006

57. Karunakaran C, Jayas DS, White, NDG. Mass determination of wheat kernels from X-ray images. American Society of Agricultural and Biological Engineers. Conference: 2004, DOI: 10.13031/2013.16742

58. Haff RP, Slaughter DC. Real-time X-ray inspection of wheat for infestation by the granary weevil, Sitophilus Granarius (L.). Transactions of the ASAE. 2004; 47(2): 531-537. DOI: 10.13031/2013.16022

59. Karunakaran C, Jayas DS, White, NDG. Soft X-ray inspection of wheat kernels infested by Sitophilus oryzae. Transactions of the ASAE. 2003; 46(3): 739-745. DOI: 10.13031/2013.13576

60. Shahin MA, Tollner EW, Gitaitis RD, Sumner DR, Maw BW. Apple classification based on surface bruises using image processing and neural networks. Transactions of the ASAE. 2002; 45 (5): 1619-1627. DOI: 10.13031/2013.11047

61. Kim S, Schatzki TF. Apple watercore sorting system using X-ray imagery. I. Algorithm development. Transactions of the ASAE. 2000; 43 (6): 1695-1702. DOI: 10.13031/2013.3070

62. Patel KK, Kar A, Khan MA. Potential of reflected UV imaging technique for detection of defects on the surface area of mango. Journal of Food Science and Technology. 2019; 56(2): 1295-1301. DOI: 10.1007/s13197-019-03597-w

63. Cosentino A. Practical notes on ultraviolet technical photography for art examination. Conservar Património. 2015; 21: 53-62. DOI: 10.14568/cp2015006

64. Özlüoymak ÖB. Development of a UV-based Imaging System for Real-Time Detection and Separation of Dried Figs Contaminated with Aflatoxins. Journal of Agricultural Sciences. 2014; 20: 302-316. DOI: 10.15832/tbd.87873

65. Nagle M, Intani K, Mahayothee B, Sardsud V, Muller J. Nondestructive mango quality assessment using image processing: inexpensive innovation for the fruit handling industry. In: Conference on international research on food security, natural resource management and rural development (Tropentag), Gottingen (Germany), 1-4. 2012

66. Özlüoymak ÖB. A Research on Separation System Design of Aflatoxin Contaminated Dried Fig [Ph.D.]. Adana, Turkey: Çukurova University; 2012

67. Slaughter DC, Obenland DM, Thompson JF, Arpaia ML, Margosan DA. Nondestructive freeze damage detection in oranges using machine vision and ultraviolet fluorescence. Postharvest Biology and Technology. 2008; 48(3): 341-346. DOI: 10.1016/j.postharvbio.2007.09.012

68. Blasco J, Aleixos N, Gomez J, Molto E. Citrus sorting by identification of the most common defects using multispectral computer vision. Journal of Food Engineering. 2007; 83(3): 384-393. DOI: 10.1016/j.jfoodeng.2007.03.027

69. Karami MA, Mirabolfathy M. Neural Network to Separate Aflatoxin Contaminated Pistachio Nuts. Acta Horticulturae. 2006; 726 (726): 605-610. DOI: 10.17660/ActaHortic.2006.726.103

70. Kleynen O, Destain MF. Detection of defects on fruits by machine vision and unsupervised segmentation. In: AgEng conference, September 12-16. Technological Institute, Leuven, Belgium, 2004; 1006-1007. http://hdl.handle.net/2268/81214

71. Steiner WE, Rieker RH, Battaglia R. Aflatoxin Contamination in Dried Figs: Distribution and Association with Fluorescence. Journal of Agricultural Food Chemistry. 1988; 36(1): 88-91. DOI: 10.1021/jf00079a022

72. Yabe K, Ando Y, Ito M, Terakado N. Simple Method for Screening Aflatoxin-Producing Molds by UV Photography. Applied and Environmental Microbiology. 1987; 53(2): 230-234. DOI: 10.1128/AEM.53.2.230-234.1987

73. Robertson GL. Food packaging principles and practice, 3rd ed. CRC Press, Taylor and Francis group, UK. 2012. ISBN 9781439862414

74. Jha SN. Non-destructive evaluation of food quality. Ist ed. Heidelberg, Springer; 2010. 18-22p. DOI: 10.1007/978-3-642-15796-7

75. ElMasry G, Sun DW. Principles of hyperspectral imaging technology. In: Da-Wen Sun, editors. Handbook of Hyperspectral imaging for food quality analysis and control. 1st ed. Academic Press, San Diego, California, USA, 2010. 3-43p. DOI: 10.1016/B978-0-12-374753-2.10001-2

76. Lou W, Nakai S. Application of artificial neural networks for predicting the thermal inactivation of bacteria: a combined effect of temperature, pH and water activity. Food Research International. 2001; 34(47): 573-579. DOI: 10.1016/S0963-9969(01)00074-6

77. Brosnan T, Sun DW. Inspection and grading of agricultural and food products by computer vision systems - a review. Computers and Electronics in Agriculture. 2002; 36 (2–3): 193-213. doi.org/10.1016/S0168-1699(02)00101-1

78. Wu D, Sun DW. Colour measurements by computer vision for food quality control-a review. Trends in Food Science and Technology. 2013; 29(1): 5-20. DOI: 10.1016/j.tifs.2012.08.004

79. Chen YR, Chao K, Kim MS. Machine vision technology for agricultural applications. Computers and Electronics in Agriculture. 2002; 36(2): 173-191. DOI: 10.1016/S0168-1699(02)00100-X

80. Hong H, Yang X, You Z, Cheng F. Visual quality detection of aquatic products using machine vision. Aquacultural Engineering. 2014; 63: 62-71. DOI: 10.1016/j.aquaeng.2014.10.003

81. Payne A, Walsh K, Subedi P, Jarvis D. Estimating mango crop yield using image analysis using fruit at 'stone hardening' stage and night time imaging. Computers and Electronics in Agriculture. 2014; 100: 160-167. DOI: 10.1016/j.compag.2013.11.011

82. Wijethunga P, Samarasinghe S, Kulasiri D, Woodhead I. Digital Image Analysis Based Automated Kiwifruit Counting Technique. 23rd International Conference Image and Vision Computing, New Zealand, Christchurch, Publisher: IEEE, 2008. 1-6. DOI: 10.1109/IVCNZ.2008.4762149

83. Song Y, Glasbey CA, Horgan GW, Polder G, Dieleman JA, van der Heijdenc GWAM. Automatic fruit recognition and counting from multiple images. Biosystems Engineering. 2014; 118: 203-215. DOI: 10.1016/j.biosystemseng.2013.12.008

84. Payne AB, Walsh KB, Subedi PP, Jarvis D. Estimation of mango crop yield using image analysis – Segmentation method. Computers and Electronics in Agriculture. 2013; 91: 57-64. DOI: 10.1016/j.compag.2012.11.009

85. Linker R, Cohen O, Naor A. Determination of the number of green apples in RGB images recorded in orchards. Computers and Electronics in Agriculture. 2012; 81: 45-57. DOI: 10.1016/j.compag.2011.11.007

86. Narendra VG, Hareesh KS. Quality inspection and grading of agricultural and food products by computer vision—a review. International Journal of Computer Applications. 2010; 2(1): 975-8887. DOI: 10.5120/612-863

87. Safren O, Alchanatis V, Ostrovsky V, Levi O. Detection of green apples in hyperspectral images of apple-tree foliage using machine vision. Transactions of the ASABE. 2007; 50(6): 2303-2313. DOI: 10.13031/2013.24083

88. Nyalala I, Okinda C, Nyalala L, Makange N, Chao Q, Chao L, Yousaf K, Chen K. Tomato volume and mass estimation using computer vision and machine learning algorithms: Cherry tomato model. Journal of Food Engineering. 2019; 263: 288-298. DOI: 10.1016/j.jfoodeng.2019.07.012

89. Weijun X, Wang F, Yang D. Research on Carrot Grading Based on Machine Vision Feature Parameters. IFAC PapersOnLine. 2019; 52(30): 30-35. DOI: 10.1016/j.ifacol.2019.12.485

90. Rodriguez-Pulido FJ, Ferrer-Gallego R, Gonzalez-Miret ML, Rivas-Gonzalo JC, Song Y, Glasbey CA, Horgan GW, Polder G, Dieleman JA, van der Heijden GWAM. Automatic fruit recognition and counting from multiple images. Biosystems Engineering. 2014; 118(1), 203-215. DOI: 10.1016/j.biosystemseng.2013.12.008

91. Aggarwal AK, Mohan R. Aspect ratio analysis using image processing for rice grain quality. International Journal of Food Engineering. 2010; 6(5): 1-14. DOI: 10.2202/1556-3758.1788

92. Dubosclard P, Larnier S, Konik H, Herbulot A, Devy M. Automated visual grading of grain kernels by machine vision. In: Twelfth International Conference on Quality Control by Artificial Vision. 2015; 9534: 1-8. DOI: 10.1117/12.2182793

93. Zareiforoush H, Minaei S, Alizadeh MR, Banakar A, Samani BH. Design, development and performance evaluation of an automatic control system for rice whitening machine based on computer vision and fuzzy logic. Computers and Electronics in Agriculture. 2016; 124:14-22. DOI: 10.1016/j.compag.2016.01.024

94. Bhat S, Panat S, Arunachalam N. Classification of rice grain varieties arranged in scattered and heap fashion using image processing. In: Ninth International Conference on Machine Vision (ICMV 2016). 2017; 10341: 1-6. DOI: 10.1117/12.2268802

95. Kiliç K, Boyaci IH, Köksel H, Küsmenoglu I. A classification system for beans using computer vision system and artificial neural networks. Journal of Food Engineering. 2007; 78(3): 897-904. DOI: 10.1016/j.jfoodeng.2005.11.030

96. Venora G, Grillo O, Ravalli C, Cremonini R. Identification of Italian landraces of bean (Phaseolus vulgaris L.) using an image analysis system. Scientia Horticulturae. 2009; 121(4): 410-418. DOI: 10.1016/j.scienta.2009.03.014

97. Laurent B, Ousman B, Dzudie T, Carl MFM, Emmanuel T. Digital camera images processing of hard-to-cook beans. Journal of Engineering and Technology Research. 2010; 2(9): 177-188. DOI: 10.5897/JETR.9000027

98. Araújo SA, Alves WAL, Belan PA, Anselmo KP. A computer vision system for automatic classification of most consumed Brazilian beans. In: Bebis G. et al. (eds) Advances in Visual Computing. ISVC 2015. Lecture Notes in Computer Science, Springer, 2015. p. 45-53. DOI: 10.1007/978-3-319-27863-6_5

99. Belan PA, De Macedo RAG, Pereira MA, Alves WAL, Araújo SA. A fast and robust approach for touching grains segmentation. In: Campilho A., Karray F., ter Haar Romeny B. (eds) Image Analysis and Recognition. Lecture Notes in Computer Science, vol 10882. Springer; 2018. p. 482-489. DOI: 10.1007/978-3-319-93000-8_54

100. Belan PA, de Macedo RAG, Alves WAL, Santana JCC, Araújo SA. Machine vision system for quality inspection of beans. The International Journal of Advanced Manufacturing Technology. 2020; 111 (11-12), 1-15. DOI: 10.1007/s00170-020-06226-5

101. Girolami A, Napolitano F, Faraone D, Braghieri A. Measurement of meat color using a computer vision system. Meat science. 2013; 93(1): 111-118. DOI: 10.1016/j.meatsci.2012.08.010

102. Jackman P, Sun DW, Allen P. Recent advances in the use of computer vision technology in the quality assessment of fresh meats. Trends in Food Science and Technology. 2011; 22(4), 185-197. DOI: 10.1016/j.tifs.2011.01.008

103. Chen K, Sun X, Qin C, Tang X. Color grading of beef fat by using computer vision and support vector machine. Computers and Electronics in Agriculture. 2010; 70(1): 27-32. DOI: 10.1016/j.compag.2009.08.006

104. Mancini RA, Hunt MC. Current research in meat color. Meat Sci. 2005; 71(1):100-121. DOI: 10.1016/j.meatsci.2005.03.003

105. Zheng C, Sun DW, Zheng L. Correlating color to moisture content of large cooked beef joints by computer vision. Journal of Food Engineering. 2006; 77(4): 858-863. DOI: 10.1016/j.jfoodeng.2005.08.013

106. Teimouri N, Omid M, Mollazade K, Mousazadeh H, Alimardani R, Karstoft H. On-line separation and sorting of chicken portions using a robust vision-based intelligent modeling approach. Biosystems Engineering. 2018; 167(4): 8-20. DOI: 10.1016/j.biosystemseng.2017.12.009

107. Gu J, He N, Wu X. A New Detection Method for Fish Freshness. In Computational Intelligence and Design (ISCID), 2014 Seventh International Symposium 2. 2014; 555-558. DOI: 10.1109/ISCID.2014.153

108. Vatansever F, Hamblin MR. Far infrared radiation (FIR): its biological effects and medical applications, Photonics Lasers Med. 2012; 1(4): 255-266. DOI: 10.1515/plm-2012-0034

109. Jain RK, Hoffman AJ, Jepsen PU, LIU PQ, Turchinovich D, Vitiello MS. Mid-infrared, long-wave infrared, and terahertz photonics: introduction. Optics Express. 2020; 28(9): 14169-14175. DOI: 10.1364/OE.395165

110. Bae TW. Spatial and temporal bilateral filter for infrared small target enhancement. Infrared Physics and Technology. 2014; 63: 42-53. doi.org/10.1016/j.infrared.2013.12.007

111. Kim DK. Flame detection using region expansions and on-line variances in infrared image. Journal of Korea Multimedia Society. 2009; 12(11): 1547-1556

112. Huang H, Yu H, Xu H, Ying Y. Near infrared spectroscopy for on/in-line monitoring of quality in foods and beverages: A review. Journal of Food Engineering. 2008; 87(3): 303-313. DOI: 10.1016/j.jfoodeng.2007.12.022

113. Peng Y, Wenxiu W. Application of near-infrared spectroscopy for assessing meat quality and safety. In: Theophanides T, editors. Handbook of Infrared Spectroscopy - Anharmonicity of Biomolecules, Crosslinking of Biopolymers, Food Quality and Medical Applications. Ist ed. IntechOpen; 2015. p. 137-163. DOI: 10.5772/58912

114. Scotter C. Use of near infrared spectroscopy in the food industry with particular reference to its applications to on/in-line food processes. Food Control. 1990; 1(3): 142-149. DOI: 10.1016/0956-7135(90)90006-X

115. Workman J, Weyer L. Practical guide and spectral atlas for interpretive near-infrared spectroscopy. 2nd ed. CRC Press, Taylor & Francis Group; 2012. 326 p. DOI: 10.1201/b11894

116. Aenugu HPR, Kumar DS, Parthiban SN, Ghosh S, Banji D. Near Infra-Red Spectroscopy—An Overview. International Journal of ChemTech Research. 2011; 2011, 3: 825-836. ISSN: 0974-4290

117. Genot V, Bock L, Dardenne P, Colinet G. Use of near-infrared reflectance spectroscopy in soil analysis. A review. Biotechnology, Agronomy, Society and Environment. 2014; 18: 247-261

118. Liu Y, Chen YR, Kim MS, Chan DE, Lefcourt AM. Development of simple algorithms for the detection of fecal contaminants on apples from visible/near infrared hyperspectral reflectance imaging. Journal of Food Engineering. 2007; 81(2): 412-418. DOI: 10.1016/j.jfoodeng.2006.11.018

119. Park B, Lawrence KC, Windham WR, Smith D. Performance of hyperspectral imaging system for poultry surface fecal contaminant detection. Journal of Food Engineering. 2006; 75(3): 340-348. DOI: 10.1016/j.jfoodeng.2005.03.060

120. Ariana D, Lu R, Guyer D.E. Hyperspectral reflectance imaging for detection of bruises on pickling cucumbers. Computers and Electronics in Agriculture. 2006; 53(1): 60-70. DOI: 10.1016/j.compag.2006.04.001

121. Xiong Z, Sun DW, Xie A, Han Z, Wang L. Potential of hyperspectral imaging for rapid prediction of hydroxyproline content in chicken meat. Food Chemistry. 2015; 175: 417-422. DOI: 10.1016/j.foodchem.2014.11.161

122. Gowen AA, O'Donnell CP, Cullen PJ, Downey G, Frias JM. Hyperspectral imaging e an emerging process analytical tool for food quality and safety control. Trends in Food Science and Technology. 2007; 18(2): 590-598. DOI: 10.1016/j.tifs.2007.06.001

123. Monteiro S, Minekawa Y, Kosugi Y, Akazawa T, Oda K. Prediction of sweetness and amino acid content in soybean crops from hyperspectral imagery. ISPRS Journal of Photogrammetry and Remote Sensing. 2007; 62(1): 2-12. DOI: 10.1016/j.isprsjprs.2006.12.002

124. Smail V, Fritz A, Wetzel D. Chemical imaging of intact seeds with NIR focal plane array assists plant breeding. Vibrational Spectroscopy. 2006; 42(2): 215-221. DOI: 10.1016/j.vibspec.2006.02.004

125. Uno Y, Prasher S, Lacroix R, Goel P, Karimi Y, Viau A, Patel R.M. Artificial neural networks to predict corn yield from Compact Airborne Spectrographic Imager data. Computers and Electronics in Agriculture. 2005; 47(2): 149-161. DOI: 10.1016/j.compag.2004.11.014

126. Lewis E, Schoppelrei J, Lee E, Kidder L. Near-infrared chemical imaging as a process analytical tool. In: Bakeev K, editor, Handbook of Process analytical technology. 1st ed. Oxford: Blackwell Publishing; 2005. p. 187-225. DOI: 10.1002/9780470988459.ch7

127. Koehler F, Lee E, Kidder L, Lewis N. Near infrared spectroscopy: the practical chemical imaging solution. Spectroscopy Europe. 2002; 14(3): 12-19

128. Mahesh S, Manickavasagan A, Jayas DS, Paliwal J, White, NDG. Feasibility of near-infrared hyperspectral imaging to differentiate Canadian wheat classes. Biosystems Engineering. 2008; 101(1): 50-57. DOI: 10.1016/j.biosystemseng.2008.05.017

129. Singh CB, Jayas DS, Paliwal J, White NDG. Identification of insect-damaged wheat kernels using short-wave nearinfrared hyperspectral and digital colour imaging. Computers and Electronics in Agriculture. 2010; 73(2): 118-125. DOI: 10.1016/j.compag.2010.06.001

130. Ariana DP, Lu R. Quality evaluation of pickling cucumbers using hyperspectral reflectance and transmittance imaging: Part II. Performance of a prototype. Sensing and Instrumentation for Food Quality and Safety. 2008; 2(3): 152-160. DOI: 10.1007/s11694-008-9058-9

131. Shaw G, Manolakis D. Signal processing for hyperspectral image exploitation. IEEE Signal Processing Magazine. 2002; 19(1): 12-16. DOI: 10.1109/79.974715

132. Lu RF, Chen YR. Hyperspectral imaging for safety inspection of food and agricultural products. In: SPIE Conference on Pathogen Detection and Remediation for Safe Eating, Boston. 1998; DOI: 10.1117/12.335771

133. Siche R, Vejarano R, Aredo V, Velasquez L, Saldana E, Quevedo R. Evaluation of Food Quality and Safety with Hyperspectral Imaging (HSI). Food Engineering Reviews. 2016; 8(3): 306-322. DOI: 10.1007/s12393-015-9137-8

134. Cheng JH, Sun D.W. Hyperspectral imaging as an effective tool for quality analysis and control of fish and other seafoods: Current research and potential applications. Trends in Food Science and Technology. 2014; 37(2): 78-91. DOI: 10.1016/j.tifs.2014.03.006

135. Huang H, Liu L, Ngadi MO. Recent developments in hyperspectral imaging for assessment of food quality and safety. Sensors. 2014; 14(4): 7248-7276. DOI: 10.3390/s140407248

136. Dumont J, Hirvonen T, Heikkinen V, Mistretta M, Granlund L, Himanen K, Fauch L, Porali I, Hiltunen J, Keski-Saari S, Nygren M, Oksanen E, Hauta-Kasari M, Keinänen M. Thermal and hyperspectral imaging for Norway spruce (Picea abies) seeds screening. Computers and Electronics in Agriculture. 2015; 116: 118-24. DOI: 10.1016/j.compag.2015.06.010

137. Manley M, Mcgoverin CM, Engelbrecht P, Geladi P. Infuence of grain topography on near infrared hyperspectral images. Talanta. 2012; 89(2): 223-230. DOI: 10.1016/j.talanta.2011.11.086

138. Ambrose A, Kandpal LM, Kim MS, Lee WH, Cho BK. High speed measurement of corn seed viability using hyperspectral imaging. Infrared Physics and Technology. 2016; 75: 173-179. DOI: 10.1016/j.infrared.2015.12.008

139. McGoverin CM, Engelbrecht P, Geladi P, Manley M. Characterisation of non-viable whole barley, wheat and sorghum grains using nearinfrared hyperspectral data and chemometrics. Analytical and Bioanalytical Chemistry. 2011; 401(7): 2283-2289. DOI: 10.1007/s00216-011-5291-x

140. Polak A, Coutts FK, Murray P, Marshall S. Use of hyperspectral imaging for cake moisture and hardness prediction. IET Image Processing. 2019; 13(7): 1152-1160. DOI: 10.1049/iet-ipr.2018.5106

141. Chen H, Qiao H, Feng Q, Xu L, Lin Q, Cai K. Rapid Detection of Pomelo Fruit Quality Using Near-Infrared Hyperspectral Imaging Combined With Chemometric Methods. Frontiers in Bioengineering and Biotechnology. 2021; 8: 616943. DOI: 10.3389/fbioe.2020.616943

142. Gao Z, Shao Y, Xuan G, Wang Y, Liu Y, Han X. Real-time hyperspectral imaging for the in-field estimation of strawberry ripeness with deep learning. Artificial Intelligence in Agriculture. 2020; 4: 31-38. DOI 10.1016/j.aiia.2020.04.003

143. Pu YY, Sun DW, Buccheri M, Grassi M, Cattaneo TMP, Gowen A. Ripeness classification of Bananito fruit (Musa acuminata, AA): a comparison study of visible spectroscopy and hyperspectral imaging. Food Anal. Methods. 2019; 12: 1693-1704. doi.org/10.1007/s12161-019-01506-7

144. González-Cabrera M, Domínguez-Vidal A, Ayora-Cañada M.J. Hyperspectral FTIR imaging of olive fruit for understanding ripening processes. Postharvest Biology and Technology. 2018; 145: 74-82. DOI: 10.1016/j.postharvbio.2018.06.008

145. Munera S, Amigo JM, Blasco J, Cubero S, Talens P, Aleixos N. Ripeness monitoring of two cultivars of nectarine using VIS-NIR hyperspectral reflectance imaging. Journal of Food Engineering. 2017; 214: 29-39. DOI: 10.1016/j.jfoodeng.2017.06.031

146. Zhang C, Guo C, Liu F, Kong W, He Y, Lou B. Hyperspectral imaging analysis for ripeness evaluation of strawberry with support vector machine. Journal of Food Engineering. 2016; 179: 11-18. DOI: 10.1016/j.jfoodeng.2016.01.002

147. Lu Y, Huang Y, Lu R. Innovative hyperspectral imaging-based techniques for quality evaluation of fruits and vegetables: a review. Applied Sciences 2017; 7(2): 1-36. doi:10.3390/app7020189

148. Zeng X, Miao Y, Ubaid S, Gao X, Zhuang S. Detection and classification of bruises of pears based on thermal images. Postharvest Biology and Technology. 2020; 161(1): 111090. DOI: 10.1016/j.postharvbio.2019.111090

149. Zhu X, Lia G. Rapid detection and visualization of slight bruise on apples using hyperspectral imaging. International journal of food properties. 2019; 22(1): 1709-1719. DOI: 10.1080/10942912.2019.1669638

150. Fan S, Li C, Huang W, Chen L. Detection of blueberry internal bruising over time using NIR hyperspectral reflectance imaging with optimum wavelengths. Postharvest Biology and Technology. 2017; 134: 55-66. DOI: 10.1016/j.postharvbio.2017.08.012

151. Zhang BH, Li JB, Fan SX, Huang WQ, Zhao CJ, Liu CL, Huang D.F. Hyperspectral Imaging Combined with Multivariate Analysis and Band Math for Detection of Common Defects on Peaches (Prunus Persica). Computers and Electronics in Agriculture. 2015; 114: 14-24. DOI: 10.1016/j.compag.2015.03.015

152. Qiang L, Mingjie T. Detection of Hidden Bruise on Kiwi fruit Using Hyperspectral Imaging and Parallelepiped Classification. Procedia Environmental Sciences. 2012; 12: 1172-1179. DOI: 10.1016/j.proenv.2012.01.404

153. Li J, Rao X, Ying Y. Detection of common defects on oranges using hyperspectral reflectance imaging. Computers and Electronics in Agriculture. 2011; 78(1): 38-48. DOI: 10.1016/j.compag.2011.05.010

154. Xing J, Baerdemaeker JD. Bruise detection on 'Jonagold' apples using hyperspectral imaging. Postharvest Biology and Technology. 2005; 37(2): 152-162. DOI: 10.1016/j.postharvbio.2005.02.015

155. Mehl PM, Chen YR, Kim MS, Chan DE. Development of hyperspectral imaging technique for the detection of apple surface defects and contaminations. Journal of Food Engineering. 2004; 61(1): 67-81. DOI: 10.1016/S0260-8774(03)00188-2

156. Kandpal LM, Cho BK. A review of the applications of spectroscopy for the detection of microbial contaminations and defects in agro foods. Journal of Biosystems Engineering. 2014; 39(3): 215-226. DOI: 10.5307/JBE.2014.39.3.215

157. Kim G, Kim GH, Park J, Kim DY, Cho BK. Application of infrared lock-in thermography for the quantitative evaluation of bruises on pears. Infrared Physics and Technology. 2014; 63: 133-139. DOI: 10.1016/j.infrared.2013.12.015

158. Lu R, Chen YR, Park B, Choi K.H. Hyperspectral imaging for detecting bruises in apples. In: ASAE Annual International Meeting, Paper No. 993120, St. Joseph, Michigan. 1999

159. Lu R. Detection of bruise on apples using near-infrared hyperspectral imaging. Transactions of the ASAE. 2003; 46(2): 1-8. DOI: 10.13031/2013.12941

160. Changyeun M, Giyoung K, Moon SK, Jongguk L, Kangjin L, Wang-Hee L, Byoung-Kwan C. On-line fresh-cut lettuce quality measurement system using hyperspectral imaging. Biosystems engineering. 2017; 156: 38-50. DOI: 10.1016/j.biosystemseng.2017.01.005

161. Yang B, Gao Y, Yan Q, Qi L, Zhu Y, Wang B. Estimation Method of Soluble Solid Content in Peach Based on Deep Features of Hyperspectral Imagery. Sensors. 2020; 20(18): 1-12. DOI: 10.3390/s20185021

162. Pu Y, Sun D, Riccioli C, Buccheri M, Grassi M, Cattaneo TM, Gowen A. Calibration transfer from micro nir spectrometer to hyperspectral imaging: A case study on predicting soluble solids content of bananito fruit (Musa acuminata). Food Analytical Methods. 2018; 11(11): 1021-1033. DOI: 10.1007/s12161-017-1055-3

163. Ma T, Li X, Inagaki T, Yang H, Tsuchikawa S. Noncontact evaluation of soluble solids content in apples by near-infrared hyperspectral imaging. Journal of Food Engineering. 2017; 224: 53-61. DOI: 10.1016/j.jfoodeng.2017.12.028

164. Pu H, Liu D, Wang L, Sun DW. Soluble Solids Content and pH Prediction and Maturity Discrimination of Lychee Fruits Using Visible and Near-Infrared Hyperspectral Imaging. Food Analytical Methods. 2016; 9(1): 235-244. DOI: 10.1007/s12161-015-0186-7

165. Li J, Peng YK, Chen LP, Huang WQ. Near-infrared hyperspectral imaging combined with cars algorithm to quantitatively determine soluble solids content in "Ya" pear. Spectroscopy and Spectral Analysis. 2014; 34(5): 1264-1269. DOI: 10.3964/j.issn.1000-0593(2014)05-1264-06

166. Leiva-Valenzuela GA, Lu R, Aguilera JM. Prediction of firmness and soluble solids content of blueberries using hyperspectral reflectance imaging. 2013; Journal of Food Engineering. 115(1): 91-98. DOI: 10.1016/j.jfoodeng.2012.10.001

167. Baiano A, Terracone C, Peri G, Romaniello R. Application of hyperspectral imaging for prediction of physico-chemical and sensory characteristics of table grapes. Computers and Electronics in Agriculture. 2012; 87: 142-151. DOI: 10.1016/j.compag.2012.06.002

168. Rajkumar P, Wang N, EImasry G, Raghavan GSV, Gariepy Y. Studies on banana fruit quality and maturity stages using hyperspectral imaging. Journal of Food Engineering. 2012; 108(1): 194-200. DOI: 10.1016/j.jfoodeng.2011.05.002

169. Mendoza F, Lu R, Ariana D, Cen H, Bailey B. Integrated spectral and image analysis of hyperspectral scattering data for prediction of apple fruit firmness and soluble solids content. Postharvest Biology and Technology. 2011; 62(2): 149-160. DOI: 10.1016/j.postharvbio.2011.05.009

170. Peng Y, Lu R. Analysis of spatially resolved hyperspectral scattering images for assessing apple fruit firmness and soluble solids content. Postharvest Biology and Technology. 2008; 48(1): 52-62. DOI: 10.1016/j.postharvbio.2007.09.019

171. ElMasry G, Wang N, ElSayed A, Ngadi M. Hyperspectral imaging for nondestructive determination of some quality attributes for strawberry. Journal of Food Engineering. 2007; 81(1): 98-107. DOI: 10.1016/j.jfoodeng.2006.10.016

172. El Fakir C, Poffo L, Billiot B, Besnard P, Goujon JM. Active Hyperspectral Mid-Infrared Imaging Based on Widely Tunable QCL Laser. Conference: 21st International Conference on Transparent Optical Networks (ICTON), Angers, France. 2019; 1-4. DOI: 10.1109/ICTON.2019.8840448

173. Buitrago MF, Groen TA, Hecker CA, Skidmore AK. Spectroscopic determination of leaf traits using infrared spectra. International journal of applied earth observation and geoinformation. 2018; 69: 237-250. DOI: 10.1016/j.jag.2017.11.014

174. Ribeiro da Luz B, Crowley JK. Spectral reflectance and emissivity features of broad leaf plants: Prospects for remote sensing in the thermal infrared (8.0-14.0 μm). Remote Sensing of Environment. 2007; 109(4): 393-405. DOI: 10.1016/j.rse.2007.01.008

175. Ribeiro da Luz B. Attenuated total reflectance spectroscopy of plant leaves: a tool for ecological and botanical studies. New phytologist. 2006; 172(2): 305-318. DOI: 10.1111/j.1469-8137.2006.01823.x

176. Chrzanowski K. Testing thermal imagers: Practical guidebook. Warsaw: Military University of Technology. 2010; INSB 978-83-61486-81-7

177. Minkina W, Dudzik S. Handbook of Infrared thermography: Errors and uncertainties. 1st ed. Chichester: Wiley; 2009. 212 p. ISBN-10: 0470747188

178. Ball M, Pinkerton H. Factors affecting the accuracy of thermal imaging cameras in volcanology. Journal of Geophysical Research Atmospheres. 2006; 111(B11): 1-14. DOI: 10.1029/2005JB003829

179. Chelladurai V, Jayas DS, White NDG. Thermal imaging for detecting fungal infection in stored wheat. Journal of Stored Products Research. 2010; 46(3): 174-179. DOI: 10.1016/j.jspr.2010.04.002

180. Baranowski P, Mazurek W, Witkowska-Walczak B, Sławiński C. Detection of early apple bruises using pulsed-phase thermography. Postharvest Biology and Technology. 2009; 53(3): 91-100. DOI: 10.1016/j.postharvbio.2009.04.006

181. Baranowski P, Mazurek W, Wozniak J, Majewska U. Detection of early bruises in apples using hyperspectral data and thermal imaging. Journal of Food Engineering. 2012; 110(3): 345-355. DOI: 10.1016/j.jfoodeng.2011.12.038

182. Kuzy J, Jiang Y, Li C. Blueberry bruise detection by pulsed thermographic imaging. Postharvest Biology and Technology. 2018; 136: 166-177. DOI: 10.1016/j.postharvbio.2017.10.011

183. Takacs R, Jovicic V, Rasic AZ, Geier D, Delgado A, Becker T. Evaluation of baking performance by means of mid-infrared imaging. Innovative Food Science and Emerging Technologies. 2020; 61: 102327. DOI: 10.1016/j.ifset.2020.102327

184. Lopez-Sanchez J, Fortuny-Guasch J. 3-D radar imaging using range migration techniques. IEEE Transactions on Antennas and Propagation. 2000; 48(5): 728-737. DOI: 10.1109/8.855491

185. Sheen D, McMakin D, Hall T. Three-dimensional millimeter-wave imaging for concealed weapon detection. IEEE Transactions on Microwave Theory and Techniques. 2001; 49(9): 1581-1592. DOI: 10.1109/22.942570

186. Cook BS, Tehrani B, Cooper JR, Tentzeris MM. Multilayer inkjet printing of millimeter-wave proximity-fed patch arrays on flexible substrates. IEEE Antennas and Wireless Propagation Letters. 2013; 12: 1351-1354. DOI: 10.1109/LAWP.2013.2286003

187. Scott S. Three-Dimensional Microwave Imaging for Indoor Environments [Ph.D.]. University of California; 2017

188. Fear EC, Meaney PM, Stuchly MA. Microwaves for breast cancer detection? IEEE Potentials. 2003; 22(1): 12-18. DOI: 10.1109/MP.2003.1180933

189. Fear EC, Hagness SC, Meaney PM, Okonieweski M, Stuchluy MA. Enhancing breast tumor detection with near-field imaging. IEEE Microwave Magazine. 2002; 3(1): 48-56. DOI: 10.1109/6668.990683

190. Ambrosanio M, Scapaticci R, Crocco L. A Simple Quantitative Inversion Approach for Microwave Imaging in Embedded Systems. International Journal of Antennas and Propagation. 2015; 12: 1-18. DOI: 10.1155/2015/129823

191. Manickavasagan A, Jayasuriya H, editors. Handbook of Imaging with Electromagnetic Spectrum. 1st ed. (eds.). Berlin, Heidelberg: Springer; 2014. 204 p. DOI: 10.1007/978-3-642-54888-8

192. Pastorino M, editor. Handbook of Microwave Imaging. 1st ed. John Wiley & Sons; 2010. 304 p. ISBN: 978-0-470-60247-8

193. Donelli M, Franceschini D, Rocca P, Massa A. Three-dimensional microwave imaging problems solved through an efficient multiscaling particle swarm optimization. IEEE Transactions on Geoscience and Remote Sensing. 2009; 47(5):1467-1481. DOI: 10.1109/TGRS.2008.2005529

194. Benedetti M, Donelli M, Lesselier D, Massa A. A two-step inverse scattering procedure for the qualitative imaging of homogeneous cracks in known host media-preliminary results. IEEE Antennas and Wireless Propagation Letters. 2007; 6: 592 - 595. DOI: 10.1109/LAWP.2007.910954

195. Huang T, Mohan AS. A microparticle swarm optimizer for the reconstruction of microwave images. IEEE Transactions on Antennas and Propagation. 2007; 55(3 I): 568-576. DOI: 10.1109/TAP.2007.891545

196. Donelli M, Massa A. Computational approach based on a particle swarm optimizer for microwave imaging of two-dimensional dielectric scatterers. IEEE Transactions on Microwave Theory and Techniques. 2005; 53(5), 1761-1776. DOI: 10.1109/TMTT.2005.847068

197. Bort E, Donelli M, Martini A, Massa A. An adaptive weighting strategy for microwave imaging problems. IEEE Transactions on Antennas and Propagation. 2005; 53(5): 1858-1862. DOI: 10.1109/TAP.2005.846811

198. Caorsi S, Massa A, Pastorino M, Donelli M. Improved microwave imaging procedure for non-destructive evaluations of two-dimensional structures. IEEE Transactions on Antennas and Propagation. 2004; 52(6): 1386-1397. DOI: 10.1109/TAP.2004.830254

199. Giordano A. Microwave Imaging Technology for Food Contamination Monitoring [MSc]. College of Electronic Engineering, Telecommunications and Physics, Politecnico Di Torino; 2018

200. Ragni L, Al-Shami A, Mikhaylenko G, Tang J. Dielectric characterization of hen eggs during storage. Journal of Food Engineering. 2007; 82(4): 450-459. DOI: 10.1016/j.jfoodeng.2007.02.063

201. Guo W, Trabelsi S, Nelson SO, Jones DR. Storage effects on dielectric properties of eggs from 10 to 1800 MHz. Journal of Food Science. 2007; 72: 335-340. DOI: 10.1111/j.1750-3841.2007.00392.x

202. Tai TC, Wu HW, Hung CY, Wang YH. Food Security Sensing System Using a Waveguide Antenna Microwave Imaging through an Example of an Egg. Sensors. 2019; 20(3): 699. DOI:10.3390/s20030699

203. Abdullah MZ, Guan LC, Lim KC, Karim AA. The applications of computer vision system and tomographic radar imaging for assessing physical properties of food. Journal of Food Engineering. 2004; 61(1): 125-135. DOI: 10.1016/S0260-8774(03)00194-8

204. Sándor K, Zsolt P, Ádám C, György B, Tamás M, Tamás D. Non-destructive imaging and spectroscopic techniques to investigate the hidden-lifestyle arthropod pests: a review. Journal of Plant Diseases and Protection. 2020; 127: 283-295. doi.org/10.1007/s41348-020-00300-6

205. Singh V, Sharma N, Singh S. A review of imaging techniques for plant disease detection. Artificial Intelligence in Agriculture. 2020; 4: 229-242. DOI: 10.1016/j.aiia.2020.10.002

206. Hart AG, Bowtell RW, Köfckenberger W, Wenseleers T, Ratnieks FLW. Magnetic resonance imaging in entomology: a critical review. Journal of Insect Science. 2003; 3(5),1-9. DOI: 10.1673/031.003.0501

207. Joyce DC, Hockings PD, Mazucco RA, Shorter AJ. 1H-Nuclear magnetic resonance imaging of ripening 'Kensington Pride' mango fruit. Functional Plant Biology. 2002; 29(7): 873-879. DOI: 10.1071/PP01150

208. Thybo AK, Szczypinski PM, Karlsson AH, Dønstrup S, Stødkilde-Jørgensen HS, Andersen HJ. Prediction of sensory texture quality attributes of cooked potatoes by NMR-imaging (MRI) of raw potatoes in combination with different

image analysis methods. Journal of Food Engineering. 2004; 61(1): 91-100. DOI: 10.1016/S0260-8774(03)00190-0

209. Koizumi M, Naito S, Ishida N, Haishi T, Kano H. A dedicated MRI for food science and agriculture. Food Science and Technology Research. 2008; 14 (1): 74 - 82. DOI: 10.3136/fstr.14.74

210. Taglienti A, Massantini R, Botondi R, Mencarelli F, Valentini M. Postharvest structural changes of Hayward kiwifruit by means of magnetic resonance imaging spectroscopy. Food Chemistry. 2009; 114(4): 1583-1589. DOI: 10.1016/j.foodchem.2008.11.066

211. Ciampa A, Dell'Abate MT, Masetti O, Valentini M, Sequi P. Seasonal chemical-physical changes of PGI Pachino cherry tomatoes detected by magnetic resonance imaging. Food Chemistry. 2010; 122(4): 1253-1260. DOI: 10.1016/j.foodchem.2010.03.078

212. Zhang L, McCarthy MJ. Measurement and evaluation of tomato maturity using magnetic resonance imaging. Postharvest Biology and Technology. 2012; 67, 37-43. DOI: 10.1016/j.postharvbio.2011.12.004

213. Melado-Herrerosa A, Munoz-Garcíab MA, Blancoc A, Valc J, Fernández-Valled M, Barreiro P. Assessment of watercore development in apples with MRI: Effect of fruit location in the canopy. Postharvest Biology and Technology. 2013; 86: 125-133. DOI: 10.1016/j.postharvbio.2013.06.030

214. Divya S, Thyagarajan D, Sujatha G. Magnetic resonance imaging technology for process control and quality maintenance in food quality operation. International Journal of Engineering and Technology (IJET). 2013; 4(6): 441-449. ISSN: 0975-4024

215. Van As H, Van Duynhovenc J. MRI of plants and foods. Journal of Magnetic Resonance. 2013; 229: 25-34. DOI: 10.1016/j.jmr.2012.12.019

216. Chizhik VI, Chernyshev YS, Donets AV, Frolov VV, Komolkin AV, Shelyapina MG. Basic principles of detection of nuclear magnetic resonance. In: Chizhik VI, Chernyshev YS, Donets AV, Frolov VV, Komolkin AV, Shelyapina MG. Handbook of Magnetic resonance and its applications. 1st ed. Chizhik, Cham: Springer; 2014. P. 93-162. DOI: 10.1007/978-3-319-05299-1_2

217. Herremansa E, Melado-Herrerosb A, Defraeyea T, Verlindenc B, Hertoga M, Verbovena P, Vald J, Fernández-Vallee ME, Bongaersf E, Estradeg P, Wevers h M, Barreirob P, Nicolaïa BM. Comparison of X-ray CT and MRI of watercore disorder of different apple cultivars. Postharvest Biology and Technology. 2014; 87: 42-50. DOI: 10.1016/j.postharvbio.2013.08.008

218. Patel KK, Khan MA, Kar A. Recent developments in applications of MRI techniques for foods and agricultural produce—An overview. Journal of Food Science and Technology. 2015; 52(1): 1-26. DOI: 10.1007/s13197-012-0917-3

219. Pfugfelder D, Metzner R, van Dusschoten D, Reichel R, Jahnke S, Koller R. Non-invasive imaging of plant roots in diferent soils using magnetic resonance imaging (MRI). Plant Methods. 2017; 13(102): 1-9. DOI: 10.1186/s13007-017-0252-9

220. Kamal T, Cheng S, Khan IA, Nawab K, Zhang T, Song Y, Wang S, Nadeem M, Riaz M, Khan MAU, Zhu BW, Tan M. Potential uses of LF-NMR and MRI in the study of water dynamics and quality measurement of fruits and vegetables. Journal of Food Processing and Preservation. 2019; 43(474): 1-21. DOI: 10.1111/jfpp.14202

221. Baek S, Lim J, Lee JG, McCarthy MJ, Kim SM. Investigation of the maturity changes of cherry tomato using magnetic resonance imaging. Applied Sciences. 2020; 10(15): 5188. DOI: 10.3390/app10155188

Index

A

Aflatoxins 201
Algorithms 187
Alimentos 158
Allergenic 83
Allergies 83
Analysis 15, 18, 49, 125, 186, 188-190, 194, 206, 219, 231, 239
Antennas 241
Articles 12-14, 49, 86
Ascorbic 153
Assessment 14
Atmosphere 122
Aureus 62
Autoclave 143

B

Bacillus 65
Bacteria 62, 65
Bagasse 11, 17-19, 40, 42-43, 47-51, 55
Banana 78
Beetroot 101
Bioactive 116, 120, 128, 133
Biology 237, 243
Botulism 65
Butter 80

C

Cactus 123
Cadmium 227
Cereals 11
Cheese 217
Chemistry 111
Chocolate 75, 79-82, 84, 86-91
Components 59, 120
Consumers 48
Cornflour 101
Criteria 11, 49
Cylinder 164, 169, 172, 179

D

Detector 192
Digital 187
D-value 142
Dynamics 159-161

E

Emission 194
Energy 40, 192, 207

F

Flavanols 72
Flavonoids 63
Fruits 81, 101, 125, 203, 222

G

Gamma-ray 192-194
Glucose 62
Glycerol 101
Grading 232
Grains 199, 204

H

Hardness 210
Harvesting 77
Hazards 66-67
Health 47-48, 52, 64, 71-72, 86-87
Humidity 60
Hydrolysis 166-167, 183

I

Infrared 205-206, 234, 236, 238

Intensity 187, 222

Internal 222

J

Journal 52-57, 69-70, 107-109, 111-112, 114, 129-132, 134-135, 150, 152-159, 161-162, 181-183, 224, 226-228, 230, 232-234, 237-240, 242-243

L

Larvae 64

M

Machines 202

Magnetic 218, 243

Matrix 43, 45

Microscopy 124

Microwave 214-217, 241-242

Minerals 46

N

Nectar 63

Nonthermal 5

Nutrition 52

O

Overtones 207

Oxidase 62

P

Pericarp 222

Pesticides 82

Phenolic 64

Plastic 95-96

Platform 79

Poisoning 81

Pollen 63

Polymers 106, 108, 114, 129

Proteins 62, 123, 126

Q

Quality 6, 18, 20, 23, 42, 53, 109, 122, 130-131, 145, 150, 185, 191, 195-196, 202, 204, 207, 220, 222-224, 231-232, 234, 236

R

Radiation 191, 200, 205, 211

Rayleigh 164

Renewable 116-117

Research 11, 18, 49, 129, 157, 219

Resistance 142

S

Safety 66, 90-91, 185, 223-224, 236

Sensory 13, 19, 22, 44-45, 50-51, 219

Signal 214, 219, 222

Simulation 149, 159, 161

Software 148, 190, 215

Solvent 98

Spectrum 188, 191, 202, 206

Stomach 63

Sugarcane 10-11, 17-19, 40, 42-43, 47-51, 55

Surface 200

Systems 151, 202, 211, 213, 223-224

T

Tensor 190

Texture 17, 22, 44

Thermal 5, 139-140, 142, 144, 146-152, 154-155, 158, 160-162, 166, 179, 212-213, 240

Thiamine 177

Tomography 194, 216, 227

V

Vegetables 81, 134, 222

Viscosity 166

Vitamin 60

W

Wavelength 191, 200, 202

X

X-rays 195

Z

Zamora 151

Z-value 142

www.ingramcontent.com/pod-product-compliance
Ingram Content Group UK Ltd.
Pitfield, Milton Keynes, MK11 3LW, UK
UKHW021327190726
13845UKWH00004B/67

9 789349 964266

Basics of Baking

Basics of Baking

SANDEEP MALIK

IMSAR, M.D.UNIVERSITY, ROHTAK, (HARYANA)

AMAN PUBLICATIONS

NEW DELHI

Published by
Rajiv Jain for Aman Publications
LG-4A, Ganpati Bhawan,
4675-B/21, Ansari Road,
Darya Ganj, Delhi - 110002
PH: 23255012, 23282127
E-mail: sshjbs@yahoo.com

First Published – 2007
2nd Edition - 2023

ISBN–81–8204–033-7

SOLE DISTRIBUTOR
JINDAL BOOK SERVICES
LG-4A, Ganpati Bhawan,
4675-B/21, Ansari Road,
Darya Ganj, Delhi - 110002

Jacket Designed by :
S.M. Graphics

Printed by:
NCT Printers , Delhi, India

PREFACE

Baking is the technique of cooking food in an oven by dry heat applied evenly throughout the oven. Breads, desserts, and meat are often baked, and baking is the primary cooking technique used to produce cakes and pastry-based goods such as pies, tarts, and quiches. Such items are sometimes referred to as baked goods.

Baking is a fine art of dedication, challenges, hardships and triumphs. It is as much a science as an art. So, in addition to mastering basic techniques and recipes, every baker must also learn about the science that underlies the baking craft. This book offers an in-depth coverage of the basics of baking. It explains the art and science of contemporary baking techniques and practices.

Written in a clear, easy-to-understand style, this book gives readers knowledge that they can apply to their own baking-whether it's selecting the right flour, understanding how different leavening agents work, or learning about using new baking ingredients and additives to enhance favourite recipes. This insightful book will be an essential key to mastering skills and effectively adapting to today's quickly evolving trends. It will be an essential companion for professional bakers, pastry chefs, beginning students and anyone who is serious about baking.

CONTENTS

	Preface	v
1.	Introduction to Baking	1
2.	Baking Ingredients	12
3.	Baking of Bread	37
4.	Preparation of Cakes	76
5.	Baking of Pies	114
6.	Making of Cookies	138
7.	Baking of Hot Rolls and Buns	161
8.	Preparation of Icings	167
9.	Measuring Ingredients	178
10.	Properties and Functions of Baking Powder	194
11.	Yeast-free Baking	205
12.	The Physics of Baking	210
13.	Baking Appliances	235
14.	Functions of Baking Ovens	261
	Bibliography	293
	Index	295

1

INTRODUCTION TO BAKING

Baking is the technique of cooking food in an oven by dry heat applied evenly throughout the oven. The person that does the baking is called a baker. Breads, desserts, and meat are often baked, and baking is the primary cooking technique used to produce cakes and pastry-based goods such as pies, tarts, and quiches. Such items are sometimes referred to as "baked goods," and are sold at a bakery.

The dry heat of baking gelatinises starch and causes the outside of the food to brown or char, giving it an attractive appearance and taste, as well as partially sealing in the food's moisture. The browning is caused by caramelisation of sugars and the Maillard reaction. Moisture is never really entirely "sealed in," however; over time, an item being baked will become drier and drier. This is often an advantage, especially in situations where drying is the desired outcome, for example in drying herbs or in roasting certain types of vegetables. The most common baked item is bread. Variations in the ovens, ingredients and recipies used in the baking of bread result in the wide variety of breads produced around the world.

To compensate for moisture loss, some items (usually meats) are basted on the surface with butter or oil to slow the loss of moisture through the skin. Some foods are replenished with moisture during baking by placing a small amount of liquid (such as water or broth) in the bottom of the pan, and letting it steam up into or around the food.

The term baking is not usually associated with the cooking of meats in this manner, it is instead termed roasting. Over time baked goods become hard in a process known as going stale, this is not primarily due to moisture being lost from the baked products but a reorganisation of the way in which the water and starch are associated over time, a process similar to recrystallisation.

Role of Baker

A baker is someone who primarily bakes and sells bread. Cakes and similar foods may also be produced as the traditional boundaries between what is produced by a baker as opposed to a pastry chef have blurred in recent decades. The place where a baker works is called a bakehouse, bakeshop, or bakery.

Today bakers work in varying environments both as employees and sometimes owning their own stores. Bakers can be found working in:

— *Large factories*. These produce bread and related products which are then transported to numerous selling points throughout a region. These normally include supermarkets, convenience stores, and the like. Bakers in these environments are largely there for quality control as machines take care of much of the labour intensive aspect of the job.

— *Small Independent bakeries*. These are largely family-run businesses. They may specialise in particular types of products, such as sourdough. In Australia many such businesses are owned by members of the Vietnamese community.

— *Chain stores*. Recent years has seen the rise of chain stores selling the same range of products. For example, Bakers Delight is an Australia-wide operation well known for its range of products. Bakers in these stores bake according to a pre-determined recipe book. This can lead to fustration as some bakers do not agree with techniques used by the franchising model.

Baker Percentage

Baker percentage is a way of indicating the proportion of ingredients when making bread. Contrary to the usual way of expressing percentages, the total does not add up to 100. All the ingredients are measured by their weight compared to the flour's. Thus, the flour accounts always for 100% and all the other ingredients make the total higher than 100. For example, if a recipe calls for 10 pounds of flour and 5 pounds of water, the corresponding percentages will be 100% and 50%.

The main advantage of this formulation is that it allows bakers to easily resize a recipe. With a percentage formula for a bread recipe, it becomes exceedingly easy to make the same batch of bread for 2 loaves as it is for 50 loaves, with a bit of simple math. Common formulations for bread include 100% flour; 60% water/ liquid, 1% yeast 2% salt, and 1% oil, lard, or butter.

Moreover, bread recipes are better done this way than by volume measurements especially in large batches, because measurements by weight are very precise, while measurements by volume can actually vary a good amount, depending on how the flour is packed into the measuring cup, how the flour is ground, etc.

The Art of Baking

Baking is basic, but it can be tricky. It's one of the more advanced kitchen arts, but you can learn it. Be forewarned right now: Baking can be addicting. Baking is awesome. And it can be easier, once you learn the basics. Unlike other types of cooking, you can't just add a little bit here or substitute a little bit there, until you are pretty advanced.

In all branches of cooking, the application of heat changes one thing into another. In baking, this mystery is even more pronounced. You make a dough or a batter, put a wet, sticky thing into the oven and eventually remove a mouthwatering solid object that smells so good people will think Betty Crocker has been incarnated in your kitchen.

The one big rule is you must measure. This is probably another factor bakers love. They are kitchen chemists, more than any other kinds of cooks, using measuring spoons and cups instead of test tubes and beakers. You must measure accurately and have the right tools, for it all to work. You need different measuring cups for liquids and dry ingredients. Flat-topped measuring cups are made to be filled to the top with dry stuff. Liquid measures are usually glass, with more space at the top to eliminate spills, and a pour spout.

A 2-cup Pyrex liquid measure is more useful than the 1-cup; a 4-cup measure is handy because it can also be used as a small mixing bowl. Add liquids to the cup, set it on the counter, and look at it sideways, on its own level. If you hold it up in front of your face, the liquid sloshes around and is hard to read accurately. One cup of flour weighs four ounces. You can buy a kitchen scale and weigh it, or you can properly measure flour: Stir the flour in a bag. Gently spoon flour into the measuring cup that you hold over a sheet of waxed paper or the flour canister.

Do not tap it or shake it to level the flour. Let the flour heap up. When the flour is over the top, use the flat side of a knife or spatula to level off the top of the cup. If you tamp flour into the cup it could weigh as much as six ounces. The only dry ingredient you pack into a cup is brown sugar. Solid shortening is packed, too, unless you purchase the stick form pre-marked in measured amounts. Do you need to sift? Only if the recipe recommends doing so, because today's all-purpose flour is pre-sifted.

If you do have to sift and you don't have a sifter, use a sieve or strainer. After you get your fix of measuring, you will arrive at the crux of baking magic. The alchemy all depends on leavening, the agents that make dough rise. Leavenings can be yeast, eggs, baking powder, baking soda combined with an acid such as vinegar, or other more arcane things, such as sourdoughs and the artisan baker's "old dough" saved from previous loaves.

Yeast can be a tricky beast, we concentrate on the quicker, easier baked goods made with baking powder. Yes, there is every kind of mix these days, but you should know how to make basic muffins, biscuits, cornbread, brownies and a loaf of banana bread.

Preheating

Why do recipes for baked goods always start out "preheat oven to...?" It's because of "oven spring," which is the initial leap that leavenings make when put into a hot oven. If your oven's not hot when you put the pans in, the mixture won't rise like it's supposed to. Most ovens will preheat in 10 to 20 minutes. You should also prepare baking pans according to the recipe. Cookies with a high fat content may not need to go on a greased surface. You can use a paper towel to smear shortening or butter into pans, and remember to coat those pesky corners well, or you can spray with PAM, one of the greatest inventions of the modern world.

Substitutions

Whole wheat is heavier. You can substitute half of the flour amount in an all-white-flour recipe with whole-wheat flour, but, more than that, and you're risking your teeth. One of the most frequent baking mistakes is the use of breakfast spread instead of butter or margarine. This is a guaranteed way to wind up throwing your results in the trash can. "Spreads" that are less than 60 percent fat have a lot of water included and will make cookies spread too thin or otherwise mess up recipes. If the first ingredient on the label is water, don't use it for baking.

Stick margarine that is at least 80 percent fat can be substituted for butter. For best results, use butter if the recipe calls for butter. Eat smaller pieces. Baking powder cannot be substituted for baking soda. They are not the same thing. Bottom line: Don't mess with the basic ingredients, the flour, liquid, salt, fat, leavening. In one area you may unleash your creativity. It's in the add-ons. Nuts and dried fruits can be substituted freely. Out of almonds? Use pecans. Or use dried red cherries instead of apricots in a scone. Coconut counts as a dried fruit.

However, substitute dry for dry. Fresh fruit cannot be used in place of dried fruit because the extra moisture in the fruit will change the finished product. Spices and extracts can often substitute for each other, too. Try nutmeg for a change instead of cinnamon, or use almond extract instead of vanilla. Remember, however, that too much of any spice will overwhelm instead of complement. Go for subtlety. In general, don't make more than one substitution per baked recipe.

Muffins Making

Why are the breakfast muffins in your local java joint so delicious? They're cake, that's why. Check the fat and calorie content if it comes in a wrapper. Any cake batter, just about, can be baked in muffin tins. But homemade muffins for breakfast and snacks don't have to be laden with fat and sugar. They actually can be healthful treats. Cakes and their cousins, such as muffins, have to be measured more accurately than breads.

If you don't measure carefully, they can turn out dry or heavy. Muffins are mixed easily by hand, with one bowl of stirred-together dry ingredients and one bowl of liquid ingredients. Most recipes call for muffin cups to be filled two-thirds full, but that's not a hard and fast rule. Oven spring, the initial leap that leavenings make when put into a hot oven, pushes muffins over the top of the cup.

Quick Breads

Like muffins, quick breads are fast and simple to make. You don't have to be quite as careful about overmixing, but all that's needed is a light combining of wet and dry, Moore advises. Overmixing can cause bread to crumble. A crack down the middle of the loaf is characteristic of these breads. Don't worry about it. You can omit salt from quick breads if you want, Moore said, but "to me, it just doesn't taste right." Omitting salt affects the rising of yeast doughs, but not quick ones.

Quick breads often are flavoured with fruits, such as pumpkin, or are great vehicles for using an excess of bananas or zucchini. Oil is most often the fat of choice in quick breads. Applesauce can be substituted directly for the same amount of oil in fruit quick breads, said Sharon Hoelscher Day, extension educator at the Maricopa County Cooperative Extension. If the recipe calls for one cup of oil, substitute one cup of applesauce. The texture may be a little heavier, but still good. In most recipes for quick breads, you can reduce the fat amount by one-fourth or one-third, Hoelscher Day said. Or substitute two egg whites for one egg.

Brownies

Brownies are a chocolate cake with very little flour, which is what makes them so chewy and good. The commercial fat substitute based on prunes can be used in most recipes for brownies, because it's dark-coloured and the flavour of dried plums blends well with chocolate.

Cornbread

Learn to make good cornbread and you never will be without a perfect companion for a bowl of chili or beans or stew. Where you were raised will determine whether you like your cornbread sweeter (the South) or less sweet (anywhere else). All cornmeals—white, yellow and the Southwestern blue—are enough alike that they all can be substituted for each other. The latest Cooks Illustrated magazine, known for its exhaustive tests to perfect recipes, notes that yellow has the most "corn" flavour.

Cornbread is basically the same as muffin batter, but with more liquid, Moore said. For recipes calling for two cups white flour to make a dozen muffins, you can substitute up to one cup of the flour with whole-wheat flour or cornmeal, adding perhaps a little more liquid for a cornmeal version. Cornbread often is baked in cast-iron pans for a crisp crust. To make it even crispier, preheat a greased or oiled cast-iron skillet or pan in the oven before pouring in the cornbread batter.

Biscuits

Shortening or butter is "cut in" to biscuit doughs and piecrusts, instead of being melted or softened and mixed in. The intact bits of fat separate the flour into layers during baking, creating the biscuit's (or piecrust's) characteristic flakiness. To "cut in," you can use a pastry cutter, which dices up the fat, or you can use two knives, sawing at the shortening in a crisscross motion. Soon, the fat and flour will bind to form crumbles. The recipe should say what size these crumbles should be: "the size of peas" or whatever. When the crumbles are that size, stop. Add liquid, and knead very lightly. Biscuits need what our foremothers called "a light hand." Think about having "a light hand," and it will be easier every time you make a new batch.

For strawberry season, you can add extra sugar to biscuit dough to create classic shortbread biscuits. Just a couple of tablespoons in a batch is enough. Cut the biscuits into 3-inch rounds, split with a fork when they're just barely cooled, and layer on sliced strawberries, sweetened a bit ahead of time so they'll make their own juice. Dollop on some whipped cream. You then can taste the real reason that you should know how to make biscuits.

Importance of Sanitation in Baking

All bakery personnel to maintain extreme cleanliness and sanitation in the bakery to protect the health of the consumer of bakery foods. It is the responsibility of the bakery production manager or the person in charge to make daily inspections of the bakery and personnel. It is advisable to set up a cleaning frequency chart to provide for the daily cleaning of every area of baking equipment that actually touches the baked bread or pastry.

Cream-filled pies and whipped cream products have been involved in food poisoning outbreaks. Protection against Staphyloccus poisoning can be controlled by good sanitation practices and by holding of prepared foods under refrigeration until they are served. At a temperature of 42 degrees F. or lower, the growth of staphylocci is completely inhibited. This temperature is merely a safeguard, but it is not completely reliable, since the bacteria merely becomes dormant at the low temperature and resume their active growth as soon the temperature is raised to a suitable level.

Insects and Animal Pests

Breeding of insects and animal pests such as rats and mice will cause unsanitary conditions. Rodents not only eat costly ingredients and manufactured products, but they also create a potential danger to consumers of bakery foods. Rodents are dangerous transmitters of Yellow Jaundice, Typhus, Food Poisoning, and Tapeworms or other internal parasites. To prevent the entrance of rodents, seal the bakery tightly. When preventive measures fail, and rodents get inside the bakery, use traps or call a Pest Control Agency. Clean all flour bins, conveyors and flour sifters. All dead flour which is flour that collects at the bottom, sides, and inside of the flour handling equipment should be removed to prevent insect eggs from hatching.

Government regulations for food products list legal requirements for the manner in which food can be processed and they have the support of law. These regulations are covered in good manufacturing practices which are part of the food, drug and cosmetic act. The act covers bakery, plants and grounds, maintence, sanitary facilities, and production and processing. Good manufacturing practices are legal requirements, therefore food processors can be prosecuted for not complying with it's provisions.

Salmonellosis

Salmonella has been known to be present in eggs, dried milk, and cocoa. Any of these ingredients, when incorporated in foods could be a source of infection. Baking usually inactivate these organisms, however, if infected marerials exist in the bakery, re-infecting the baked product is possible. When Staphylococci infects an individual, it may be dead, but the poison produced by this bacterium can still make one sick. The poison is not destroyed by heat. However, Salmonella does not produce any poison and must be alive to make one sick.

Personal Cleanliness

All bakery personnel should observe and practice rules of cleanliness such as keeping the body clean. The bakery should provide hot showers for bakery personnel. Clean white uniforms including headgear should be worn while on duty. Arm pits should always be covered. Hands should be washed with soap and hot water when returning from work after any absence, particularly after using the toilet. Personnel having skin eruptions or infectious cuts should never be allowed to work with food.

Clean bakery equipment should be mandatory in bakery production, because unsanitary bakery equipment may cause the two most dreaded deseases in bakery foods-mold and rope.

Role of Food Inspectors

Food inspectors, state health department inspectors, and city health department inspectors are authorised to perform inspections. Also inspectors from the American institute of baking which developed the consolidated sanitation standards are recognised by the government and are authorised to make legal inspections. These inspections are authorised to be made at reasonable working hours. When inspectors arrive at the bakery, he or she presents credentials before presenting the notice of inspection form to the plant manager. At that point the inspector has legal authority to begin the inspection.

BAKING INGREDIENTS

Ingredients play special role in baking and cooking process. Some functions are critical to the success of the finished product. Some of these ingredients can be substituted with other ingredients and produce the same outcome but many times substitutions will have some affect on the end product.

DRY INGREDIENTS

Baking Powder

A combination of baking soda, an acid and a moisture absorber, which together form a leavening agent. Just the right amount of acid is added to the baking soda so that when the baking powder is added to moist ingredients, there is a reaction that releases carbon dioxide bubbles. This reaction is what makes the batter or dough rise.

Baking Soda

An ingredient used as a leavening agent. When it is mixed with an acidic ingredient and then comes in contact with liquid ingredients, a chemical reaction occurs. The chemical reaction releases carbon dioxide gas and causes the rising action in dough or batter. Some of the acidic ingredients that cause the reaction with baking soda are vinegar, lemon juice, orange juice, buttermilk, sour milk, sour cream, honey, chocolate and molasses. The chemical reaction begins as soon as the liquid is added, so the dough or batter should be baked as soon as possible after it has been mixed. If too much time passes before baking, the gases can escape and the product will not rise properly when baked.

Bran

Bran is the outer layer of a kernel of grain. Bran is added to cereals and baked goods to increase the nutritional value and to add fiber.

Bread-crumbs

Breadcrumbs are used to add texture and body to foods. They are added in with other ingredients or they can be used as a coating for meats, fish and vegetables. Breadcrumbs are used fresh or dried, and are used in different degrees of coarseness. Fresh breadcrumbs are handmade. Dried breadcrumbs can be made by hand but are also sold in food stores already dried and prepackaged.

Brown Sugar

Granulated sugar that has molasses syrup added to it. The molasses syrup softens the texture of the sugar. Brown sugar is available in a light and dark variety. The darker has a more intense flavour. Use light brown sugar unless dark brown is specifically specified in the recipe.

Canning & Pickling Salt

A fine-grained salt that does not contain any additives,

which prevents the brine from becoming cloudy when canning and pickling food. It is used when canning and pickling vegetables and meats. It can also be used in the same manner as table salt but it may start to clump when exposed to excessive humidity.

Cocoa Powder

An unsweetened powder made from the leftover chocolate liquor that has had the cocoa butter removed. The chocolate liquor is made from cacao beans and about 75% of it is extracted as cocoa butter and the remainder is dried and ground into cocoa powder. Cocoa powder is used to add chocolate flavouring to baked goods.

Corn Meal/Flour

Made from ground dried corn. It is available coarse, medium or fine ground. Corn flour is ground finer than fine ground corn meal. Corn meal or flour will not form gluten so products made from them have a more coarse, dense texture than products made from some of the other flours. Corn meal and corn flour provide a corn flavour to products it is used in.

Cornstarch

A thickening agent used in sauces, gravies, soups and desserts. It is finely ground from the heart of the corn kernel and has twice the thickening power of flour.

Tartar Cream

A white powder made from the acid deposits that form on the inside of the wine barrels during wine making. It is added to egg whites to give them more volume and stability. It also helps to stop sugar syrups from crystallising and makes creamier frostings and candies when it is one of their ingredients. This acidic powder is also added to baking soda to create commercial baking powder.

Flour

Generally, flour is the ingredient that gives baked goods their form and structure. When the flour is mixed with water, the proteins in flour interact with each other to form gluten. Gluten is what gives the dough its elasticity and the ability to stretch as the leavening agent produces the carbon dioxide gases that cause the dough to rise. Different types of wheat flour contain different amounts of proteins for forming gluten. Soft wheat have a lower protein content than hard wheat. A flour with higher protein content is more desirable for yeast breads, whereas lower protein content is better for cakes, pastries, and quick breads to produce a more tender texture.

All-Purpose Flour

All-purpose flour is a multipurpose flour made from hard and soft wheat. It can be used for all baking purposes. It is available bleached and unbleached.

Arrowroot Flour

Arrowroot flour is a fine powder ground from the arrowroot plant. It does not have any flavour and its texture is similar to cornstarch. It is used as a thickener for sauces, gravies, puddings and dessert fillings. Arrowroot flour has approximately 50% more thickening power than flour.

Bread Flour

Bread flour is a hard wheat flour that is higher in gluten

forming protein. It is used with yeast breads because the higher gluten content produces bread that is lighter, has more volume and is less crumbly.

Cake Flour

Cake flour is a soft-wheat flour with a fine-texture and a high starch content. It contains a lower level of protein than bread flour. When mixed with moisture, the protein in flour turns to gluten that gives the baked goods strength and structure. Pastries and cakes need only a limited amount of gluten. Since the pastry flour contains less protein it creates tender pastries and cakes. Cake flour is also referred to as pastry flour.

Oat Flour

Oat flour is ground from rolled oats. It provides a nutlike flavour and has a high protein content. Oat flour does not form gluten like the wheat flour, so it must be used in combination with a wheat flour so that the bread will rise. When using oat flour, about 1/3 of the flour can be oat flour and 2/3 should be wheat flour.

Pastry Flour

Pastry flour is a soft-wheat flour with a fine-texture and a high starch content. It contains a lower level of protein than bread flour. When mixed with moisture, the protein in flour turns to gluten that gives the baked goods strength and structure. Pastries and cakes need only a limited amount of gluten. Since the pastry flour contains less protein it creates tender pastries and cakes. The higher level of protein found in bread flour is required to create the structure of the bread crumb. Pastry flour is also referred to as cake flour.

Potato Starch Flour

Potato starch flour is a non-gluten forming flour that is ground from dried, cooked potatoes. It is a fine textured flour that has a mild potato flavour. Potato starch flour must be used in combination with a gluten forming flour, such as a all-purpose flour.

Rice Flour

Rice flour is a flour made from finely ground long or short grain rice. It is used as a thickening agent to thicken puddings, desserts, and sauces. It is also used to make noodles. It does not form gluten, so it can be substituted for wheat flour in some recipes to be used by people that do not tolerate gluten.

Rye Flour

Rye flour is ground from grains of rye grass and has a slightly sweet-sour flavour. It does not contain enough protein to develop the gluten necessary for the bread to rise sufficiently, so it must be combined with wheat flour. Rye flour produces a denser and darker loaf of bread and is available in light, medium and dark varieties. The darker the flour the darker the loaf of bread will be. When using light rye flour in combination with wheat flour, you can use about 40% rye flour. When using the medium or dark rye flour, the amount of rye flour must be decreased.

Self-rising Flour

Self-rising flour is an all-purpose flour into which baking powder (a leavening agent) and salt have been added. When used in a recipe in place of all-purpose flour, the baking powder and salt are omitted from the other ingredients.

Semolina Flour

Semolina flour is ground from hard durum winter wheat. Semolina flour produces stronger dough because it contains a high level of gluten, which adds elasticity to the dough. Dough made from semolina flour is too hard to handle when rolling out by hand because of its stickiness but when mixed with unbleached white flour it produces a strong dough that is easier to handle. Semolina flour is used when making pasta and some breads.

Soy Flour

Soy flour is flour ground from hulled and roasted soybeans. Soy flour cannot form gluten and contains no starch, but it is very high in protein. It is generally used to increase the protein in baked goods but you must be cautious as to how much you use because it will affect the taste and cause increased browning if too much is used. The flour is produced as natural (defatted), which has the oils removed, or full-fat, which retains the soybean oils.

Triticale Flour

Triticale flour is ground from the man-made grain triticale. Triticale is a high protein man-made grain produced by crossbreeding wheat and rye. The man-made grain was produce in effort to duplicate the protein and bread-making qualities of wheat and the durability and high lysine content of rye. Triticale flour provides a nutty flavour similar to wheat and rye, and generally contains more protein than wheat flour. Because of its durability, it handles better than rye flour but not as well as wheat flour. It can be substituted in equal portions for recipes requiring either wheat or rye. Yeast dough made with triticale flour should be fermented for less time than dough made with wheat flour.

Wheat Flour

Whole wheat four is ground from the full wheat kernel, including the bran, germ, and the endosperm. Whole wheat flour is brownish in colour and is much more nutritious than refined white flour. It is often added to white flour to improve the nutrition of baked goods. Whole wheat flour is available in fine, medium and coarse ground. As the coarseness of the flour increases, the coarseness of the baked good's texture increases.

Due to the higher unsaturated oil content in whole wheat flour, it has a shorter shelf life. Food products made with whole wheat flour have a higher potential of becoming rancid if kept for long periods of time. The life of the food product can be extended if stored in the refrigerator. If substituting whole wheat flour for part of the white flour when making yeast and quick breads, each cup of white flour that you replace should be replace by 1 cup plus 2 tablespoons of whole wheat flour. Bread made with whole wheat flour will produce a smaller, denser loaf.

Granulated Sugar

Fine or extra-fine sugar crystals made from refined beet or cane sugar. It serves several purpose in baking. It provides sweetness, adds tenderness, and helps in the browning of baked goods. When added to yeast dough, it helps the yeast begin producing the gas for the dough to rise and helps to retain the moisture in the bread. It is also referred to as white sugar.

Herbs

Herbs are plants that have fragrant leaves and stems that are used to season foods. Herbs can be used fresh or dried, but the quantity you should use in a recipe may vary greatly depending on which type is used. Be sure to read the recipe carefully to see if different quantities are indicated for dried versus fresh. There are many different types of herbs, but some that are commonly available are: angelica, basil, bay leaves, burnet, chives, chervil, cilantro/ coriander, dill, fennel, garlic, hyssop, lavender, lemon balm, lemon grass, lemon verbena, marjoram, mint, oregano, parsley, pepper, rosemary, sage, savory, sorrel, tarragon, thyme, watercress, and wintergreen.

Kosher Salt

A coarse grained salt that does not contain any additives. It is preferred for its texture and flavour. Because it is coarser grained, it takes up more space when measuring. When using it in place of regular salt, you will need to increase the amount called for in the recipe. Add an additional 1/8 tsp. of kosher salt when the recipe calls for 1 tsp. of salt.

Oatmeal

A cereal ground from hulled and cleaned oats. It can be cooked to eat as a breakfast cereal or it can be used as an ingredient for baking. Oatmeal is available as old fashioned rolled oats, which are larger flakes that take approximately 15 minutes to cook or as quick-cooking oats, which are smaller cut flakes that take approximately 5 minutes to cook. Both types can be interchanged when baking. It is also available as an instant oatmeal, which is oatmeal that is made up of very small cut flakes that are precooked and dried. They require minimal cooking but are not recommended for use in baking.

Pearl Sugar

Large, smoothly polished crystals of sugar, which are used for decorating purposes.

Pepper

A savory spice that has been ground from peppercorns. Pepper is available as whole peppercorns that can be freshly ground and can also be purchased cracked or in ground form.

Powdered Sugar

Granulated sugar that has been finely ground into a powder. It is used to make creamy frostings and to add a decorative coating to baked goods. Cornstarch is sometimes added to the powdered sugar to prevent it from absorbing moisture and forming clumps. Also referred to as confectioner's sugar.

Salt

Salt is a seasoning that sharpens the flavour of other ingredients. If the amount of salt called for in a receipe is reduced, other seasonings should be increased to compensate for the flavour loss. Salt will also enhance the sweetness of a food. The salt in yeast dough slows the fermentation of the yeast, so reducing the amount of salt used will affect the finished product. The loaves of bread that will rise to quickly, which will affect the flavour and shape of the loaves.

Salt Substitute

Generally a salt substitute consists of potassium chloride. It is bitter in taste but provides for a lower sodium intake. It should not be used in baking. There are also salt substitute mixtures available which consists of a mix of potassium chloride and sodium chloride. The mixture provides for a lower sodium intake but its taste is similar to regular salt.

Sea Salt

Sea salt is produced through the evaporation of sea water. It is available in fine and coarse grained crystals. It is preferred for its fresh, distinctive taste, which will vary depending on the location it is from.

Spices

Spices are taken from a variety of plants. Spices are found in the stems, roots, buds, bark, seeds, and fruit of plants. They have strong aromas and strong flavours that are used to add flavour to baked and cooked foods. Most spices are available in both whole and ground form. Ground spices will lose their potency faster than whole spices. Some of the spices you will find are allspice, aniseeds, caraway, cayenne pepper, chili powder, cinnamon, cloves, coriander seeds, cumin, curry powder, fennel seeds, ginger, juniper berries, mace, nutmeg, paprika, pepper, peppercorns, saffron, Sichuan pepper, star anise, sumac, and turmeric.

Superfine Sugar

Granulated sugar that has been ground into extra-fine granules. Superfine sugar melts easily in liquid and works well in most baking recipes.

Table Salt

A fine-grained salt that contains additives that prevent caking, which allows the salt to flow freely. It is also available with iodine added. Table salt is used as an "at the table" seasoning that is added to food when it is eaten at a meal but is also used as an ingredient when baking and cooking. It should not be used in canning and pickling because the additives in it can cause the water to turn cloudy and pickles to darken in colour.

Unflavoured Gelatin

Unflavoured gelatin is made from the collagen protein found in the bone, skin, and cartilage of animals. It is added to sweet and savory foods to cause the liquid in the dishes to jell, which thickens or gives shape to the dish. One tablespoon of gelatin is generally enough to jell two cups of liquid.

Wheat Germ

Refers to the embryo of the wheat kernel, which is removed for the production of refined white flour, but remains as part of whole-wheat flour. The wheat germ is a good source of protein, vitamins, and minerals. It is used to add fiber and flavour, and can be substituted for up to 1/3 of the flour called for in a recipe.

Yeast

Tiny microorganisms that grow and multiply when activated by the addition of warm water. As the microorganisms increase in size and number, they give off carbon dioxide, which causes the dough to rise and produce a light and airy loaf of bread.

The addition or removal of other ingredients, such as sugar and salt, affect the action of the yeast, and if less yeast is used than called for, it will increase the amount of time it takes the dough to reach the proper volume. There are three types of yeast available, which include active dry yeast sold in granule form; compressed fresh yeast sold in cake form, which is found in the refrigerator section of a food store; and quick-rising yeast, which decreases the rising time by approximately 1/3 in comparison to active dry yeast.

Solid Ingredients

Almond Paste

A paste made from blanched almonds, sugar and glycerine. The paste is used as a flavouring agent for pastries. Almond extract is sometimes added to the paste to bring out the almond flavour. The paste is also used for decorating baked goods. Almond paste is similar to marzipan but they are not interchangeable in recipes.

Baking or Bitter Chocolate

Unsweetened chocolate that contains no other ingredients other than chocolate liqueur and 50 to 58% cocoa butter. Do not substitute any of the sweetened chocolates in a recipe calling for unsweetened chocolate.

Caramel

Sugar that has been slowly heated until it melts and becomes slightly burned. Once the sugar has melted, it takes on varying degrees of colour and develops a distinct flavour. If it is cooked too long it will become bitter and quite dark in colour. It is used in candies, desserts and savory dishes to add colour and flavour.

Caramel Candies

Candies made form butter, sugar and cream. They are generally sold as small square candies that are individually wrapped. These squares are melted and used as a filling or topping for desserts or baked goods. They are also melted and used to coat apples to make caramel apples.

Chocolate

Cocoa beans are ground and then processed to separate

the cocoa butter from the chocolate liqueur. The chocolate liqueur is then further processed and mix with other ingredients to create different varieties of chocolate. Some of the different varieties are dark chocolate, milk chocolate, white chocolate, unsweetened, bittersweet, and semi-sweet chocolate.

Chocolate Chips

Chocolate chips are small round droplets of chocolate. They are available in semisweet, bittersweet, milk, mint and white chocolate. You will also find them in mini, standard and large sizes. Chocolate chips will hold their shape when melted unless they are stirred together. They are popular to use in cookies, bars, muffins, and quick breads because when heated they will hold their shape. They add texture and bursts of flavour without being overbearing.

Cream Cheese

A smooth white cheese that is made from cream and milk. Cream cheese is a fresh cheese so it does not keep for a long period of time. It has a sweet tangy flavour, and is available in several varieties: regular, which contains about 33% milkfat; light, which is softer and contains less calories and fat than regular; nonfat, which is made from skim milk and contains no fat; and whipped, which is regular cream cheese that has been whipped to make it fluffy and easier to spread. It is also available in several flavours. Cream cheese is used as a spread but is also used to make dips, desserts and frostings.

Dark Chocolate

Dark chocolate includes sweet, semisweet and bittersweet chocolate. They are made up of chocolate liqueur, sugar and vanilla. Bittersweet contains more chocolate liqueur than sweet and semisweet chocolate. Generally the different varieties can be interchanged in recipes with little affect on the finished product.

Flavoured Chips

Baking chips that are shaped the same as chocolate chips but are different flavours. They are available in a variety of flavours such as vanilla, cherry, peanut butter, and butterscotch. They can be used in the same manner as chocolate chips to add texture and flavour to baked goods.

Fruits

There are many different types of fruit that are used in baking and cooking, such as apples, bananas, oranges, lemons, limes, pineapple, strawberries, peaches, pears, apricots, and berries. They are used fresh, cooked, and dried. Check the recipe for the type of fruit needed and how it needs to be prepared.

Milk Chocolate

Chocolate that is enriched with milk powder. It contains less chocolate liqueur and is lighter in colour than any of the dark chocolates.

Raisins

Raisins are dried grapes, which can be eaten plain or used as an ingredient in both sweet and savory dishes. They are available in two varieties, dark and golden. The dark raisin is dark in colour, has a drier, more shriveled texture, and has a sweet flavour. The golden raisin is a light golden colour, is more moist and plump, and has a sweet, tangy flavour. Raisins add flavour and texture to salads, baked goods, and cooked dishes.

Sour Cream

Sour cream is a dairy product produced from sweet pasteurised cream. It has a thick consistency and a flavour that is slightly tangy. It is fairly acidic, which activates the leavening action of baking soda. It is used as a topping and for making dips. It is also used as an ingredient in both sweet and savory dishes. Sour cream is also available in low-fat and nonfat varieties.

Vegetables

There are many different types of vegetables that can be used when cooking, such as potatoes, beans, peas, carrots, corn, onions, broccoli, cauliflower, and many more. They are used fresh, cooked, and dried. Check the recipe for the type of vegetables needed and to see how they need to be prepared. Vegetables are used on their own to create a vegetable dish or with other ingredients to create a one-dish meal.

White Chocolate

White chocolate is not actually a true chocolate because it does not contain any chocolate liqueur. It contains pure cocoa butter, sugar, milk solids, lecithin, and vanilla. It is can be eaten on its own or it can be used in cookies, cakes, bars, and desserts. It is also used as a coating on snacks, such as pretzels and nuts.

Liquid Ingredients

Buttermilk

Traditionally buttermilk was the liquid that was left after whole milk has been churned into butter, but today commercial buttermilk is produced by adding lactic-acid bacteria to low-fat or nonfat milk. Buttermilk has a thick consistency and adds a tangy flavour to baked goods, sauces, and dressings. It can be used in many recipes to replace yogurt and sour milk. Buttermilk is also available in powdered form.

Corn Syrup

A thick, sweet syrup extracted from corn. It is available in a light and dark variety. The light syrup is clear in colour and sweet but does not have much flavour. The dark syrup is brown in colour and has a sweet molasses or caramel-like flavour. Corn syrup works well in frostings, candies and jams because it hinders crystallisation. It produces a denser, chewier dough when used in baked goods. Corn syrup can also be used in place of honey or maple syrup but it is not as sweet so it may have a slight affect on the finished product.

Eggs

Eggs can be cooked and eaten as a food themselves but they also serve several purposes as an ingredient. They give cooked dishes and baked goods structure. They also add flavour, colour, liquid, protein, and fat. The egg yolks add to the tenderness and provide consistent flavour and texture to the product. Egg whites add volume to the recipe. Two egg whites can be substituted for one whole egg if concerned about cholesterol. Grade A large eggs are the best size to use for most recipes calling for eggs.

Evaporated Milk

Whole, low-fat, or nonfat milk that has had approximately 60% of the water removed. It has been heat-treated to sterilize it in the can. This allows the milk to be stored at room temperature. Evaporated milk is used undiluted in sauces, soups and custards to give them a rich creamy texture. It is also used in baked goods and other dishes. Do not interchange with sweetened condensed milk, which is very sweet and sticky. Evaporated milk can also be reconstituted by adding water and it can then be added to recipes calling for milk.

Extracts

Extracts are produced by extracting the oils from fruits, nuts, and plants and then dissolving the in an alcohol base. Extracts are used to add flavouring to baked goods, candies, desserts, sweet fillings, and frostings. There are imitation flavourings also available that are used in the same manner as extracts but to provide the best flavour to your foods use pure extracts. Some of the common extracts available are vanilla, almond, maple, anise, lemon, orange, and peppermint.

Food Colouring

A colouring solution used to add colour to foods. Food colouring is generally added to frostings or batters for decorating purposes but can also be added to other foods to enhance their colour. Food colouring adds colour but does not affect flavour. It is generally available in the primary colours and they can be mixed to create additional colours. Start with a small amount and add more to darken to the desired colour. Food colouring is also available in a paste form.

Half milk and Half Light Cream

A commercially produced product that consists of half milk and half light cream. It is used on cereal, as a coffee creamer and as an ingredient in many recipes.

Heavy Cream

A dairy product that has a butterfat content of at least 36%. It is also known as whipping cream and double cream. Heavy cream provides a rich, creamy texture to the foods it is added to. It is used in sweet and savory sauces, baked goods, desserts and dessert toppings. It is also used as an ingredient in cocktail drinks.

Honey

Honey is the nectar from flowers and other plants that is collected by bees. Honey is used as a sweetener and has a distinctive flavour that is influenced by the type of flowers and plants that the nectar is taken from. A product containing honey is likely to be fairly moist because it contains fructose, which absorbs moisture from the air around it. When using honey in a recipe in place of sugar, use 3/4 cup plus 1 tablespoon of honey per cup of sugar being replace.

The liquid in the recipe should also be decreased by 2 tablespoons to help offset the additional moisture that will be created by the honey. If too much honey is used in a recipe, excessive browning may occur. Honey is also used as a spread similar to jam or jelly. It is available in three basic forms—liquid honey (liquid containing none of the comb), chunk-style honey (liquid with pieces of the comb), and comb honey (a square or round piece of the honeycomb containing the honey inside).

Juices

Juices are liquids extracted from food items that are used to make beverages and sauces. The fruit juices are also used as liquid ingredients in recipes for some baked goods. Fruit juices are acidic, so when combined with baking soda in a recipe, they cause the chemical reaction that results in the baking soda acting as a leavening agent.

Light Cream

A dairy product containing 18 to 30% milkfat, also referred to as table cream. It is used in making soups, sauces, desserts and baked goods. When used in place of heavy cream, the foods texture will not be as creamy. Light cream will separate or curdle easier than heavy cream so when using in soups and sauces, use a lower heat, heat slowly, and do not let it boil.

Low-fat Milk

Milk that has had most of the fat removed. It is available in varieties containing 1% or 2% milkfat. It can be used in recipes as a substitution for whole milk or fat-free milk.

Milk

When called for in a recipe it is referring to cow's milk. It is available as whole milk, low-fat milk, and skim (fat-free) milk. Also available powdered and condensed. Most often they are interchangeable in a recipe. The milk adds nutrition and browning capabilities. When heating milk, heat slowly and watch carefully to prevent it from scorching.

Molasses

A thick, strong flavoured syrup, produced as a byproduct of sugar refining. Molasses is used as a sweetener but is not as sweet as sugar. It is available in three varieties: light, which is from the first boiling and the sweetest of the three; dark, which is from the second boiling and is a little less sweet but has a deeper flavour; and blackstrap, which is from the third boiling and is very dark in colour with a bitter flavour. Blackstrap is rarely used in baking. Light and dark molasses are interchangeable in recipes, resulting in the finished product having a slightly deeper flavour when the dark molasses is used.

Skim Milk

A fat-free variety of milk that contains no more than .5% of milkfat. Skim milk can be interchanged with low-fat or whole milk in a recipe without a noticeable difference in the end product. Depending on the amount of milk called for in the recipe, you may notice slightly less flavour and richness when skim milk is substituted whole milk.

Soy Milk

A milk substitute made from pressed, cooked soybeans. Plain soy milk can be used in place of cow's milk in puddings, sauces and soups. Using soy milk in place of cow's milk will affect the taste of the end product. You may want to start by only replacing half of the cow's milk with soy milk. In addition to plain soy milk, there are flavoured varieties, and low-fat, fat-free, and fortified versions. Soy milk is high in protein, low in saturated fats, and cholesterol free. People who are lactose intolerant can use soy milk as a replacement for milk in some of their recipes. Soy milk has a tendency to curdle so you have to be careful when adding to acidic foods.

Sweetened Condensed Milk

Whole milk that has had 60% of the water removed and then sugar added. The amount of sugar added makes up 40% of the sweetened condensed milk's volume. It is very sweet and sticky, and adds a richness to the foods it is added to. It is used to make candies, bars, pies, and puddings. Sweetened condensed milk turns a golden colour and takes on a caramel like flavour when heated.

It also thickens as it is heated. Sweetened condensed milk cannot be used as a substitution for other milk product.

Vinegar

An acidic solution resulting from allowing wine or an alcoholic liquid to ferment a second time. Vinegar adds a sour, pungent taste to the foods it is added to. It has a tendency to lose its pungency when heated so if you desire a stronger flavour from the vinegar when adding it to a heated dish, you should add it after the dish is removed from the heat. The two most common types of vinegar are apple cider vinegar, which has a tart fruity flavour and distilled white vinegar, which has a more pungent flavour that is too harsh for regular cooking. Apple cider vinegar is most often used in recipes for foods such as salads, dressings, sauces, marinades and condiments. The distilled white is used most often for pickling and cleaning purposes. Other varieties include balsamic vinegar, malt vinegar, wine vinegar, and rice vinegar.

Vinegar-Balsamic

A vinegar made from sweet, white Trebbiano grapes. Traditional balsamic vinegar is aged for at least 12 years and is very expensive. Most commercially manufactured balsamic vinegar are aged for a shorter period of time. They are aged in stainless steel tanks for 6 months to a year and in wooden barrels for 2 to 12 years. It adds a sweet-sour taste to the foods it is added to and it has a dark colour. It can be used for salad dressings, sauces, marinades and gravies.

Vinegar-Malt

Malt vinegar is made from the fermentation of malt liquor. It has a natural caramel colour but is sometimes distilled to make it clear. It has a strong distinctive taste and is popular to use for pickling and sprinkled on fish and chips.

Vinegar-Rice

Vinegar made from fermented rice wine. It has a clean, mild flavour that is slightly sweet, which goes well with herbs, spices and fruits. Rice vinegar is light in colour, which helps maintain the colour of the ingredients it is added to. It can be sprinkled on salads, fruits, vegetables and stir fry dishes to give them a spark of flavour. It is popular in Japanese and Chinese dishes.

Vinegar-Wine

Vinegar made from fermented red or white wine. Red wine vinegar is dark red in colour and has a sharp flavour. White wine vinegar is clear or has a very light gold tint to it. It is acidic in flavour. They are used to add a spark of flavour to fruits, salad dressings, marinades, sauces, and glazes.

Fat Ingredients

Butter

The semi-solid product that is produce from churning cream. It is at least 80% milkfat and is available in salted or unsalted varieties. The unsalted variety is often used when cooking because it allows the cook to have more control over the seasoning. Butter creates flaky pastries, and it provides flavour, tenderness, crispness, and a golden-brown colour to many foods. It is used when frying and sautéing food, as a spread for baked goods such as breads and rolls, and as a flavouring for cooked vegetables. When eaten, it does not leave a greasy feeling in your mouth because it melts at body temperature. It is an ingredient in hundreds of recipes including breads, pastries, cakes, entrees, and sauces.

Lard

Lard is pork fat that has been rendered. In rendering, the fat is heated, causing the connective tissues to brown and turn crispy. They are then strained from the fat to purify the lard. Lard provides a slightly nutty flavour and because of its richness, it creates very tender, flaky pastries and pie crusts. It can be used in place of butter in baking recipes, but the amount should be reduced by approximately 25 percent.

Margarine

Margarine is made from partially hydrogenated vegetable oil, water, milk solids, salt and generally colouring. It is similar to butter in appearance and texture. Its flavour is also similar to butter but is not as rich. In addition to regular margarine (containing 80% fat), it is available in several other varieties, such as reduced fat (containing 60% fat), low-fat (containing 40% or less fat), and fat-free (containing less than half a gram of fat). It can be used in most recipes in place of butter but there will be a slight loss of flavour. For best results when substituting margarine, use regular margarine and not one of the reduced fat varieties. Margarine will create a softer dough but butter should be used when making pastries to provide the most flakiness.

Oil

Oils are extracted from seeds and fruits such as corn, cottonseed, soybean, sesame, sunflower, safflower, olive, and avocado. Some oils are extracted from nuts such as peanut, walnut, almond, and hazelnut. Oils are used to fry and sauté foods, for making salad dressings, and as an ingredient in baking. Some of the common oils used for these purposes are vegetable oil, corn oil, canola oil, olive oil, sunflower oil, soybean oil, and peanut oil. Oil is used in recipes for breads, cakes, cookies and muffins. Oil can be used in place of butter or margarine in most recipes by using 7/8 cup of oil for each cup of butter or margarine that is called for. Substituting the oil will result in the finished product having a heavier texture. Because oil has a higher smoke point than butter or margarine, it works better for sautéing, frying, deep-frying and stir-frying.

Alternate Ingredients

There are a variety of alternate ingredients that can be used instead of oil or butter to reduce the fat content when cooking or baking. Substitute products exist with reduced fat and no fat and in different forms such as sprays, spreads and liquids. Fruit purees or applesauce can be used as oil substitutes for baking purposes. When using reduced fat or fat- free substitutes, be cautious as to when and how they are used. Some recipes require a certain amount of fat to be successful.

3

Baking of Bread

Bread is the most important article of food. Many processes have been employed in making and baking; and as a result, from the first flat cake has come the perfect loaf. Bread is made from flour of wheat, or other cereals, by addition of water, salt, and a ferment. Wheat flour is best adapted for bread making, as it contains gluten in the right proportion to make the spongy loaf. But for its slight deficiency in fat, wheat bread is a perfect food; hence arose the custom of spreading it with butter. It should be remembered, in speaking of wheat bread as perfect food, that it must be made of flour rich in gluten.

Next to wheat flour ranks rye in importance for bread making; but it is best used in combination with wheat, for alone it makes heavy, sticky, moist bread. Corn also needs to be used in combination with wheat for bread making, for if used alone the bread will be crumbly. The miller, in order to produce flour which will make the white loaf, in the process of grinding wheat has been forced to remove the inner bran coats, so rich in mineral matter, and much of the gluten intimately connected with them.

A grain of wheat consists of:

— an outer covering or husk, which is always removed before milling;

— bran coats, which contain mineral matter;

— gluten, the proteid matter and fat; and

— starch, the centre and largest part of the grain.

Wheat is distinguished as white and soft, or red and hard. The former is known as winter wheat, having been sown in the fall, and living through the winter; the latter is known as spring wheat, having been sown in the spring. Wheat is milled for converting into flour by processes producing essentially the same results, all requiring cleansing, grinding, and bolting. Entire wheat flour has only the outer husk removed, the remainder of the kernel being finely ground. Graham flour, confounded with entire wheat, is too often found to be an inferior flour, mixed with coarse bran. Grinding is accomplished by one of four systems:

— low milling;

— Hungarian system, or high milling;

— roller milling; and

— by a machine known as distintegrator.

In low milling process, grooved stones are employed for grinding. The stones are enclosed in a metal case, and provision is made within case for passage of air to prevent wheat from becoming overheated. The lower stone being permanently fixed, the upper stone being so balanced above it that grooves may exactly correspond, when upper stone rotates, sharp edges of grooves meet each other, and operate like a pair of scissors. By this process flour is made ready for bolting by one grinding.

In high milling process, grooved stones are employed, but are kept so far apart that at first the wheat is only bruised, and a series of grindings and siftings is necessary. This process is applicable only to the hardest wheats, and is partially supplanted by roller-milling. In roller-milling, wheat is subjected to action of a pair of steel or chilled-iron horizontal rollers, having toothed surfaces. They revolve in opposite directions, at different rates of speed, and have a cutting action. Porcelain rollers, with rough surfaces, are sometimes employed. In this system, grinding is accomplished by cutting rather than crushing.

The disintegrator consists of a pair of circular metal disks, set face to face, studded with circles of projecting bars so arranged that circles of bars on one disk alternate with those of the other. The disks are mounted on the same centre, and so closely set to one another that projecting bars of one disk come quite close to plane surface of the other. They are inclosed within an external casing. The disks are caused to rotate in opposite directions with great rapidity, and the grain is almost instantaneously reduced to a powder.

After grinding comes bolting by which process the different grades of flour are obtained. The ground wheat is placed in octagonal cylinders (covered with silk or linen bolting-cloth of different degrees of fineness), which are allowed to rotate, thus forcing the wheat through. The flour from first siftings contains the largest percentage of gluten. Flour is branded under different names to suit manufacturer or dealer. In consequence, the same wheat, milled by the same process, makes flour which is sold under different names.

Essential Ingredients

Flour

Wheat flour is essential because it is milled from the only cereal grain known to contain the proteins glutenin and gliadin which when combined with water form gluten, the elastic material which holds the gas produced by the chemical reaction of the yeast enzymes on sugar. As in building a house, the frame must be built. Gluten forms the framework of bread and also the cell structure of the interior of the loaf. Rye flour contains both glutenin and gliadin but is incapable of forming gluten because there are substances in rye flour that interferes with it's ability to form gluten. That is the reason why wheat flour must be included as a large percentage of the flour in rye bread formulas.

Separation of gluten from wheat flour

First you must wash out all the starches from a mixture of flour and water as follows:

— Mix a small amount of flour (about 8 ounces) with just enough water to form a stiff ball of dough.

— Soak the ball of dough in water for about 30 minutes.

— Over a fine mesh sieve, and under running water from a faucet, wash out all the starch. When all the starch has been removed the water will run clear. Warning, if you try to wash gluten from soft wheat flour (cake flour), you will be able to get only a tiny bit of gluten from it. Hard wheat flour (bread flour) will yield a large amount of gluten. Rye flour will yield zero amount of gluten.

If you bake the ball of wet gluten at about 400 degrees F. until all the mositure has evaporated and cut the dry gluten in half.

Separation of glutenin and gliadin from gluten.

Once you have washed out the starches, you end up with raw wet gluten. If you soak the ball of gluten in pure ethyl alcohol, the glutenin and the gliadin will separate out. The gliadin is the sticky part and will form long tiny silky looking strands when touched with the finger. The glutenin on the other hand will look and feel like tough raw rubber.

Types of Wheat Used to Producing Bread Flour

Hard spring wheat and hard winter wheat are the best types of wheat for producing quality breads. Spring wheats are grown in the Northwest where there is less rainfall than in other wheat growing areas. This results in a higher percentage of protein and a lower percentage of starch than wheats grown in the Southwest where more rain falls. Hard spring wheat generally produce loaves with greater volume than winter wheats, but with slightly more open crumb texture. Millers combine the two types of wheat in their blend to improve interior loaf characteristics.

Hard White Wheat grown in some Western States, is also used in bread production. It is slightly lower in protein than spring and winter wheats. Bakers add Vital Wheat Gluten and/or Dough Strengtheners to doughs to make up for the deficiency. Duram Wheats which contain a higher percentage of proteins than does most other types of wheats are used primarily in making spaghetti and macaroni products. Milling of wheat into flour. Basically, after the wheat has been cleaned and tempered, and the percentage of moisture within the grain has been adjusted, the wheat passes through several reduction rollers.

The grounded middlings are separated into several streams of flour by sieving and bolting. The bran particles which have been removed are used in cereals. Some are also used for animal feed. 100 pounds of cleaned wheat generally yield only about 70 to 76 pounds of middlings which can be ground into various grades of flour.

Flour grades

Whole wheat flour is flour produced from 100 percent of the wheat. By adding Vital Wheat Gluten to their formulas bakers are producing high quality White Whole Wheat Bread.

Straight grade of flour is all of the flour produced after the outer portion (bran and germ) have been removed. 100 pounds of wheat will produce about 72 pounds of straight flour. Straight flour is simular to all purpose flour found in grocer stores.

Patent flour is a highly refined flour which remains after all the clear flours have been removed. Patent flours are produced from the intermost part of the wheat where the best quality proteins are located. 100 pounds of wheat will produce only about 40 to 60 pounds of patent flour, and the middlings that remain go into clear flours.

Clear flour is the portion of the flour remaining after the patent flour streams have been separated. Clear flour generally contain a higher percentage of protein than the other grades, but the quality of the protein is lower. This type pf flour is generally used to produce French bread. It is also mixed with wheat flour and rye flour by the baker to produce loaves with greater volume.

In order to produce quality breads from fresh milled flour it must be allowed to age or mature in storage for about a month. Bleaching and maturing agents are used to artifically age and whiten the flour. This results in flour that can be used immediately after milling to produce a quality product.

Strength of flour is its ability to be made into well piled loaves. The flour should have a high content of quality protein to retain gas and contain enough natural sugars and diastatic enzymes to produce enough sugar from the starch for uniform gas production.

Tolerance is the ability to produce a quality product for some time after the optimum fermentation time has elapsed. and in the event that the dough was overmixed. Enrichment of flour is the process of replacing the vitamins and minerals removed during the milling process.. Most of the vitamins and minerals are located in the outer portions of the wheat, and since most of the outer portions of the wheat are not used to produce the best grades of flour, these essential elements must be replaced. The exception is whole wheat flour.

The two most important enzymes in flour are protease and diastase. Protease conditions the gluten, improving its elasticity and its ability to retain the gas produced during fermentation. Diastase changes some of the starch in flour to dextrins and maltose sugar. Some flours do not contain a sufficient amount of diastase enzyme due to poor climatic conditions under which the wheat was grown. Sproted wheat can be added to the flour by the miller or the baker can add a specially prepared diastatic malt in his formula.

Rye flour is a finely ground flour prepared from the rye grain. It contains about the same amount of protein as wheat flour. However, the flour contains gummy substances which prevent formation of gluten, accounting for the small compact loaves of bread made made from a dough containing only rye flour. To produce rye bread with acceptable volume, not more than 20 percent dark rye flour, 30 percent medium rye flour, and 40 percent light rye flour can be used in the formula. White bread flour must be used in the formula so enough gluten can be formed to hold the gas during fermentation, and to form the structure of the texture.

Water

Water is a basic ingredient in bread baking. It would be impossible to produce a loaf of bread without water in some form. There are several types of water. Hard water produces better quality bread than any type of water. Soft water weekens the gluten during mixing and fermentation. This can be corrected to some degree by increasing the percentage of salt in the formula slightly and by using mineral yeast food in the formula. Alkaline water is the most harmful, because it doesn't only weaken the gluten, but retards fermentation. Yeast likes a slightly acid medium to perform at its best. The weakening of the gluten and retarding effect on yeast can be corrected by using an acid ingredient such as vinegar (acedic acid) or lactic acid. Special types of mineral yeast food has been developed to correct this problem.

Yeast

Yeast is one of the essential ingredients in bread production. It is a one cell plant which multiplies by a process known as budding. Under the right conditions of water, sugars, warmth, and dissolved minerals, yeast causes fermentation. Yeast is available in compressed form which must be kept under refrigeration until it is ready to be used and active dry yeast which need not be refrigerated. Active dry yeast has an extended shelf life. Both will generally produce satisfactory results. However, a majority of large Commercial Bread Bakeries use compressed yeast in their formulas. When using active dry yeast half as much active dry yeast is required as compared with compressed yeast.

Enzymes

Enzymes in yeast consist of invertase, maltase, zymase and protease. There are others, but they are not important in bread production. The protease enzyme in yeast is only active if the cell wall of the yeast is damaged in some way. That is the only time it can penetrate the cell wall of the yeast. There are always a few damaged yeast cells especially in dry yeast where some cells are damaged during the drying process, or in yeast that has gotten a little old. If the protease enzyme does penetrate the cell wall of the yeast, it will weaken the gluten.

Invertase enzyme is an entercelluar enzyme. Sucrose, (cane or beet sugar) when dissolved enter the cell wall and are changed to dextrose and fructose, which are later changed by the enzyme zymase. Maltase attacks malt sugar. The malt sugar is changed to two molecules of the simple sugars dextrose. Zymase is the enzyme which changes the simple sugars, dextrose and fructose into carbon dioxide gas and alcohol, and several esters which result in producing the unique aroma flavour and taste of breads. Also, the gas produced causes the dough to rise.

Salt

Salt is another essential ingredient in quality bread production. It has several functions. It regulates fermentation. Too little salt causes the dough to ferment too fast and wild, while too much salt slows down the fermentation process. Salt is a very powerful ingredient and it must be scaled very carefully when weighing the ingredients for the dough. In the proper amount, salt produces good grain and texture, because it strengthens the gluten allowing it to fully mature during fermentation. It also prevents the growth of wild yeast and bacteria. A whiter crumb is produced and last but not least, it brings out the taste and flavour of bread.

Enriching Ingredients

Sugar

Cane or beet sugar (sucrose) are basically the same. They are both compound sugars and must be changed to simple sugars by the yeast enzyme invertase before the yeast can convert them to carbon dioxide gas and alcohol by the yeast enzyme zymase. Dextrose also known as corn sugar is produced by converting corn starch into sugar. It is interchangeable with cane and beet sugar in bread production. However, its sweetening value is only about 72 as compared to a sweetness value of 100 for cane or beet sugar.

Fructose or levulose, found in fruits, molases, and honey is a left handed sugar which means it reflects polarized light to the left whereas dextrose reflects polarized light to the right. High fructose has in recent years been produced from corn starch by a special process which was unknown just a few years ago. It has a sweetening value considerably higher than any other type of sugar (about 172). Lactose sugar is sugar found in milk, and is not fermentable by bakers yeast. At the end of the baking process it remains as lactose sugar in the baked bread. Its sweetening value is very low compared to the other sugars.

Sugar is not considered an essential ingredient in bread baking. This is because flour contains a small amount of natural sugars and some starch is converted to sugar during the fermentation process. It does have many important functions. Some of the added sugar is converted to carbon dioxide and alcohol by the yeast enzymes. It provides necessary sweetness, helps to produce a golden brown colour of the crust, improves the texture of the crumb, helps to retain moisture in the crumb and adds to the nutritional value of the bread. High percentage of sugar as used in sweet doughs and Danish pastry doughs retard yeast activity. That is the reason why when increasing the percentage of sugar in doughs, the percentage of yeast must be increased by the same proportion.

Milk

Several years ago non fat dry milk was the type of milk generally used in bread baking, but today it has become too expensive to use by most commercial bread bakeries.

In it's place they use milk subsitutes such as milk whey in combination with other ingredients such as soyflour. It has a stabalising effect on fermentation, preventing wild fermentation. It improves crust colour because of the lactose sugar it contains. The lactose sugar is not fermentable by baker's yeast. It also improves texture, crumb colour, flavour and taste, and keeping quality of the baked loaf.

If non fat dry milk is used in bread it must be heated to a high enough temperature during the drying process to destroy bacteria which weakens the gluten in the dough. Milk dried by the vacuum drying process must be properly heat treated prior to being dried otherwise considerable difficulty can be expected during mixing and fermentation of the dough.

Shortenings

Shortenings are available from many sources. Any type of shortening can be used in bread production with satisfactory results. Shortening acts as a lubricant in the dough, making the dough more pliable, prevents stickeness, and reduces the amount of dusting flour necessary during the make-up process.. When shortening is used, the dough expands more easily and smoothly. In the baked product it makes the crust more tender, improves the keeping quality and produces a crumb that is soft and chewey. Because of the cutting effect on the bran in whole wheat flour, it is almost impossible to produce a loaf of whole wheat bread with acceptable volume without using shortening in the formula.

Optional Ingredients

To produce a variety of quality breads, optional ingredients are required. They result in uniformity of products and efficiency of operations, assist in increasing the tolerances of doughs due to production variables, and help to satisfy the demand for variety in the flavour and taste of breads. Mold inhibitors prevent bread from becoming molded.

Mineral yeast food was developed to stabilise water by adding mineral salts, which are essential in dough fermentation and conditioning. Mineral salts condition soft water, and acid salts neutralise the alkalinity of alkaline water which is harmful to yeast and gluten. Doughs made with mineral yeast foods are not sticky and are very soft. They go through machines with less trouble and will mold up free of air pockets. Following are some of the improved loaf characteristics resulting from the use of mineral yeast foods:—

> Greater loaf volume and ovenspring, improved texture and crumb colour, better crust colour, and greater uniformity.

Wheat Flour contain between 2 and 3 percent Pentosans which are the residual cell wall material that is left as a result of milling wheat into flour. When an enzyme named Xylanases is added to the dough, it breaks down the pentosans resulting in strengthening doughs and improving the baked product. Xylanase enzymes also strengthen multi-grain doughs and doughs for rolls and buns. This a very powerful enzyme therefore only 100 to 200 parts per million can be used to achieve the desired results.

Malt cereal syrup is available with low diastatic activity, high diastatic activity, and non diastatic activity depending upon the amount of diastase enzymes they contain. Low diastatic malt is produced for use with flour which has a fair amount of diastatic enzymes. High diastatic malt is used with flours which have a relatively small amount of diastatic enzymes. Non diastatic malt would be used with flours having a relatively high amount of diastatic enzymes. The amount of enzymes contained in malt is controlled by the manufacturer by heating the malt to various temperatures during production.

Malt syrups are concentrated products made by evaporating the water extract of malted barley and other cereal grains. In addition to containing diastatic enzymes, diastatic malt also contain enzymes which assist in adjusting proteolytic action of the dough to compensate for flours of different proteolytic content. High diastatic malts will also be high in proteolytic activity and vice versa. In addition to enzyme functions, malted cereal syrup contain a high percentage of fermentable maltose sugar as well as soluble proteins and natural salts which serve as valuable yeast nutrients.

Rye flavouring is a concentrated flavoring ingredient in powdered form. It gives rye bread an acetic taste as well as a spiced flavour. A preliminary sour dough is not required to enhance the flavour and aroma of rye bread.

Poppy seeds are used as a topping for French breads and rolls. They produce a nutty flavour in the baked product. Just before the loaves or rolls are loaded in the oven, they are washed with a cooked cornstarch wash which is thin enough to brush on or spray on. Egg wash can also be used. The seeds are then sprinkled on. The wash keeps the seeds from falling off the loaf or rolls.

Sesame seeds are small oval-shaped seeds of the sesamum family. The seeds are pearly white in colour and produce a rich, toasted nut flavour. They are, like poppy seeds, sprinkled on washed French breads and rolls.

Staling of Bakery Items

There are two ways in which bakery products stale. They are Starch Retrogradation (Firming of the Crumb) and by getting infected by molds and rope. In simple terms, Staling of crumb (firming of crumb)-is the process the starch molecules go through when they shrink upon cooling. Starch molecule consists of a very long chain of Carbon, Hydrogen and Oxygen that are stretched out when warm and feel soft. Upon cooling, the chain shrinks and thus become firm which is called staling. You have probably experienced that when a stale product is warmed, it becomes soft. The starch chain has stretched again. Upon cooling, it shrinks again and become firm.

Anti-Staling Ingredients

Emulsifiers. For the past several years bakers used emulsifiers called bread softeners to produce bread that will remain soft for a longer period of time. It is added to the dough during mixing. Some of the more common ones are monoglycrides, calcium steroyl lactylate, and sodium steroyl lactylate. The softening action takes place after the bread is baked. Also, Potato bread will reist staling because potatoes act as anti-staling ingredients to some degree. Some anti-staling ingredients also perform as dough conditioners or dough strengtheners.

Enzymes. Enzyme manufacturers are hard at work on generic engineering and protein engineering producing enzymes to extend the shelf life of bread many fold.

Uses of Stale Bread

All pieces of bread should be saved and utilised. Large pieces are best for toast. Soft stale bread, from which crust is removed, when crumbed, is called stale bread crumbs, or raspings, and is used for puddings, griddle-cakes, omelets, scalloped dishes, and dipping food to be fried. Remains of bread, from which crusts have not been removed, are dried in oven, rolled, and sifted. These are called dry bread crumbs, and are useful for crumbing croquettes, cutlets, fish, meat, etc.

Mold and Mold Inhibitors

Sanitation plays a very important role in preventing mold in bread. Mold spores do not survive baking temperatures. The interior of the loaf, when it comes out of the oven is about 210 to 212 degrees F. which will destroy any mold spores which may be present in the dough. Therefore, bread and other bakery products can only be contaminated after they leave the oven. Some of the more dangerous areas for mold contmination are storage rooms, and slicing machine blades which come in direct contact with the interior of the loaf where there is an abundant supply of food and moisture. Mold spores also thrive in dark places.

You can extend the length of time that it takes bakery products to mold by several days by using Mold Inhibitors such as Calcium Propionate for yeast raised doughs and sodium propionate in chemically leavened products. Propionates are present in many foods, but in very small amounts. Swiss Cheese, however is an exception. For this reason, Swiss Cheese rarely molds, unless it is improperly developed. Propionates may be obtained by the oxidisation of propyl alcohol, forming propionic acid. The propionic acid is in turn combined with other chemicals to form the well known Sodium and Calcium propionates sold under different Trades Names.

Mold Inhibitors react as an alkaline in doughs, and since yeast doesn't like an alkaline condition, Mineral Yeast Foods containing monocalcium phosphate are added to the dough. Monocalcium Phosphate reacts as an acid in doughs therefore counteracting the alkaline which is formed by the propionates. Also, vinegar can be used at the rate of about 1 pint per 100 pounds of flour. Inhibitors are called inhibitors, because not enough is used to kill the mold. They only retard the growth of molds. Bread will mold eventually if kept in a warm moist invironment. The amounts of Calcium Propionates to use in bread varies with the climate, season of the year, or type of product.

Dark Breads require more than White Breads. For average climates, 2.5 to 3.5 ounces are used per 100 pounds of flour in White Breads and 4.0 to 5.0 ounces are used in Dark Breads.

Types and Colour of Molds. There are many different types of molds and they have different colours. Mold spores are practically everywhere, because they are very tiny and are carried in the air. They are so tiny that they can only be seen under a microscope. Mold Spores are like seeds that you plant in the garden. When they come in contact with the proper food, moisture and warmth, the spores produce mold plants which you can see with the naked eye.

Rope

Rope is a bread disease caused by the bacteria, Bacillus mesentericus. This disease breaks down the cells of the bread and leaves a sticky, pasty mass. When the crumb is pressed together, and pulled apart, it will stretch into long, sticky, web-like strands. The product will have the odor of overripe cantalope. The rope bacteria are too small to see with the naked eye, but they can be seen with a microscope. The bacteria can be present in the ingredients, especially flour and yeast.

Unlike mold, rope spores are not destroyed by baking temperatures. Calcium propionate, sodium diacetate or one pint of vinegar per 100 pounds of flour can be used in bread doughs to increase the shelf life of the product. If the bakery is contaminated, thorough cleaning with special chemicals will be necessary and/ or the bakery may have to be steam cleaned.

Bread Making

Fermented bread is made by mixing tc a dough, flour, with a definite quantity of water, milk, or water and milk, salt, and a ferment. Sugar is usually added to hasten fermentation. Dough is then kneaded that the ingredients may be thoroughly incorporated, covered, and allowed to rise in a temperature of 68° F., until dough has doubled its bulk. This change has been caused by action of the ferment, which attacks some of the starch in flour, and changes it to sugar, and sugar in turn to alcohol and carbon dioxide, thus lightening the whole mass.

Dough is then kneaded a second time to break bubbles and distribute evenly the carbon dioxide. It is shaped in loaves, put in greased bread pans (they being half filled), covered, allowed to rise in temperature same as for first rising, to double its bulk. If risen too long, it will be full of large holes; if not risen long enough, it will be heavy and soggy. If pans containing loaves are put in too hot a place while rising, a heavy streak will be found near bottom of loaf.

To shape bread dough in loaves, divide dough in parts, each part large enough for a loaf, knead until smooth, and if possible avoid seams in under part of loaf. If baked in brick pan, place two loaved in one pan, brushed between with a little melted butter. If baked in long shallow pan, when well kneaded, roll with both hands to lengthen, care being taken that it is smooth and of uniform thickness. Where long loaves are baked on sheets, shape and roll loosely in a towel sprinkled with corn meal for last rising. To shape bread dough in biscuits, pull or cut off as many small pieces (having them of uniform size) as there are to be biscuits.

Flour palms of hands slightly; take up each piece and shape separately, lifting, with thumb and first two fingers of right hand, and placing in palm of left hand, constantly moving dough round and round, while folding towards the centre; when smooth, turn it over and roll between palms of hands. Place in greased pans near together, brushed between with a little melted butter, which will cause biscuits to separate easily after baking. For finger rolls, shape biscuits and roll with one hand on part of board where there is no flour, until of desired length, care being taken to make smooth, of uniform size, and round at ends.

Biscuits may be shaped in a great variety of ways, but they should always be small. Large biscuits, though equally good, never tempt one by their daintiness. Bread is often brushed over with milk or butter before baking, to make a darker crust. Where bread is allowed to rise over night, a small piece of yeast cake must be used; one-fourth yeast cake to one pint liquid is sufficient, one-third yeast cake to one quart liquid. Bread mixed and baked during the day requires a large quantity of yeast; one yeast cake, or sometimes even more, to one pint of liquid. Bread dough mixed with a large quantity of yeast should be watched during rising, and cut down as soon as mixture doubles its bulk. If proper care is taken, the bread will be found most satisfactory, having neither "yeasty" nor sour taste. Fermented bread was formerly raised by means of leaven.

Bread Baking

Bread is baked;

— to kill ferment,

— to make soluble the starch,

— to drive off alcohol and carbon dioxide, and

— to form brown crust of pleasant flavour.

Bread should be baked in a hot oven. If the oven be too hot the crust will brown quickly before the heat has reached the centre, and prevent further rising; loaf should continue rising for first fifteen minutes of baking, when it should begin to brown, and continue browning for the next twenty minutes. The last fifteen minutes it should finish baking, when the heat may be reduced. When bread is done, it will not cling to sides of pan, and may be easily removed.

Biscuits require more heat than loaf bread, should continue rising the first five minutes, and begin to brown in eight minutes. Experience is the best guide for testing temperature of oven. Various oven themometers have been made, but none have proved practical. Bread may be brushed over with melted butter, three minutes before removal from oven, if a more tender crust is desired.

Remove loaves at once from pans, and place side down on a wire bread or cake cooler. If a crisp crust is desired, allow bread to cool without covering; if soft crust, cover with a towel during cooling. When cool, put in tin box or stone jar, and cover closely. Never keep bread wrapped in cloth, as the cloth will absorb moisture and transmit an unpleasant taste to bread. Bread tins or jars should be washed and scalded twice a week in winter, and every other day in summer; otherwise bread is apt to mould.

As there are so many ways of using small and stale pieces of bread, care should be taken that none is wasted. Unfermented bread is raised without a ferment, the carbon dioxide being produced by the use of soda (alkaline salt) and an acid. Soda, employed in combination with cream of tartar, for raising mixtures, in proportion of one-third soda to two-thirds cream of tartar, was formerly used to a great extent, but has been generally superseded by baking powder. Soda bicarbonate ($NaHCO_3$) is manufactured from sodium chloride (NaCl), common salt or cryolite.

Baking powder is composed of soda and cream of tartar in definite, correct proportions, mixed with small quantity of dry material (flour or cornstarch) to keep action from taking place. If found to contain alum or ammonia, it is impure. In using baking powder, allow two teaspoons baking powder to each cup of flour, when eggs are not used; to egg mixtures allow one and one-half teaspoons baking powder.

When a recipe calls for soda and cream of tartar, in substituting baking powder use double amount of cream of tartar given. Soda and cream of tartar, or baking powder mixtures, are made light by liberation of gas in mixture; the gas in soda is set free by the acid in cream of tartar; in order to accomplish this, moisture and heat are both required. As soon as moisture is added to baking powder mixtures, the gas will begin to escape; hence the necessity of baking as soon as possible.

If baking powder only is used for raising, put mixture to be cooked in a hot oven. Cream of tartar ($HKC_4O_6H_4$) is obtained from argols found adhering to bottom and sides of wine casks, which are ninety per cent cream of tartar. The argols are ground and dissolved in boiling water, colouring matter removed by filtering through animal charcoal, and by a process of recrystallisation the cream of tartar of commerce is obtained.

The acid found in molasses, sour milk, and lemon juice will liberate gas in soda, but the action is much quicker than when cream of tartar is used. Fermented and unfermented breads are raised to be made light and porous, that they may be easily acted upon by the digestive ferments. Some mixtures are made light by beating sufficiently to enclose a large amount of air, and when baked in a hot oven air is forced to expand. Aerated bread is made light by carbon dioxide forced into dough under pressure. The carbon dioxide is generated from sulphuric acid and lime. Aerated bread is of close texture, and has a flavour peculiar to itself. It is a product of the baker's skill, but has found little favour except in few localities.

Water Bread

2 cups boiling water	2 1/2 teaspoons salt
1 tablespoon butter dissolved in	1/4 yeast cake
1 tablespoon lard water	1/4 cup lukewarm
2 tablespoon sugar	6 cups sifted flour

Put butter, lard, sugar, and salt in bread raiser, or large bowl without a lip; pour on boiling water; when lukewarm, add dissolved yeast cake and five cups of flour; then stir until thoroughly mixed, using a knife or mixing-spoon. Add remaining flour, mix, and turn on a floured board, leaving a clean bowl; knead until mixture is smooth, elastic to touch, and bubbles may be seen under the surface. Some practice is required to knead quickly, but the motion once acquired will never be forgotten. Return to bowl, cover with a clean cloth kept for the purpose, and board or tin cover; let rise over night in temperature of 65° F.

In morning cut down : this is accomplished by cutting through and turning over dough several times with a case knife, and checks fermentation for a short time; dough may be again raised, and recut down if it is not convenient to shape into loaves or biscuits after first cutting. When properly cared for, bread need never sour. Toss on board slightly floured, knead, shape into loaves or biscuits, place in greased pans, having pans nearly half full. Cover, let rise again to double its bulk, and bake in hot oven. This recipe will make a double loaf of bread and pan of biscuit. Cottolene, crisco, or beef drippings may be used for shortening, one-third less being required. Bread shortened with butter has a good flavour, but is not as white as when lard is used.

Milk and Water Bread

- 1 cup scalded milk
- 1 cup boiling water
- 1 tablespoon lard
- 1 tablespoon butter
- 1 yeast cake dissolved in
- 1/4 cup lukewarm water
- 6 cups sifted flour, or one cup white flour and enough entire wheat flour to knead
- 21/2 teaspoon salt

Prepare and bake as Water Bread. When entire wheat flour is used add three tablespoons molasses. Bread may be mixed, raised, and baked in five hours, by using one yeast cake. Bread made in this way has proved most satisfactory. It is usually mixed in the morning, and the cook is able to watch the dough while rising and keep it at uniform temperature. It is often desirable to place bowl containing dough in pan of water, keeping water at uniform temperature of from 95° to 100° F. Cooks who have not proved themselves satisfactory bread makers are successful when employing this method.

Entire Wheat Bread

- 2 cups scalded milk
- 1/4 cup sugar or
- 1/3 cup molasses
- 2 teaspoons salt
- 1 yeast cake dissolved in
- 1/4 cup lukewarm water
- 42/3 cups coarse entire wheat flour

Add sweetening and salt to milk; cool, and when lukewarm add dissolved yeast cake and flour; beat well, cover, and let rise to double its bulk. Again beat, and turn into greased bread pans, having pans one-half full; let rise, and bake. Entire Wheat Bread should not quite double its bulk during last rising. This mixture may be baked in gem pans.

German Caraway Bread

Follow recipe for Milk and Water Bread , using rye flour in place of entire wheat flour, and one tablespoon sugar for sweetening. After first rising while kneading add one-third tablespoon caraway seed. Shape, let rise again, and bake in a loaf.

Entire Wheat and White Flour Bread

Use same ingredients as for Entire Wheat Bread, with exception of flour. For flour use three and one-fourth cups entire wheat and two and three-fourths cups white flour. The dough should be slightly kneaded, and if handled quickly will not stick to board. Loaves and biscuits should be shaped with hands instead of pouring into pans, as in Entire Wheat Bread.

Graham Bread

2 cups hot liquid (water, or milk and water)	1/4 yeast cake dissolved in
1/4 cup lukewarm water	
1/3 cup molasses	3 cups flour
21/2 teaspoons salt	3 cups Graham flour

Prepare and bake as Entire Wheat Bread. The bran remaining in sieve after sifting Graham flour should be discarded. If used for muffins, use two and one-half cups liquid.

Third Bread

2 cups lukewarm water	1 cup rye flour

1 yeast cake	1 cup granulated corn meal
1/2 tablespoon salt	
1/2 cup molasses	3 cups flour

Dissolve yeast cake in water, add remaining ingredients, and mix thoroughly. Let rise, shape, let rise again, and bake as Entire Wheat Bread.

Rolled Oats Bread

2 cups boiling water dissolved in	1/2 yeast cake
1/2 cup molasses	1/2 cup lukewarm water
1/2 tablespoon salt	1 cup rolled oats
1 tablespoon butter	5 cups flour

Add boiling water to oats and let stand one hour; add molasses, salt, butter, dissolved yeast cake, and flour; let rise, beat thoroughly, turn into buttered bread pans, let rise again, and bake. To make shaping of biscuits easy, take up mixture by spoonfuls, drop into plate of flour, and have palms of hands well covered with flour before attempting to shape, or drop from spoon into buttered muffin tins.

Rye Biscuit

2 cups hot liquid 1 cup boiling water	1/4 yeast cake 11/2 teaspoons salt
1 cup rye flakes	1 yeast cake dissolved in
2 tablespoons butter	1 cup lukewarm water
1/3 cup molasses	Flour

Make same as Rolled Oats Bread.

Rye Bread

- 1 cup scalded milk
- 1 cup boiling water
- 1 tablespoon lard
- 1 tablespoon butter
- 1/3 cup brown sugar
- 1 1/2 teaspoons salt
- 1/4 yeast cake dissolved in
- 1/4 cup lukewarm water
- 3 cups flour
- Rye meal

To milk and water add lard, butter, sugar, and salt; when lukewarm, add dissolved yeast cake and flour, beat thoroughly, cover, and let rise until light. Add rye meal until dough is stiff enough to knead; knead thoroughly, let rise, shape in loaves, let rise again, and bake.

Date Bread

Use recipe for Health Food Muffins . After the first rising, while kneading, add two-thirds cup each of English walnut meats cut in small pieces, and dates stoned and cut in pieces. Shape in a loaf, let rise in pan, and bake fifty minutes in a moderate oven. This bread is well adapted for sandwiches.

Boston Brown Bread

- 1 cup rye meal
- 1 cup granulated corn meal
- 1 cup Graham flour
- 2 cups sour milk, or
- 3/4 tablespoon soda
- 1 teaspoon salt
- 3/4 cup molasses
- 1 3/4 cups sweet milk or water

Mix and sift dry ingredients, add molasses and milk, stir until well mixed, turn into a well-buttered mould, and steam three and one-half hours. The cover should be buttered before being placed on mould, and then tied down with string; otherwise the bread in rising might force off cover. Mould should never be filled more than two-thirds full.

A melon-mould or one-pound baking-powder boxes make the most attractive-shaped loaves, but a five-pound lard pail answers the purpose. For steaming, place mould on a trivet in kettle containing boiling water, allowing water to come half-way up around mould, cover closely, and steam, adding, as needed, more boiling water.

New England Brown Bread

11/2 cups stale bread	11/2 Rye meal
31/4 cups cold water	11/2 Granulated corn meal
3/4 cup molasses	11/2 Graham flour
11/2 teaspoons salt	3 teaspoons soda

Soak bread in two cups of the water over night. In the morning rub through colander, add molasses, dry ingredients mixed and sifted, and remaining water. Stir until well mixed, fill buttered one-pound baking-powder boxes two-thirds full, cover, and steam two hours.

Indian Bread

11/2 cups Graham flour	1 teaspoon salt
1 cup Indian meal	1/2 cup molasses
1/2 tablespoon soda	12/3 cups milk

Mix and steam same as Boston Brown Bread.

Steamed Graham Bread

3 cups Arlington meal	1 teaspoon salt

1 cup flour	1 cup molasses (scant)
31/2 teaspoons soda	21/2 cups sour milk

Mix same as Boston Brown Bread and steam four hours. This bread may often be eaten when bread containing corn meal could not be digested.

Parker House Rolls

2 cups scalded milk	2 teaspoons salt
3 tablespoons butter	1 yeast cake dissolved in
2 tablespoons sugar	1/4 cup lukewarm water

Add butter, sugar, and salt to milk; when lukewarm, add dissolved yeast cake and three cups of flour. Beat thoroughly, cover, and let rise until light; cut down, and add enough flour to knead (it will take about two and one-half cups). Let rise again, toss on slightly floured board, knead, pat, and roll out to one-third inch thickness. Shape with biscuit-cutter, first dipped in flour. Dip the handle of a case knife in flour, and with it make a crease through the middle of each piece; brush over one-half of each piece with melted butter, fold, and press edges together.

Place in greased pan, one inch apart, cover, let rise, and bake in hot oven twelve to fifteen minutes. As rolls rise they will part slightly, and if hastened in rising are apt to lose their shape. Parker House Rolls may be shaped by cutting or tearing off small pieces of dough, and shaping round like a biscuit; place in rows on floured board, cover, and let rise fifteen minutes. With handle of large wooden spoon, or toy rolling-pin, roll through centre of each biscuit, brush edge of lower halves with melted butter, fold, press lightly, place in buttered pan one inch apart, cover, let rise, and bake.

Dinner Rolls

Use same ingredients as for Parker House Rolls, allowing one-fourth cup butter. Shape in small biscuits, place in rows on a floured board, cover with cloth and pan, and let rise until light and well puffed. Flour handle of wooden spoon and make a deep crease in middle of each biscuit, take up, and press edges together. Place closely in buttered pan brushing with butter between biscuits, cover, let rise, and bake twelve to fifteen minutes in hot oven. From this same mixture crescents, braids, twists, bow-knots, clover leaves, and other fancy shapes may be made.

Sticks

1 cup scalded milk	1 yeast cake dissolved in
1/4 cup butter	1/4 cup lukewarm water
11/2 tablespoons sugar	White 1 egg
1/2 teaspoon salt	33/4 cups flour

Add butter, sugar, and salt to milk; when lukewarm, add dissolved yeast cake, white of egg well beaten, and flour. Knead, let rise, shape, let rise again, and start baking in a hot oven, reducing heat, that sticks may be crisp and dry. To shape sticks, first shape as small biscuits, roll on board (where there is no flour) with hands until eight inches in length, keeping of uniform size and rounded ends, which may be done by bringing fingers close to, but not over, ends of sticks.

Salad Sticks

Follow recipe for Sticks. Let rise, and add salt to dough, allowing two teaspoons to each cup of dough. Shape in small sticks, let rise again, sprinkle with salt, and bake in a slow oven. If preferred glazed, brush over with egg yolk slightly beaten and diluted with one-half tablespoon cold water.

Swedish Rolls

Use recipe for Salad Rolls. Roll to one-fourth inch thickness, spread with butter, and sprinkle with two tablespoons sugar mixed with one-third teaspoon cinnamon, one-third cup stoned raisins finely chopped, and two tablespoons chopped citron; roll up like jelly roll, and cut in three-fourths inch pieces. Place pieces in pan close together, flat side down. Again let rise, and bake in a hot oven. When rolls are taken from oven, brush over with white of egg slightly beaten, diluted with one-half tablespoon water; return to oven to dry egg, and thus glaze top.

Sweet French Rolls

1 cup milk
1 yeast cake dissolved in
1/4 cup lukewarm water
Flour
1/4 cup sugar
1 teaspoon salt
1 egg
Yolk one egg
1/8 teaspoon mace
1/4 cup melted butter

Scald milk; when lukewarm, add dissolved yeast cake and one and one-half cups flour; beat well, cover, and let rise until light. Add sugar, salt, eggs well beaten, mace, and butter, and enough more flour to knead; knead, let rise again, shape, and bake same as Salad Rolls, or roll in a long strip to one-fourth inch in thickness, spread with butter, roll up like jelly roll, and cut in one-inch pieces. Place pieces in pan close together, flat side down. A few gratings from the rind of a lemon or one-half teaspoon lemon extract may be substituted in place of mace.

Luncheon Rolls

1/2 cup scalded milk	2 tablespoons melted butter
2 tablespoons sugar	1 egg
1/2 teaspoon salt	Few gratings from rind of lemon
1/2 yeast cake dissolved in	
2 tablespoons lukewarm water	Flour

Add sugar and salt to milk; when lukewarm, add dissolved yeast cake and three-fourths cup flour. Cover and let rise; then add butter, egg well beaten, grated rind of lemon, and one and one-fourth cups flour. Let rise again, roll to one-half inch thickness, shape with small biscuitcutter, place in buttered pan close together, let rise again, and bake. These rolls may be ready to serve in three hours if one and one-half yeast cakes are used.

French Rusks

2 cups scalded milk	Flour
1/4 cup butter	1 egg
1/4 cup sugar	Yolks 2 eggs
1 teaspoon salt	Whites 2 eggs
1 yeast cake dissolved in	3/4 teaspoon vanilla
1/4 cup lukewarm water	

Add butter, sugar, and salt to scalded milk; when lukewarm add dissolved yeast cake and three cups flour. Cover and let rise; add egg and egg yolks well beaten, and enough flour to knead. Let rise again, and shape as Parker House Rolls. Before baking, make three parallel creases on top of each roll. When nearly done, brush over with whites of eggs beaten slightly, diluted with one tablespoon cold water and vanilla. Sprinkle with sugar.

Rusks (Zweiback)

- 1/2 cup scalded milk
- 1/2 teaspoon salt
- 2 yeast cakes
- 1/4 cup sugar
- 1/4 cup melted butter
- 3 eggs

Add yeast cakes to milk when lukewarm; then add salt and one cup flour, and let rise until very light. Add sugar, butter, eggs unbeaten, and flour enough to handle. Shape as finger rolls, and place close together on a buttered sheet in parallel rows, two inches apart; let rise again and bake twenty minutes. When cold, cut diagonally in one-half inch slices, and brown evenly in oven.

German Coffee Bread

- 1 cup scalded milk
- 1/3 cup butter, or butter and lard
- 1/4 cup sugar
- 1/2 teaspoon salt
- 1 egg
- 1/3 yeast cake dissolved in
- 1/4 cup lukewarm milk
- 1/2 cup raisins stoned and cut in pieces

Add butter, sugar, and salt to milk; when lukewarm, add dissolved yeast cake, egg well beaten, flour to make stiff batter, and raisins; cover, and let rise over night; in morning spread in buttered dripping-pan one-half inch thick. Cover and let rise again. Before baking, brush over with beaten egg, and cover with following mixture : Melt three tablespoons butter, add one-third cup sugar and one teaspoon cinnamon. When sugar is partially melted, add three tablespoons flour.

Coffee Cakes

- 1 cup scalded milk
- 4 yolks of eggs
- 1/2 cup sugar
- 2 yeast cakes

3 eggs	1/2 teaspoon extract lemon or
2/3 cup butter	2 pounded cardamon seeds
42/3 cups flour	

Cool milk; when lukewarm, add yeast cakes, and when they are dissolved add remaining ingredients, and beat thoroughly with hand ten minutes; let rise six hours. Keep in ice-box over night; in morning turn on floured board, roll in long rectangular piece one-fourth inch thick; spread with softened butter, fold from sides toward centre to make three layers. Cut off pieces three-fourths inch wide; cover and let rise. Take each piece separately in hands and twist from ends in opposite directions, coil and bring ends together at top of cake. Let rise in pans and bake twenty minutes in a moderate oven; cool and brush over with confectioners' sugar, moistened with boiling water to spread, and flavoured with vanilla.

Coffee Rolls

2 cups milk	1 egg
11/2 yeast cakes	1/2 teaspoon cinnamon
Butter	1 teaspoon salt
Lard 1/2 cup each	Melted butter
Sugar	Confectioners' sugar
Flour	Vanilla

Scald milk, when lukewarm add yeast cakes, and as soon as dissolved add three and one-half cups flour. Beat thoroughly, cover, and let rise; then add butter, lard, sugar, egg unbeaten, cinnamon, salt, and flour enough to knead. Knead until well mixed, cover, and let rise. Turn mixture on a floured cloth. Roll into a long, rectangular piece one-fourth inch thick. Brush over with melted butter, fold from ends toward centre to make three layers and cut off pieces three-fourths inch wide. Cover and let rise.

Take each piece separately in hands and twist from ends in opposite directions, then shape in a coil. Place in buttered pans, cover, again let rise, and bake in a moderate oven twenty minutes. Cool slightly, and brush over with confectioners' sugar moistened with boiling water and flavoured with vanilla.

Swedish Bread

21/2 cups scalded milk	2/3 cup sugar
1 yeast cake	1 egg, well beaten
Flour	1/4 teaspoon salt
1/2 cup melted butter	1 teaspoon almond extract

Add yeast cake to one-half cup milk which has been allowed to cool until lukewarm; as soon as dissolved add one-half cup flour, beat thoroughly, cover, and let rise. When light, add remaining milk and four and one-half cups flour. Stir until thoroughly mixed, cover, and again let rise; then add remaining ingredients and one and one-half cups flour. Toss on a floured cloth and knead, using one-half cup flour, cover, and again let rise. Shape as Swedish Tea Braid or Tea Ring I or II, and bake. Swedish Tea Braid. Cut off three pieces of mixture of equal size and roll, using the hands, in pieces of uniform size; then braid.

Put on a buttered sheet, cover, let rise, brush over with yolk of one egg, slightly beaten, and diluted with one-half tablespoon cold water, and sprinkle with finely chopped blanched almonds. Bake in a moderate oven. Swedish Tea Ring I. Shape as tea braid, form in shape of ring, and proceed as with tea braid, having almonds blanched and cut in slices crosswise. Swedish Tea Ring II. Take one-third Swedish Bread mixture and shape, using the hands, in a long roll. Put on an unfloured board and roll, using a rolling-pin, as thinly as possible. Mixture will adhere to board but may be easily lifted with a knife. Spread with melted butter, sprinkle with sugar and chopped blanched almonds or cinnamon. Roll like a jelly roll, cut a piece from each end and join ends to form ring. Place on a buttered sheet, and cut with scissors and shape. Let rise, and proceed as with Tea Ring I.

Dutch Apple Cake

1 cup scalded milk	23/4 cups flour
1/3 cup butter	Melted butter
1/3 cup sugar	5 sour apples
1/3 teaspoon salt	1/4 cup
1 yeast cake	1/2 teaspoon cinnamon
2 eggs	2 tablespoons currants

Mix first four ingredients. When lukewarm add yeast cake, eggs unbeaten, and flour to make a soft dough. Cover, let rise, beat thoroughly, and again let rise. Spread in a buttered dripping-pan as thinly as possible and brush over with melted butter. Pare, cut in eighths, and remove cores from apples. Press sharp edges of apples into the dough in parallel rows lengthwise of pan. Sprinkle with sugar mixed with cinnamon and sprinkle with currants. Cover, let rise, and bake in a moderate oven thirty minutes. Cut in squares and serve hot or cold with whipped cream sweetened and flavoured.

Raised Muffins

1 cup scalded milk	3/4 teaspoon salt
1 cup boiling water	1/4 yeast cake

2 tablespoons butter	1 egg
1/4 cup sugar	4 cups flour

Add butter, sugar, and salt to milk and water; when lukewarm, add yeast cake, and when dissolved, egg well beaten, and flour; beat thoroughly, cover, and let rise over night. In morning, fill buttered muffin rings two-thirds full; let rise until rings are full, and bake thirty minutes in hot oven.

Grilled Muffins

Put buttered muffin rings on a hot greased griddle. Fill one-half full with raised muffin mixture, and cook slowly until well risen and browned underneath; turn muffins and rings and brown the other side. This is a convenient way of cooking muffins when oven is not in condition for baking.

Raised Hominy Muffins

1 cup warm cooked hominy	1 teaspoon salt
1/4 cup butter	1/4 yeast cake
1 cup scalded milk	1/4 cup lukewarm water
3 tablespoons sugar	31/4 cups flour

Mix first five ingredients; when lukewarm add yeast cake, dissolved in lukewarm water and flour. Cover, and let rise over night. In the morning cut down, fill buttered gem pans two-thirds full, let rise, one hour, and bake in a moderate oven. Unless cooked hominy is rather stiff more flour will be needed.

Raised Rice Muffins

Make same as Raised Hominy Muffins, substituting one cup hot boiled rice in place of hominy, and adding the whites of two eggs beaten until stiff.

Raised Oatmeal Muffins

3/4 cup scalded milk
1/4 cup sugar
1/2 teaspoon salt
21/2 cups flour
1/4 yeast cake dissolved in
1/4 cup lukewarm milk
1 cup cold cooked oatmeal

Add sugar and salt to scalded milk; when lukewarm, add dissolved yeast cake. Work oatmeal into flour with tips of fingers, and add to first mixture; beat thoroughly, cover, and let rise over night. In morning, fill buttered iron gem pans two-thirds full, let rise on back of range that pan may gradually heat and mixture rise to fill pan. Bake in moderate oven twenty-five to thirty minutes.

Health Food Muffins

1 cup warm wheat mush
1/4 cup brown sugar
1/2 teaspoon salt
1 tablespoon butter
1/4 yeast cake
1/4 cup lukewarm water
21/4 cups flour

Mix first four ingredients, add yeast cake dissolved in lukewarm water, and flour; then knead. Cover, and let rise over night. In the morning cut down, fill buttered gem pans two-thirds full, again let rise and bake in a moderate oven. This mixture, when baked in a loaf, makes a delicious bread.

Squash Biscuits

1/2 cup squash (steamed and sifted)
1/4 yeast cake dissolved in

- 1/4 cup sugar
- 1/2 teaspoon salt
- 1/2 cup scalded milk
- 1/4 cup lukewarm water
- 1/4 cup butter
- 21/2 cups flour

Add squash, sugar, salt, and butter to milk; when lukewarm, add dissolved yeast cake and flour; cover, and let rise over night. In morning shape into biscuits, let rise, and bake.

Imperial Muffins

- 1 cup scalded milk
- 1/4 cup sugar
- 1/2 teaspoon salt
- 1/3 yeast cake dissolved in
- 1/4 cup lukewarm water
- 13/4 cups flour
- 1 cup corn meal
- 1/4 cup butter

Add sugar and salt to milk; when lukewarm add dissolved yeast cake, and one and one-fourth cups flour. Cover, and let rise until light, then add corn meal, remaining flour, and butter. Let rise over night; in the morning fill buttered muffin rings two-thirds full; let rise until rings are full and bake thirty minutes in hot oven.

Dry Toast

Cut stale bread in one-fourth inch slices. Crust may or may not be removed. Put slices on wire toaster, lock toaster and place over clear fire to dry, holding some distance from coals; turn and dry other side. Hold nearer to coals and colour a golden brown on each side. Toast, if piled compactly and allowed to stand, will soon become moist. Toast may be buttered at table or before sending to table.

Water Toast

Dip slices of dry toast quickly in boiling salted water, allowing one-half teaspoon salt to one cup boiling water. Spread slices with butter, and serve at once.

Milk Toast I

1 pint scalded milk	1/2 teaspoon salt
2 tablespoons butter	4 tablespoons cold water
21/2 tablespoons bread flour	6 slices dry toast

Add cold water gradually to flour to make a smooth, thin paste. Add to milk, stirring constantly until thickened, cover, and cook twenty minutes; then add salt and butter in small pieces. Dip slices of toast separately in sauce; when soft, remove to serving dish. Pour remaining sauce over all.

Milk Toast II

Use ingredients given in Milk Toast I, omitting cold water, and make as Thin White Sauce. Dip toast in sauce.

Brown Bread Milk Toast

Make same as Milk Toast, using slices of toasted brown bread in place of white bread. Brown bread is better toasted by first drying slices in oven.

Tomato Cream Toast

11/2 cups stewed	3 tablespoons butter
1/2 cup scalded cream	3 tablespoons flour
1/4 teaspoon soda	1/2 teaspoon salt
6 slices toast	

Put butter in saucepan; when melted and bubbling, add flour, mixed with salt, and stir in gradually tomato, to which soda has been added, then add cream. Dip slices of toast in sauce. Serve as soon as made.

German Toast

3 eggs
2 tablespoons sugar
1/2 teaspoon salt
1 cup milk
6 slices stale bread

Beat eggs slightly, add salt, sugar, and milk; strain into a shallow dish. Soak bread in mixture until soft. Cook on a hot, well-greased griddle; brown on one side, turn and brown other side. Serve for breakfast or luncheon, or with a sauce for dessert.

Brewis

Break stale bits or slices of brown and white bread in small pieces, allowing one and one-half cups brown bread to one-half cup white bread. Butter a hot frying pan, put in bread, and cover with equal parts milk and water. Cook until soft; add butter and salt to taste.

Bread for Garnishing

Dry toast is often used for garnishing, cut in various shapes. Always shape before toasting. Cubes of bread, toast points, and small oblong pieces are most common. Cubes of stale bread, from which centres are removed, are fried in deep fat and called croûstades; half-inch cubes, browned in butter, or fried in deep fat, are called croûtons.

4

Preparation of Cakes

A cake is a form of baked food, normally combine some kind of wheat byproduct, a sweetening agent (commonly sugar), a binding agent (generally egg, though gluten or starch are often used by vegetarians and vegans), fats (usually butter or margarine, although a fruit puree can be substituted to avoid using fat), a liquid (milk, water or fruit juice), flavours and some form of leavening agent (such as yeast or baking powder). Cake is often the dessert of choice for meals at ceremonial occasions, particularly weddings or birthday parties. Cake is considered delicious in many cultures.

There are two basic types of cakes, *batter type* which contains a high percentage of shortening and the *foam type* which contains little, if any shortening at all. From these two basic mixes, many varieties of cakes can be made. Quality cakes depend on many factors such as ingredients used, mixing method, batter temperature, baking, etc.

Ingredients

Each ingredient used in cake baking is employed for the specific characteristics it has and/or the result it has on the finished product. If these effects are understood, the ingredients may be selected with the assurance that the products produced will be good.

Flour

Flour is the primary structure builder in most cakes. The gluten formed during mixing coagulates during baking and assists in supporting the heavy weight of sugar and shortening. Cake flour used in cake baking is made from soft wheats. Flours milled for bread baking will be made from hard wheat. Cake flour should have a low protein content (7 to 9 percent). Cake flour should be properly bleached, because bleaching helps to carry more sugar and shortening as well as water during mixing. Cake flour should also have a P.H. of around 5.2 which is slightly acid. This acidity helps to mellow or soften the gluten.

Sugar

Sugar is used in cake as a sweetener Sugar is considered a tenderiser because of its tenderising effect resulting from the softening action on flour proteins. Sugar lowers the carmelisation point of the batter, allowing the cake crust to color at a lower temperature. Sugar helps to retain moisture left in the baked cake after baking thereby keeping the cake moist and edible for several days. A portion (about 50 percent) but not all of the sugar may be replaced with sirup. When this is done, the liquid content of the sirup must be deducted from the liquid going into the mix.

For example, if 50 pounds of cane sugar is to be replaced with a sirup containing 80 percent sugar solids, it will be necessary to divide 50 pounds of needed sugar solids 50, by 80 percent (the percentage of sugar solids contained in the sirup) equals 62.5 pounds of sirup to use. Multiply 62.5 pounds of sirup times 20 percent (the percentage of water in the sirup) equals 12.5 pounds of water in 62.5 pounds of sirup. Subtract 12 pounds of water from the amount of water called for in the formula so the water content of the formula will remain constant. The greater the percentage of sugar contained in the formula, the longer the cake will stay fresh. The amount to use will be covered later under formula balance. For optimum results, sugar must be completely dissolved in the batter. This is the reason why in balancing a cake formula, especially in high ratio cakes (cakes containing greater quantities of sugar and water than flour in the formula) water must exceed the sugar in the formula.

Salt

Salt is used primarily for taste and to tone down the sweetness.

Shortening

Shortening incorporates air in the cake batter during mixing. This air helps to obtain volume in the baked cake. It also tenderises the cake. In making high ratio cakes, an emulsified shortening must be used to form an emulsion. Regular shortening is not capable of forming an emulsion when large amounts of sugar and liquid are used. If not enough liquid is used to dissolve the sugar, the cake will collapse in the center.

Eggs

Eggs furnish structure, moisture, flavor, color, and food value to the cake. Their most valuable contribution is structure. The proteins in the eggs coagulates during baking and assists the flour as a structure builder. Egg is the only ingredient that can be used to regulate the toughening action in a cake. If a weak flour is used, the eggs can be increased. If the percentage of shortening (a tenderiser) called for in the formula is increased, the eggs must be increased also. It is important to know the percentages of fat, moisture, and protein content of eggs when balancing cake formulas.

To get the same amount of solids when using egg whites in lew of whole eggs, the amount of egg whites would have to be increased and the moisture going into the mix must be decreased. To get the same amount of moisture when using egg yolks in lew of whole eggs, the moisture content must be increased. The amount of eggs to use in a cake formula will be discussed in formula balance.

Milk Solids

Milk solids have a binding effect on the protein of the flour, thereby increasing the toughness in a cake. A portion of the total solids in milk contain lactose sugar, which carmelises at a low temperature (270 to 275 degrees F.). It is used to control crust color. It, along with the proteins in milk, adds food value and flavor to the cake, and helps to retain moisture in the cake. If liquid milk is used in lew of dry milk powder, it will be necessary to know the liquid content of the milk in order to make adjustments to the formula so the formula can be properly balanced.

The various types of milk are composed of the following percentages of components: Fresh whole milk is composed of 8.5 percent solids; 3.5 percent butterfat; 88 percent water. Milk, Dry Whole is composed of 72 percent solids; 26 percent butterfat; 2 percent water. Milk, Solids Nonfat is composed of 97 percent solids; 1.5 percent butterfat; 1.5 percent water.

Leavening

Cakes are leavened mainly in three ways. Incorporation of air during mixing, chemically leavened and vapor pressure created in the oven. The manner of leavening depends upon the type of cake being made in regard richness of formula, consistency of batter and baking temperature. Cakes low in water and high in ennriching ingredients get a larger amount of leavening during mixing and require less chemical leaveners than cakes made from lean formulas high in liquids. In addition to leavening the cake, chemical leaveners control the eating qualities of the cake. Excess soda , for example results in an undesirable "soapy taste". Baking Soda is necessary to produce the rich red color in Devils Food Cake. However, one must be careful not to use too much soda in the formula. To produce a chocolate cake having a brown crumb color rather than a red crumb color, Baking Soda must be left out.

Some types of cakes require no chemical leavening. Examples are the true Pound Cake and the Basic Sponge Cake. These two cakes are leavened entirely by physical means such as air incorporated in the batter during mixing and vapor pressure created during baking. This is the reason why these types of cakes require a longer mixing period and precise control of temperature of the batter during mixing.

Liquids

Liquids in some form is required in every cake formula. The liquid may be in the form of water, liquid milk, eggs or any other ingredient which contain water. Water has several functions in cake production. It developes the gluten, dissolves the sugar, makes the function of baking powder possible, regulates the batter consistency, and controls the temperature of the batter. It is possible to carefully regulate the water portion of the formula by figuring the liquid content of any liquid ingredient used in the batter.

The amount of water going into the cake formula is partially controlled by the type of shortening used. An emulsified type shortening will carry considerably more water in the mix, thus allowing the use of more sugar such as in High Ratio Cakes The total liquids should always equal or exceed the weight of the sugar in the formula, because all of the sugar in the formula must be dissolved to produce a quality cake.

Balancing Cake Formulas

In order to create a cake batter that will produce high quality cakes, certain amounts of the different ingredients have to be put together in a definite sequence at controlled mixing speed, time and temperature. The general relationship of ingredients that have to be brought into balance, differ according to the type of cake to be produced. In other words, the formula balance for batter cakes differs considerably from that of the foam type cake.

Batter Type Cakes

The following general rules apply to Batter type cakes:

— *Rule 1.* The weight of the sugar should equal or exceed the weight of the flour. There is a top limit, of course in the amount of sugar which can go into a cake. For White and Yellow Layer Cakes, 145 percent sugar-flour ratio seems to be about the generally accepted practical top limit. Higher sugar-flour ratios are possible in cakes containing cocoa or chocolate. The more cocoa or chocolate used in the formula, the higher the sugar-flour ratio can be. The amount of liquid also become significant in determining the amount of sugar to use.

When Rule no. 1 is applied and a specific amount of sugar is selected, both the amount of sugar and the amount of flour become fixed. To set up the formula, it then becomes necessary to consider the amount of shortening, eggs and liquid which can be used. As the percentage of shortening is increased, the percentage of eggs must be increased by the same amount. This is due to the fact that shortening is a tenderiser and to keep the cake from being over tenderised, additional structure in the form of eggs is needed.

— *Rule 2.* The weight of eggs should equal or exceed the weight of the shortening. In applying this rule, the type of cake desired must be considered. For example, a true pound cake will have equal parts of shortening or butter, sugar.flour and eggs. A high ratio layer cake will have about 50 or 60 percent as much shortening as flour and the eggs should at least equal the amount of shortening in the mix. Eggs generally exceed the shortening by 5 or 10 percent. Since shortening carries air into the batter, a cake with a high percentage of shortening will be classified as a rich formula. The air carried by the shortening will result in less chemical leavening being needed.

— *Rule 3.* The combined weight of the eggs plus the liquid, should equal or exceed the weight of the sugar. In layer type cakes, the weight of the liquids usually exceed the sugar by 20 t0 30 percent. In devils food cake, the liquids usually exceed the sugar by 40 to 50 percent. In pound cakes, best results are obtained if the liquids and sugar are nearly equal because pound cake batter needs to be slightly thicker.

Sponge Cake

The basic sponge cake is composed of flour, eggs, sugar, salt and vanilla. Following is an example of a basis spong formula in baker's percent:

Ingredients	*Percent*
Flour, cake	100
Sugar	166
Eggs	166
Salt	3

By the addition of flour to the above Basis Formula, the following rules may be applied to produce a more economical sponge cake: For each part of flour added, add the following: 0.75 to 1 part Sugar; 0.75 to 1 part Liquid Milk; 0.015 to 0.030 parts Baking Powder; 0.03 parts salt, and 0.01 part Vanilla.

Layer Cakes

Following is an example of a layer cake formula in baker's percent:

Ingredients	Minimum	Maximum
Flour, Cake	100	100
sugar	100	145
shortening, emulsified	30	60
Eggs, Whole	30	70
Salt	2	4
Milk, Dry Nonfat	7	15
Baking Powder	3	6
Liquids-water+eggs	100	125

Pound Cake

Following is an example of a pound cake formula in baker's percent:

Ingredients	Minimum	Maximum
Flour, Cake	100	100
sugar	100	125
shortening, emulsified	40	100
Eggs, Whole	40	100
Salt	2	3
Baking Powder	0	2
Liquids-water+eggs	100	125

Cake Mixing Procedures

Batter-Type Cakes

There are several ways to mix Batter-Type Cakes. The preferred way of mixing is the 2-stage method because of its simplicity and small chance for error. Regardless of the method used, the bowl must be scraped frequently to keep the batter smooth at all times. The shortening should be plastic, not too hard or too soft. The mixer should be started and run in slow speed until all the ingredients are combined to prevent splashing. Use the correct size bowl for the amount of batter being mixed. Have just enough batter in the bowl to cover the mixing paddle. The finished batter should be between 72 and 78 degrees F.

Two-stage method

— Carefully weigh all the ingredients.

— Sift all dry ingredients together including the dry milk when used. Place the shortening, dry ingredients, and about 75 percent of the water into the mixing bowl. Mix for 3 minutes at medium speed. Scrape bowl down thoroughly.

— Combine eggs, remaining water and vanilla. Add slowly to ingredients in mixing bowl while mixing in slow speed. Scrape bowl down again. Mix for 3 minutes in medium speed.

Streamlined method

— Place all ingredients, except eggs, in the mixing bowl.

— Using the wire beater attachment, beat the batter at high speed for about 1 minute.

— Add the eggs and stir in slow speed for about 1 minute. This method requires a slight reduction in baking powder (about 20 percent less) due to the greater aeration of the batter during mixing.

Sugar water method

— Place all of the sugar and slightly half the weight of the water in the mixing bowl.

— Agitate the mixture for about 30 seconds, or intil the sugar is dessolved. (Remember to reduce the baking powder by approximately 20 percent).

— Add the dry ingredients and the shortening and mix until the batter is smooth.

— Add the eggs and the remaining water, and mix until the batter is smooth.

Flour-Shortening Method (Blending Method)

— In the blending method, place the flour and shortening in the mixing bowl. Blend them together until the flour particles are thoroughly coated by the fat.

— The remaining dry ingredients (sugar, salt, baking powder, and dry milk when used used) are then added and blended well.

— About 75 percent of the liquid is added and the mixture is mixed until it is homogeneous.

— The remaining liquid, including the eggs is added in small portions and the mixing is continued for about 5 minutes, making sure to scrape the bowl occasionally to assure a smooth batter.

Foam Type Cakes

Sponge and angel food cakes are mixed differently from batter-type cakes because is necessary to incorporate airinto the mix to obtain the desired volume. The eggs are beaten with a portion of the sugar and salt to form a foamy mass. When liquid is included in the formula, it is added gradually to the beaten egg, sugar, and salt mixture. The flour, starch when used, and the baking powder are sifted together and gently folded into the whipped mixture.

Sponge cake-regular

When whole eggs or egg yolks are used in the sponge mix, the volume will be improved if they are mixed with the sugar and salt and heated over a hot water bath to a temperature of 100 to 110 degrees F. The sugar will dissolve more readily and become more evenly distributed as when making jelly rolls. Do not overheat the eggs, because if the eggs are partially cooked, the volume of the finished product will be reduced. Beat the mixture until it is light and until it will hold a crease when a finger in drawn across the top of the batter. To aid the rolling process when making jelly rolls, a portion of the sugar can be substituted with sirup.

Angel food cake

Angel food cake is prepared from egg whites, granulated sugar, salt, vanilla and cream of tartar, and flour. The mixing method is given below:

— Make sure the mixing bowl and the beater are free of any grease.

— Using a wire beater, beat the egg whites, cream of tarter, salt and vanilla until foamy. The egg whites should be fairly cool (about70 degrees F.)

— Add approximately one-half the sugar in a slow stream, and beat to a wet peak (the foam will form a peak and will feel wet to the finger) . Note: When the peak feels dry to the finger, the mixture has been overbeaten.

— Sift the flour and the remaining sugar 5 times and gently fold this mixture into the beaten mixture.

Normally the folding step is done by hand, but can be done with the machine if done very carefully. As soon as all the lumps disappears, quit folding.

— After completing the folding step, deposit the mixture into grease free Angel Food pans and bake at about 400 Degrees F.

— Place baked cake upside down on a wire cooling rack until thoroughly cooled before removing from the pan. If this is not done the cake will shrink and be low in volume.

Chiffon cakes

Chiffon cakes are another variation of foam cakes. These formulas contain a high percentage of liquid, with egg yolks and egg whites forming the greatest portion of the liquid. The aeration comes from two sources, the air beaten into the egg whites, and the baking powder incorporated as part of the other ingredients. Salad oil makes up the shortening of the mix. Except for the higher egg content and salad oil used, the formulas are very similar to high sugar layer cake formulas. The mixing procedure consists of two stages as follows:

— All the ingredients except the egg whites and about one-half the sugar are mixed to form a batter. To make a lemon or orange shiffon cake, a small amount of lemon juice or orange juice may be added, or the extract can be used.

— The egg whites, the remaining One-half of the sugar the cream of tartar, and vanilla are whipped separately to form a medium peak as in Angel Food Cake. Remember that the bowl and wire attachment should be free of grease.

— The batter which was mixed previously, is gently folded into the beaten meringue. The principle of baking is the same as for Angel Food Cake, except the oven temperature should be slightly lower, about 350- to 375 degrees F. The cake is usally baked in an Angel Food Cake pan, but can be baked in layer cake pans or sheet pans.

Tools and Equipment for Cake Making

The following are the tools and equipment that the most useful to have. They do make the job of taking a simple layer cake, filling and frosting, and turning them into a fancy dessert, much easier and less frustrating.

— Mixers
— Mixing Bowls
— Double Boiler
— Pastry Brushes
— Wax & Parchment Paper
— Round Pans
— Jellyroll Pans
— Cooling Racks
— Candy Thermometer
— Cake Boards
— Long Serrated Knife
— Revolving Cake Stand
— Spatulas: Straight & Angled
— Drywall/Construction Spatulas
— Plastic Bottles
— Pastry Triangle Comb
— Pastry Pin
— Marble Board
— Sieves

— Zester
— Grater
— Cake Boxes

Mixers

Depending on whether you're making a cake, or a cream topping, the type of mixer you use can affect the recipe. We can use the portable handheld mixer when we need to mix something over the stove, or when making a small quantity of whipping cream, topping, icings, etc., which don't require all the power of a stand-up mixer.

Mixing Bowls

These bowls can be easily cleaned in the sink or dishwasher. The best part of having the extra large bowls, is that they can be used to sift dry ingredients in them

Uses:

— Double boiler
— Mixing ingredients: batter, dough, etc.,
— Sifting flour
— Thoroughly mixing dry ingredients

Double Boiler

You can't find these everywhere; but when you do, they're expensive. You can construct a makeshift one by setting a plain stainless steel bowl over a large pot. Some recipes call for the bowl not to touch the water and others do. There are two bowls for this technique: a large shallow one and a deep narrow one. These are great, because the heat and steam that builds-up from the hot (or simmering) water underneath the bowl, gently and indirectly heat to melt, or cook, a given ingredient.

Uses:

— For melting chocolate
— Heating whipped cream to make sauces, Ganache, fillings, etc
— Heating eggs for Genoise cake

Pastry Brushes

Natural bristle brushes are synthetics. Natural brushes are more flexible and manoeuvrable when greasing pans and getting into their corners. They also hold more liquid, which makes brushing syrup onto cakes less frustrating and time consuming,

Uses:

— For greasing pans
— For brushing syrup onto layer cakes

Wax and Parchment Paper

These papers help to prevent cakes and other baked goodies from sticking to their pans. The "Wax Paper" technique doesn't always guarantee that things won't stick to the pan. The "Parchment Paper" technique, in comparison, is much more reliable and increases the guarantee that things will release from the pan easier.

Both papers can be used to make moulds for chocolate decorations, as well as, for making paper cones for piping decorations. When you peel off the wax paper, small shreds of paper adhere in the hardened chocolate, which aren't appealing. Using parchment paper for chocolate mould decorations, because nothing sticks to parchment paper. Wax paper for cones and piping, because it's less expensive than parchment paper and the paper is used to pour, not to contain and mould chocolate; there are no worries about shreds of paper. If you're in doubt, as always, then experiment with both types and see which best suits your needs.

Pans: 8", 9" and 10" rounds

These are the most common pan sizes when making a cake to serve a small gathering of 8 - 10 people. For a larger crowd, consider making two or more round cakes, or a rectangular cake, using sheet cakes made in jellyroll pans.

Jellyroll Pans

There's a lot one can do with a simple layer of sheet cake. You can make a rectangular layer cake, or a jellyroll fashion cake, or whatever you can imagine. Jellyroll pans can also be used to bake cookies and other goodies. This is a "two-in-one" investment.

Cooling Racks

Cooling racks are a must for any type of cake making. They're inexpensive and they prevent a cooling cake from developing a soggy, dampened bottom (crust).

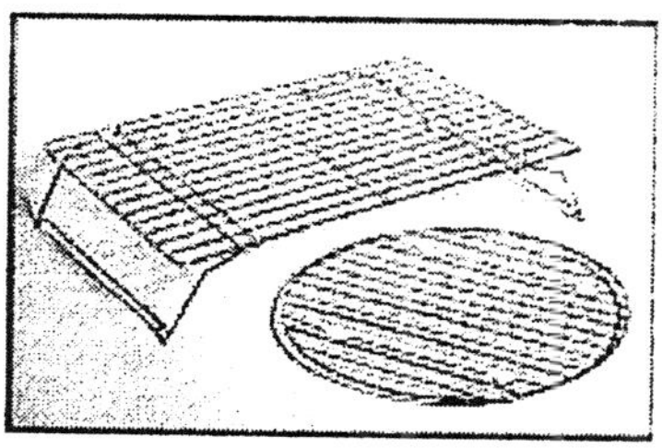

Cooling racks

You can use a cookie, cake, or any other type of rack, as long as the rack elevates and allows air to flow around and cool the cake down.

Candy Thermometer

Syrup is particular, in that, if it's above, or below, a required temperature, the recipe may not work out as intended. Instead of having to rely on the texture of the syrup; how it forms threads between your fingers, or balls in a glass of cold water, you can make better syrup, using a candy thermometer and without burning your fingers and ruining your dessert. Candy thermometers are inexpensive; but do check around for the best price.

Cake Boards

Cake boards give your cake a more professional look, as well as, make your life easier. Instead of worrying where you're going to serve the cake on, why not use a cake board instead? Select boards that are at least 1 inch larger that your cake pans. This way you get a little more space around the cake to add extra decorations.

Cake boards can be found in cake decorations and supplies stores, as well as, in cooking shops and most bakeries. They come in various sizes and shapes, and they usually sell for about a dollar or so. Next time a guest brings over a bakery made cake, save the cake board and box, if they're in good condition and re-use for home use.

Long Serrated Knife

Long serrated knives are used to split a cake into layers. Make sure that the length of the blade is at least 12 - 14 inches long (or longer than your widest cake pan), 1 - 1 1/2 inches wide and 1/8 of an inch thick.

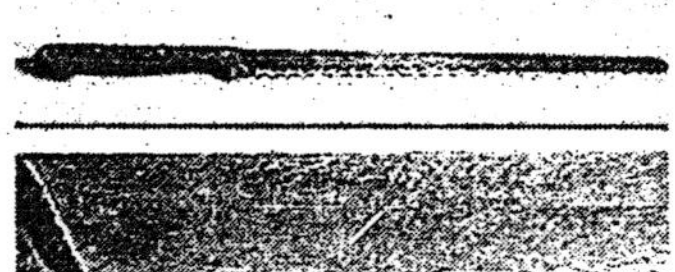

Long serrated knives

Uses:

— To split cakes into layers,

— To decorate the top of frosted cakes

— To help lift cakes and desserts off revolving cake stands; used a long spatula.

Revolving Cake Stand

This makes frosting and decorating a cake easier. These can be made of plastic or metal. Select one with a heavy based stand with a removable turntable (disk), usually metal based. The base itself should be heavy to support and prevent a dessert from toppling over. The turntable should measure about 3/16 of an inch thick and 12 inches in diameter.

Revolving Cake Stand

The pivot, or shaft holding the turntable, should fit snugly into the base opening without being "tipsy" or wobbly. When you give the turntable a good spin, it should smoothly revolve without wobbling or grabbing, and gradually come to a complete stop without jerking. The revolving stands is that you don't have to stop; to put down your pastry bag and then turn your cake around, to complete a continuous stream of piped frosting around the cake. When you try to grab and rotate the cake board with one hand, sometimes, you can easily poke the side of the decorated cake, which can be frustrating. This is one of those tools that you can do without, until you've worked with one.

Spatulas

Straight and angled types of spatulas can be used to apply and spread toppings and frostings, as well as, to slice and serve foods, depending on the nature of the dessert.

Spatulas: Straight and Angled

Angled spatulas come in very handy when applying and spreading frosting on a cake. A long and wide knife can be used instead, but; more pressure is applied to the tip of the knife, which causes tip dipping in the frosting. When an angled spatula is used, the pressure is distributed more evenly from the tip, all the way through to the handle, and thus applies even pressure to produce a more professional and smoothly frosted cake. Spatulas should not be confused with "rubber scrapers".

Drywall/Construction Spatulas

These tools are made for construction use: drywall, cement, etc., but they can be used to apply and spread frostings onto cakes, as well as, to make fancy chocolate decorations on marble slabs, which will be featured in the upcoming months.

Construction Spatulas

Plastic Bottles

Similar to plastic catsup and mustard plastic bottles, these can be used to apply syrup onto cake layers, as well as, squeesing out thinned jellies and sauces onto desserts.

Plastic bottles

— *For syrup*: Pierce cap/nozzle with a clean pin; don't snip off end; otherwise, the syrup will gush out, instead of pouring out in a thin stream. Substitution: a plain plastic water or pop bottle; it works the same.

— *For thinned jellies and sauces*: snip off a small bit of the tip/nozzle and test opening before snipping off more.

Pastry Triangle Comb

This is a small aluminum triangle, about 3 inches big, with a set of trowelled teeth on each side that vary in size and depth. This tool, when passed across a frosted cake, creates a decorative look.

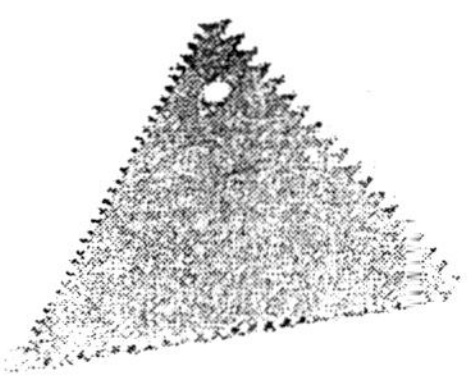

Pastry Triangle Comb

A short serrated knife can be used instead, but to a certain degree: only the side of the cake. Perfectly smoothing out frosting, on a cake, can be tedious especially when trying to eliminate "pass" marks. A pastry comb goes over these marks and turns them into decorative edges.

Pastry Bag and Accessories

These are used to pipe out and create: stars, swirls, tubes, rosettes and so on. They can also be used to pass other mixtures, such as: mashed potato mixture, choux pastry, Gnocchi, etc.

Pastry Pin

A large thumbtack, about 3 inches large, used as a mini-rotating stand that is held in between your fingers.

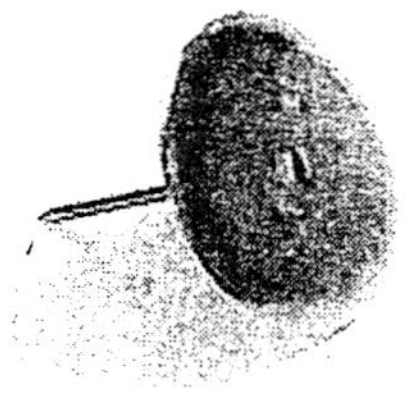

Pastry Pin

To create flower decorations and alike; you simultaneously rotate the pin, as you press out cream using a pastry bag fitted with a rosette tip. Substitution: a jar.

Marble Board

These come in various sizes and thickness. We can use a marble slab to make decorative chocolate ribbons, using drywall spatulas, which will be featured in the coming months. If you're looking to buy a marble slab, look for one that is large enough to give you plenty of space to roll out dough, yet small enough to fit into your refrigerator. Marble boards used for rolling pastry dough will be discussed, in depth, in the near future.

Sieves: Small, Medium and Large

These can be used to individually sift and eliminate clumps in dry ingredients, as well as, to thoroughly mix together dry ingredients: flour, salt and baking powder. By sifting ingredients together, as described in most cake recipes, the baking powder becomes evenly incorporated with the dry ingredients, with fewer possibilities of developing large air bubbles in the cake. If the baking powder is added directly to the batter, or is not sifted properly with the dry ingredients, then large bubbles may form in the cake during baking.

Zester

This utensil peels thin strips from the zest of citrus fruit: lemon, lime, orange, etc.

Zester

The strips can then be used for decorations.

Cake Boxes

Cake boxes are great for covering, refrigerating and transporting desserts, especially cakes. They're light and easy to assemble. A cake box, when tied with an inexpensive ribbon, keeps the cake in tact and presentable when taking it over to a friend or guest's home. Cake boxes can be found in cake decorations and supplies stores, as well as, in cooking shops and most bakeries. They come in various sizes and shapes, and they are inexpensive.

Types of Cake Pans

Almost any pan can be used for any cake recipe unless specified otherwise in the recipe, provided that baking times are adjusted.

Springform

A pan (without a non-stick coating) that has a detachable side and bottom. These are ideal for cheesecake, since they reflect heat and bake the cheesecake slowly. Most common types are made of the shiny tin; however, newer ones come in grey matt, as well as coloured exteriors.

They can also be used as a mould to create a layer or mousse type of cake.

Preparation:

Most recipes require to be greased with butter, unless otherwise stated. Most cheesecakes use cookie crumb crusts that include butter, which also acts as a releasing agent.

Angel food:

A detachable tube pan: a side part and cone with bottom part. Made especially for baking angel food cake.

For angel food cake: the pan is left ungreased. The cake needs to cling and climb onto the sides of the pan during baking. If the pan is greased, the cake won't bake high as expected; the grease will make the sides slippery and the batter won't rise. The trick to using this pan is to cool the cake in an inverted position. When cooled, run a knife between the cake and the sides of the pan to release the cake from the pan. If the cake is left to cool in the same position (upright) as it was baked, the cake will sink in.

Ring Shaped Pans

Bundt

These pans can be made from either heavy duty cast iron or plain aluminum. Both of which can be coated with a non-stick coating.

What makes these pans so unique is that the baked cake, when inverted for presentation, offers an appealing looking dessert.

Savarin

This is a uniquely shaped ring pan designed especially for Savarin. They are usually made of aluminum, with or without a non-stick coating. As with the Bundt type, when inverted for presentation, these offer an appealing looking dessert.

Ring or tube

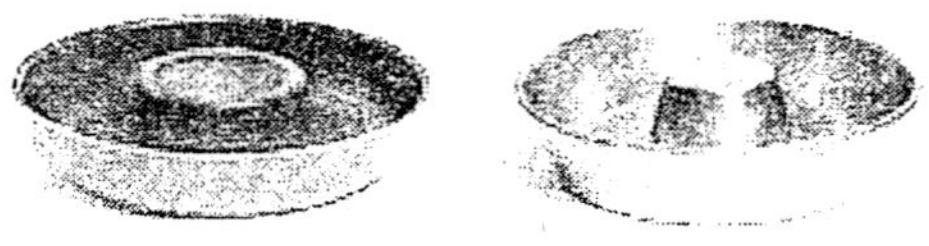

These are made of aluminum, with or without a non-stick coating and come with or without a centre opening.

Round pans:

Usually made of aluminum, with or without a non-stick coating. Ideal for layer cakes and genoise. They come in various sizes, though typical for genoise is the 9 1/2 - 10 1/2 . 8 1/2 - 9 1/2 pans are ideal for cakes made from scratch or from a commercial "box" mix.

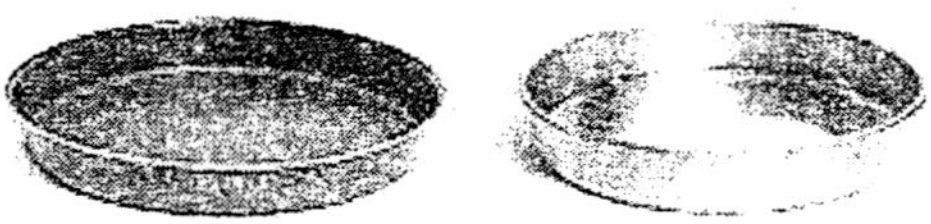

Without a non-stick coating: greased and then lined with wax or parchment paper.

Square pans:

Usually made of aluminum or tin, 9" x 9" x 2" with or without a non-stick coating. Ideal for brownies, slices, and bars. Not recommended for cakes since the centre of the cake forms a tall mound and the top looks humpy and uneven.

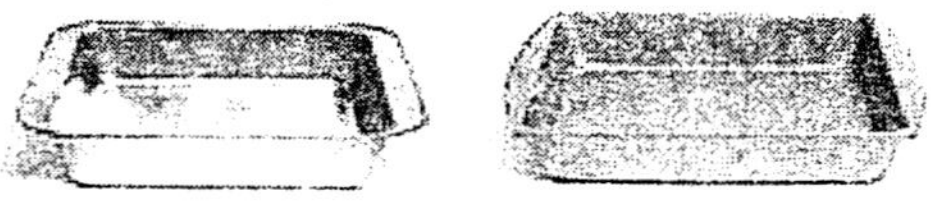

Pieces of cake cut from the sides are typically thinner than those nearest the centre, which are thicker and higher. Without a non-stick coating: greased and then lined with wax or parchment paper.

Rectangle pans:

Usually made of aluminum or tin. 9" x 13" x 2 1/2" with or without a non-stick coating. Not recommended for cakes since the centre of the cake forms a tall mound and the top looks humpy and uneven. Pieces of cake cut from the sides are typically thinner than those nearest the centre, which are thicker and higher.

These are more ideal for filo pastry desserts than cakes. Look for thick aluminum pans; thinner metal doesn't cook and brown bottom crust well.

Loaf pans:

Usually made of aluminum with a non-stick coating that are ideal for loaf breads; cakes and pound cakes.

Jellyroll pans

Made of aluminum, with or without a non-stick coating. Ideal for sheet cakes and jellyrolls, as well as for toasting bread crumbs and nuts.

They can also be used for baking cookies.

Decorative Pans:

Flan

Made of tin without a non-stick coating.

Shaped pans, sea shells, madeleines, etc.

Made of tin or thin cast iron without a non-stick coating. Straight edged pans, such as a heart, can be greased and lined with parchment paper. Intricate designed pans, such as a flan or seashells, can be sprayed with a non-stick spray, or greased and floured.

Muffin pans

Usually made of aluminum with a non-stick coating.

Better ones are made with thick metal, whereas, poorer ones are made thinner and can easily be bent out of shape.

Pie plates

Glass with or without a non-stick coating: ideal for cookie crumb crust pie, which doesn't need too much baking and browning. Non-stick coated pie plates are coated with a thin non-stick coating, which is transparent and slightly dingy to off white in colour.

Ceramic: great for custard filled pies, such as pumpkin pie.

The thickness of the ceramic absorbs and distributes heat well with excellent bottom crust browning.

Aluminum without a non-stick coating: ideal for fruit filled pies topped with a layer of pastry dough, such as: apple, cherry, etc., which require a long baking time, about 35 - 45 minutes.

These pans, if the metal is thick, can produce excellent browned bottom crusts. Thinner plates don't fare as well.

Aluminum with a non-stick coating: ideal for making pie shells to be filled with a cooked cream or custard, such as mousse or lemon meringue.

Since most pastry dough baked for these types of recipes require a short baking time, about 12 - 20 minutes, the dark non-stick coating helps to produce an excellent browned bottom crust.

Cake Decorating Ingredients

There's no trick in combining or using them. The key thing are follows:

- Walnuts
- Almonds
- Chocolate & White Chips
- Bars of Chocolate
- Cocoa Powder
- Cinnamon Powder
- Icing Sugar
- Sprinkles
- Cookies
- Maraschino or Candied Cherries

Whole Shelled Walnuts

They can be used whole, crushed or pulverised. It's best to have large pieces, since they can be used as a whole and if needed, pulverised into smaller pieces and used elsewhere.

Walnuts

Usually toast walnuts to "burn off" their oil; it's this oil that makes untoasted walnuts taste bitter.

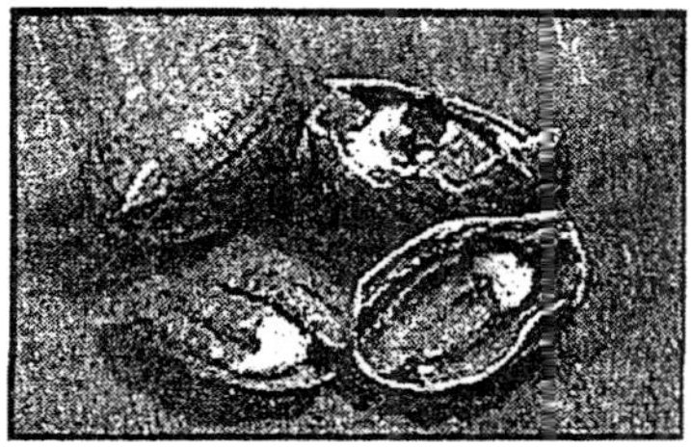

Pecans

Walnuts, in most cases, need to be toasted, unless otherwise stated.

Whole and Slivered Almonds, With or Without Skins

They can be used whole, crushed or pulverised. It's best to have whole almonds, since they can be easily pulverised into smaller pieces and used elsewhere.

Almonds

Slivered Almonds with skins

You could do a lot with these and use them in: cookies, cakes, fillings, and decorations.

— Toasted whole almonds can be used in place of cherries when studding stars or swirls of piped topping
— Toasted slivered almonds, with or without skins, can be used to cover the side of a cake or dessert
— Slivered almonds with skins have a contrasting look, which work great on cookies

Chocolate and White Chocolate Chips

Traditionally used "as is" when making cookies.

Chips can also be melted and used to make chocolate moulds and decorations.

Uses:

— Mixed in with cake, muffin batter, or even dough
— Whole chips can be randomly tossed on top of a cake
— Meticulously placed on a cake
— To coat the side of a cake; freeze chips until hard and then pulverize

Bars of Chocolate

Baking, coating, white, unsweetened and baker's semi-sweet come in bars or slabs.

Uses:

- When melted, they can be piped on cookies, cake and other desserts
- Make a fancy chocolate leaf decorations.
- They can be grated on a grater or scraped with a zester to form tid-bits and then sprinkled over desserts
- They can be made into chocolate ribbons

Cocoa Powder

A good dusting of cocoa powder over a frosted cake can produce a simple, yet fancy decoration. Cocoa powder is ground cocoa nuts, less the butter (fat) and calories. You have a choice between cocoa powder: a light and a dark one. The light one is inexpensive plain cocoa powder found everywhere. The dark one is plain cocoa powder that has been processed with alkali to mellow and darken the cocoa, which brings up its price and makes it more expensive and harder to find than plain cocoa. You can create a contrast effect simply by using these two side by side.

- Light cocoa powder.
- Dark cocoa powder.

Cinnamon Powder

You can always use cinnamon powder, if cocoa powder isn't available, to create a nice effect, as well as, to offer a different, yet slight flavour to your dessert. Cinnamon goes well with chocolate, vanilla, almonds and walnuts.

Icing Sugar

Also known as confectioners sugar, can also be used as a cake decoration when sifted over a plain unfrosted cake. The best part of having icing sugar on hand, is that you can combine it with a bit of cocoa or chocolate, vanilla, butter or shortening, and water or milk (cream), to create a quick and easy icing (or frosting). It only takes "a little bit of this and that", to turn a plain old coffee cake into an irresistible dessert.

Sprinkles - Chocolate and Multi Coloured

They break up the monotony of "all white" or "all chocolate" desserts. Note that all sprinkles are not alike. Some may bleed into the topping (dairy and non-dairy) after a few hours of refrigeration. Bleeding refers to when the colour of the sprinkles tint the topping.

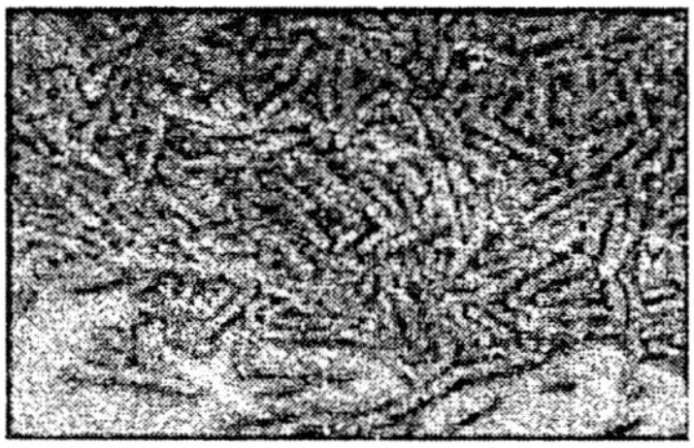

This doesn't affect the taste of the cake; though, it gives the cake a unique, yet different look.

Rainbow "Round" sprinkles

Cookies can be placed "as is" around and over of a cake. They can be soaked in syrup, with or without a liqueur.

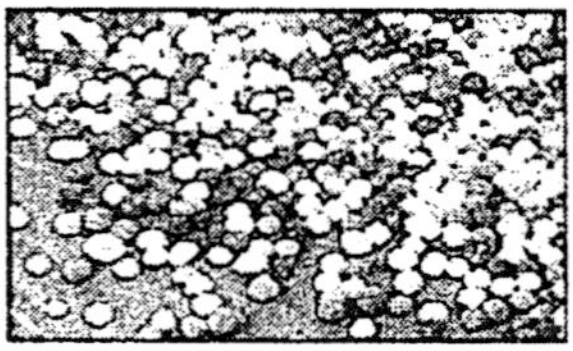

Dragees

They can also be crushed, or pulverised, and then be used in a filling or cream, or sprinkled around and over a cake.

Cherries

They can be used whole, halved, or quartered. Candied cherries come in red and green. Halves are typically used in most recipes.

If, however, you're decorating a cake and you're short on cherries, you can always quarter them instead and then add another element around them to take the focus away from their small size.

Tips for Cake Baking

Positions an oven rack on the middle leve in the oven (or otherwise indicated); place filled pan(s) centred on the middle of the rack. Allow the heat to circulate equally between walls and pan(s). Always preheat your oven as indicated in the recipe. If cake batter stands beyond ten minutes without being baked, the leavening agent in it will make the batter frothy and will produce an undesirable cake.

Colour & Crust

Position your oven's rack in the middle level of your oven.

— Baking on the lowest rack might overbake the bottom and scorch it

— Baking on the highest (near the brolier) rack might overbake the top and scorch it

If you are planning to bake two pans at once, then approximate two centres on the middle rack, allowing equal space between both pans; they should not be touching each other.

Preventing Sticking

"Wax Paper" Method:

This is a great method for greasing pans that are flat and that do not have decorative edges: 9" x 2" round cake pan for baking layer cakes, or 9" x 9" square pan for baking brownies. Measure and cut wax paper to fit bottom of pan. Brush room temperature vegetable shortening inside the bottom of the pan, and line with wax paper. Press firmly into place to adhere to pan; brush again with shortening. Then, lightly grease the sides. One exception is Angel Food Cake.

"Greased and Floured" Method:

This is a great method for greasing pans that have decorative edges: bundt, twirl, springform pan, etc. The wax paper method would be rather difficult to use, since wax paper does not adhere to curved edges. Grease pan (bottom and side), using a pastry brush, with either softened butter or shortening. Sprinkle 2 tablespoon all-purpose flour, which works the best. Tilt the pan to distribute the flour to adhere to greased areas. Finally, turn pan over and tap to remove excess flour.

Even Texture and Tops

Lightly grease the side of the pan according to methods. If you grease the side entirely, the batter will have a hard time clinging onto the pan while baking and will not rise well. Your cake will develop a lip around its edge, with a deep small indentation, and a centre hump with cracks in it.

Fill pan halfway with batter. Raise filled pan to about six inches above a hard surface and allow to drop. You eliminate any trapped air bubbles in the batter that create large holes in a finished cake. You will see that the top of the batter will have little air bubbles (this means that most of the air bubbles have risen to the top and have been eliminated). Air pockets develop during the mixing stage and the pouring of the batter into the pans (as the cake bakes, air pockets will try to escape upwards, thus creating small holes throughout the cake).

Quickly spin the pan and allow the batter to cling onto its side. Then quickly place in preheated oven and bake accordingly. This allows some batter to cling onto an ungreased surface (when cake has finished baking, the centre will have risen to the same height as its side, thus creating a flat top).

Preventing Soggy Bottoms

When cake has finished baking, remove it from the oven and allow it to cool in its pan, on a cooling rack for about ten minutes. This allows cake to keep its shape as it cools off.

To remove cake from pan, run a butter knife around the pan, between the cake and the side. Invert cake onto a cooling rack; lift off pan (and any wax paper), and return cake to an upright position; allow to cool completely on cooling rack.

When a hot surface comes into contact with a cold surface, or cold air mass, condensation might develop (water might be absorbed by the cake, causing it to have a soggy bottom, which might make it stick to its pan

Racks allow cakes to cool off at a faster and equal rate, and also prevent them from becoming soggy.

Testing for Doneness

You can determine when a cake is baked:

— When the cake's side pulls away from the pan

— And, when a knife inserted in the middle comes out clean; if traces of batter are seen, then continue baking and testing again (after 5 minutes) making sure that knife comes out clean with no traces of batter.

How to Grease a Cake Pan

There's nothing more frustrating than trying to remove a baked cake that's stuck in its pan. It's more likely that the wrong greasing technique was used, rather than the wrong pan size or type. There are a few techniques to prepare pans to prevent this from happening.

Preparation Techniques:

— Non-stick sprays

— Greased with shortening

— Greased with shortening and then dusted with all-purpose flour

— Greased with shortening and fitted with wax paper or parchment paper

Non-stick sprays are great for pans that have many crevices, bumps and valleys, such as a flan pan or Madeleine (sea shells), which are made of thin, shiny tin. This type of material reflects heat, rather than absorbing it. The typical greasing and/or flour technique may produce a gummy layer; the grease and flour doesn't cook well and the batter may stick to the pan. Non-stick sprays are more ideal for these pans, since they coat these types of pans evenly with lecithin and allow for easy removal without sticking. It's very difficult to cut wax or parchment paper and to then press them into the crevices, so a spray comes in handy.

When using shiny pans such as these, it's best to allow the baked cake to rest in its pan for about 3 - 7 minutes to retain its shape, before inverting the pan to release the cake. If the cake is left longer than that time, condensation will develop between the bottom crust of the cake and the interior bottom of the hot pan, which will moisten the crust of the cake and will cause the cake to stick to the pan, defeating the purpose of using a non-stick spray. Another thing to consider when using a non-stick spray is that during baking, the edge of the side of the cake will become a little deformed: lopsided. The baked cake will look as if someone had "cookie cut" the cake out of a larger piece of cake with the some of the edges raised, which to me isn't appealing.

5

Baking of Pies

Pies are generally classified into three main classifications as follows: Double Crust Pies such as Fruit Filled Pies, Baked Custard Pies, and Pre-Baked Shell Type Pies.

Pie Crust

Pie crusts are divided into three main classifications such as:

1. *Long flake*: To produce a long flake, the flour and shortening are mixed only until the fat is about the size of walnuts.
2. *Short flake*: To produce a short flake, the flour and shortening are mixed until the fat is the size of peas.
3. *Mealy tender crust*: To produce a mealy tender crust, the flour and fat are mixed until there are very few fat particles.

Slight variations or combinations of the three mixing methods listed above are used by some bakers. Also, the dough may be mixed by hand or by using the pastry blender attachment when using a mixing machine. If a pastry blender attachment is not available the mixing paddle can be used. When using the mixing machine, extra care must be used to prevent overmixing the flour and fat together and also after the water has been added.

Directions for Making the Pie

Of all the pastries, pies mean home and hearth. Apple pie is as much a symbol of America as baseball and bald eagles. A pie cooling on the windowsill is Americana. Not much has changed in pie baking over the centuries, except that maybe we use shortening instead of lard and the cherries come from a can instead of off the tree in the back yard. The components are basic: tender crust, filling of fruit or custard or lemon with meringue. Fill your crust with eggs and vegetables or meat and it's called a quiche.

Although real men may not eat quiche, most men will hunker down to a big piece of apple pie a la mode or a la cheese and think himself the luckiest guy on Earth. The key is the crust. It must be flaky, not tough. It must be just the right shade of golden brown. It must bake evenly. It must be thin on the bottom so it does not get soggy or stay raw. Tenderness, taste, a sculptured appearance—these are good things. Once mastered, exemplary pie baking is easy to achieve, pie after pie. Ingredients for a classic piecrust marry best when chilled—even the flour, which can be cooled in the refrigerator until needed.

Choose your own method of mixing: two knives or forks drawn against each other to "cut" ingredients; a pastry blender (arcs of plastic or wire attached to a handle); blending with your hands. The latter is most successful. Your recipe will tell you to work the mixture until it resembles coarse cornmeal. If you've never worked with cornmeal, this requires a good imagination. In general, crust should be a little crumbly before it is rolled, the flour incorporated into the fat, not too sticky or too dry.

A few attempts might be required until you can feel a properly worked dough. The less you work the dough, the more tender it will be. Overworking dough is like tanning a hide. This is a bad thing. Traditional crust ingredients include flour, salt, fat (old-fashioned lard, shortening, butter-flavour shortening or butter), milk and sometimes an acidic liquid such as vinegar or lemon juice. Acids break down proteins in the dough and produce a flakier crust. Using shortening increases flakiness, but butter will produce a richer taste.

Form the dough into a ball and then turn out on a well-floured board. Some people roll their dough between two pieces of wax paper or plastic wrap. This can be tricky because the paper can slip around. A properly dusted board will allow easy rolling. And keep flouring your rolling pin. Because there is so much shortening in a piecrust, pans usually need not be greased or floured. Roll your dough thin—about 1/8-inch thick—until you have a round of dough slightly larger than your pie pan—1/2- to 3/4-inch overhang. You want to have enough excess to seal the edges of a two-crust pie or form a decorative edge on a one-crust.

To transfer the rolled dough into the pie pan, dust it lightly with flour and fold into quarters. Place the quartered dough in the pan, with the point in the middle, and then unfold it, pressing it firmly into place but being careful not to stretch it. If a section tears slightly, it can be patched easily. With a knife or fork, poke several holes in the bottom to allow steam to escape. If you are doing a one-crust pie, make the edge attractive by crimping or rolling inward. Make sure it is even in thickness or some parts will brown too soon.

For a two-crust pie, add the filling and then top with the second round of dough. Cut several slits in the top layer, in a pattern, to allow steam to escape. Juicy fillings may ooze a little from the top. This is a good thing. But don't overfill. For another type of two-crust pie, roll out the second half of your dough (from a two-crust recipe) and cut into 1/2- to 1-inch strips to form a latticework top. You can use your own pattern. Weave the strips in and out or begin in one "corner" and alternate overlapping strips.

This is where the art of the crust comes into play. Make your pie as beautiful as it is tasty. To seal layers, moisten the edges of the bottom crust slightly and press down firmly on the edges of the top layer. Crimp or roll the edges. Place the pie pan in the centre of the oven for even baking. Bake until the crust is a beautiful golden brown. Some crusts are pre-baked and then baked some more after the filling is added. Some crusts are pre-baked and then finished with a filling that is chilled until firm. Some crusts are made of ground nuts, butter and sugar, baked until hardened and toasty flavoured.

Some crusts are not baked at all. These include cookie-crumb and graham-cracker crusts. Fillings usually are chilled. For rich, sweet pies and tarts, a crust by itself might be a kind of giant butter cookie. Your choices of pie fillings are nearly endless: fruit, custard, coconut cream, chocolate cream, mincemeat, pumpkin, sweet potato, molasses, rhubarb, lemon chiffon, lemon meringue, pecan, almond, peanut butter.

Some tips for baking the perfect pie:

— Be sure all crust ingredients, even the flour, are cold.

— For a tender crust, use low-protein flour.

— For a rich crust, substitute butter for shortening, or use half butter and half shortening.

— Work ingredients just until blended. Overworked dough results in a tough crust.

— When rolling dough, roll out from centre and don't roll off the edge.

— Cut rolled-out dough to fit pie pan with extra room to form a nice edge. Lay dough in pan and press into sides without stretching.

— Do not overfill crust. It will expand and ooze during baking.

— When baking a two-crust pie, poke holes in top crust to allow steam to escape.

— Brush top with beaten egg white mixed with a little water.

— If top crust edge browns too quickly before pie is cooked through, cover edges with a strip of aluminum foil and continue baking.

Some tips for baking the perfect pie:

— Sift flour and baking powder well to remove lumps. If sugar has lumps, sift it, too.

— Do not overmix.

— When recipe calls for beaten egg whites, use a clean, grease-free bowl. Beat whites until stiff but not dry. They should appear glossy.

— When melting chocolate, use a double boiler to prevent burning.

— Test the accuracy of your oven with a small oven thermometer set inside. Adjust temperatures accordingly.

— Never bake a cake in a microwave oven unless recipe is especially designed for microwaving.

— Ingredients usually are separated into wet (eggs, milk, liquid flavourings) and dry (flour, baking soda, baking powder, salt). Incorporate them gradually for easy mixing and to prevent lumps.

— Accuracy in measuring is essential.

Flour Preparation

Preheat the oven to 425 degrees. Measure the pour by spooning it gently from the container into a 1-cup dry measure. Don't shake the cup or tamp it. Sweep off the excess with a knife. Pour the measured flour into a 3-quart bowl. Repeat for the other 1 cup flour, then use a 1/4-cup dry measure for the remaining flour and put that into the bowl, too. Add the salt and stir with a fork to combine. Press shortening into a glass measure up to the 3/4-cup mark, packing it down into a solid mass with no gaps or air bubbles. Scrape it into the flour. Now, work the shortening and flour together, using a pastry blender, two knives, or your fingertips.

Mixing with Pastry Blender

Make small circular downward motions around the sides and bottom of the bowl with the wires of the blender, cutting through the shortening into the flour.

Occasionally, scrape down the sides of the bowl and scrape the collected flour and shortening off the wires into the bowl. When mixed enough, the mixture will have irregular granules of fat about the size of soft bread crumbs.

Mixing with Two Knives

Hold a table knife in each hand. Cut into the flour and

shortening with the knives, rapidly drawing them through the mixture, toward one another. Repeat about 15 or 20 cuts, scraping down the sides of the bowl so that all the particles of fat and flour are combined. Repeat until the mixture looks like fresh bread crumbs.

Measures: 1 tablespoon cold water, sprinkle it over the flour mixture, and stir with a fork to distribute the moisture evenly. Repeat this step four times. The mixture will now begin to look like dough. For the dough to become cohesive, you will need to add a tablespoon or more water. Sprinkle 1/2 tablespoon at a time over the dough and stir with a fork. The dough should hold together, but if it doesn't, add a few more drops of water where it seems dry and crumbly.

Dough Forming

Form half the dough into a rough round cake about 1 inch thick and 3 inches in diameter. Repeat with the other half. Refrigerate if you want, but it's not necessary. Lightly dust a rolling surface with flour and have extra flour at hand.

Place one piece of dough on the surface, sprinkle the top with a little flour, and flatten the dough by firmly pressing with the rolling pin in several places. Start in the centre of the dough and roll lightly in all directions, lifting and turning the dough after every few rollings and sprinkling on a little more flour if it sticks.

Use a flat-blade metal spatula if you need to, to help lift and turn the dough. Rub flour on the rolling pin if it becomes sticky. Try to keep the shape as round as possible; don't worry too much about ragged edges and small tears. It's easy to repair them once the dough is in the pie plate. Continue rolling until the dough is about as thick as cardboard and about inches larger than the pie pan.

Slide a flat metal spatula all around under the dough to loosen it from the rolling surface. Roll the dough up around the rolling pin, beginning at the edge nearest you, and unroll it into the pie pan, beginning at the farthest edge of the pan and unrolling toward you. Press the dough into the pan, working from the edges toward the centre, eliminating any gaps and air holes and making a snug fit. Leave about 1 inch of overhang all around the pan for a decorative edge. Trim excess overhang with scissors or sharp knife.

Top Crust

Mound filling into the dough-lined pan so that the centre of the filling rises an inch or so above the rim. Dot filling with bits of butter. Roll out the remaining half of dough just as you did the first one, making it the samē thickness but slightly larger in diameter. To ensure that the top crust is centered with neat-looking steam vents, fold the dough in half, then in half again. With a knife make three or four angular cuts about 3/4 inch long on each folded side. With your fingertips, spread a few drops of water on the rim of the dough in the pie as an adhesive for the top crust.

Now, place the point of the folded dough in the centre of the filling and unfold the dough. Trim the dough all around to only 1/2 inch of overhang. Press the edges of the two crusts together, and turn the overhanging flap under itself all around, to make a well-sealed, upstanding ridge.

Crimping the Edges

Crimp the edge to make a decorative border using one of the following methods: a) Press the tines of a fork into the edges all around; b) use your two thumbs on the inside edge to pinch together the dough, while pushing your forefingers on the outside edge, to make the dough stand up.

Adding a Glaze

For a golden glaze, use your fingers or a pastry brush to gently spread 2 tablespoons of cream or milk over the top crust. Sprinkle evenly with 1 tablespoon of sugar.

Pie Dough Ingredients

1. *Flour*: Flour is the main structure builder in pie

doughs. Pastry flour generally produces superior pie doughs, however, a combination of 60 percent bread flour and 40 percent cake flour can produce quality pies if the dough is mixed properly. If all bread flour is used, the dough will shrink excessively when baked and the crust will be tough. On the other hand, if all cake flour is used, the amount of shortening used must be reduced otherwise, the dough will be very difficult to handle during make-up.

2. *Shortening*: Shortening is responsible for flakiness, tenderness, taste, palatability and keeping quality of the crust. Shortening may be animal, vegetable or a blend of animal and vegetable. Butter, margarine, and lard produced from hog fat, are highly flavoured. Many years ago, lard was the favourite of bakers. Today, bakers use a combination of lard and vegetable shortening. Too high a percentage of shortening in the formula will result in excessive tenderness of the crust and too low a percentage will produce a tough crust.

3. *Salt*: Salt enhances the flavour of other ingredients. Salt also has a strengthening effect upon the flour proteins which is desirable.

4. *Colouring Agents*: Colouring agents are generally used to assist in developing the golden brown crust colour during baking. For example when sugars and milk are used in pie dough, they carmelize during baking. The disadvantage of using these ingredients is that they tend to destroy the flake and increase moisture absorption in the crust after baking. Some bakers use a pie wash such as a mixture of equal parts of water and eggs which has been beaten slightly with a wire whip. Milk and eggs or melted butter can also be used. The wash is painted on or sprayed on the top of the pie just before loading them into the oven. The wash containing eggs produce a shiny glazed surface to the crust, but they tend to make the crust soggy. Butter

produces a more tender crust, but the crust will not have a glaze, and will tend to be dull in colour.

4. *Water*: Water dissolves the salt, developes the flour proteins, and controls the consistency and temperature of the dough. Excessive water in the dough is undersible because it takes longer to bake the crust, and toughens the crust. Water used in the dough should be ice cold to prevent softening the fat in the dough. If time permits, the mixed dough should be chilled in the refrigerator slightly. This will assist in keeping the dough from being sticky during make-up.

Filling Ingredients

Although an attractive, tender, flaky, golden brown crust with good flavour, taste and aroma is necessary, a quality, attractive and tasty filling will in the end determine the final quality of the pie. Some well known bakery ingredient manufacturers make available to the baker high quality prepared fruit pie fillings and cream pie fillings. Bakers use these prepared mixes to save time and labour costs. For those bakers who prefer to prepare their own pie dough and pie filling formulas, the following information can be used as a guide:

Pie dough baker's percentage range showing minimum and maximum amounts of each ingredient that should be used in the formula

Ingredients	Minimum	Maximum
Flour	100	100
Shortening	50	75
Salt	3	4
Sugar	0	10
Milk, Dry Nonfat	0	5
Ice Water	20	50

Pie dough true percentage range showing minimum and maximum amounts of each ingredient that should be used in the formula

Ingredients	Minimum	Maximum
Flour	44	50
Shortening	24	32
Salt	0.5/	1.0
Sugar	0	3
Milk, Dry Nonfat	0	2
Ice Water	15	24

Fruit pie filling true percentage range showing minimum and maximum amounts of each ingredient that should be used in the formula

Ingredients	Minimum	Maximum
Sugar	15	25
Salt	0	0.25
Glucose Syrup	0	20
Cornstarch	3	4
Fruit	35	50
Juice or Water	30	45
Butter	0	1.5

Cream pie filling true percentage range showing minimum and maximum amounts of each ingredient that should be used in the formula

Ingredients	Minimum	Maximum
Sugar	15	25
Salt	0.125	0.25
Eggs or Yolk	10	20
Milk, Nonfat Dry	0	10

Chocolate	5	7
Cocoa	3	4
Cornstarch	4	5
Fruit Juice & Rind	4.5	6.5
Butter	0	3

Cream Pies

Cream for Cream Type Pies are generally poured into prebaked pie shells and topped with Meringue. There are several methods of preparing meringue, however the use of a cooked type meringue is recommended, because the meringue is more stable and will not break down or weep during storage or when chilled in the refrigerator. The meringue is made using a commercially prepared special stabaliser or with a cooked cornstarch mixture.

Custard pie filling true percentage range showing minimum and maximum amounts of each ingredient that should be used in the formula

Ingredients	Minimum	Maximum
Sugar	16	20
Salt	0.125	0.25
Eggs or Yolk	10	20
Milk, Nonfat Dry	5	8
Spices	0.125	7
Cocoa	0.125	0.5
Cornstarch	0	1.5
Pumpkin	28	36
Butter	0	1.5

Cornstarches in Pie Baking

There are several types of starches available to the pie baker for thickening the fruit juices and or milk and water.

1. Pure food powdered starch- a pure refined corn starch prepared from ordinary field corn. This type of starch is not as stable as the waxy maize starch and will break down and become watery after long storage or when used to thicken fruit juices containing a high acid content such as cherry or pineapple juices. It contains amylase and amylopectin. It can be modified to increase it's clarity and stability.
2. Waxy starch-this type of starch is refined and modified. It is prepared from waxy maize. This is an exceptionally clear, gel-producing starch with a short tender body and extreme stability. Waxy starches generally swell faster than other starches, but they also become thinner during cooking. The cooked filling will not get thicker when the pie cools as does that made with regular starch.
3. Pre-gelatinised starch (cold water starch)- this starch is generally know as instant starch. It does not require cooking. The juice is drained from the fruit. If insufficient juices are present in the fruit, water can be added to make up the shortage. All the dry ingredients (starch, sugar, salt, spiced, etc.) Of the filling are blended together thoroughly in a dry mixing bowl. Add the liquids (juices and water when used) gradually to the dry ingredients and mix until smooth. Carefully fold drained fruit into thickened mixture. Chill filling in refrigerator until ready to be used. Regardless of the type of starch used, the prepared filling should be chilled in the refrigerator before being used to reduce the possibility of the filling boiling out of the pie during baking. Generally if the oven is heated to about 450 degrees before the pie is loaded into the oven, the crust will be fully baked before the filling gets hot enough to come to a boil and spill out of the pie.

4. Tapaioca starch-this type of starch is also used in pie fillings by some bakers. One of the charaacteristics of all starches is their ability to swell when cooked in water or fruit juices. They become gelatinised and in turn cause the mixture to be thickend.

Role of Fruits in Pies

The following types of fruits will produce quality pies if processed properly:

1. Fresh fruit-wash, drain and mix into a slurry of cooked starch, sugar, salt, drained juice and water plus any spices or other ingredients such as lemon juice, and butter that may used. Refrigerate until ready to use.
2. Frozen fruit-thaw, drain the juice, cook a slurry of drained juice and water, starch, sugar and salt, etc. And mix the fruit into the slurry. If sugar has been added to the fruit, this must be taken into consideration. Excess sugar in the recipe will interfere with the ability of the starch to form a jell.
3. Canned fruit-drain the juice, if insufficient juice is present, add water to make up the shortage. Cook a slurry of drained juice, water when added, starch, sugar, salt and other ingredients when used. Then mix the fruit into the cooked mixture.
4. Dehydrated and dried fruits-reconstitute, cook into a slurry as for other types of fruits and mix the fruit into the cooked mixture. Today, dried fruits are not shriviled as they once were. They should be considered fresh fruit from which water has been removed. Years ago when dried fruits were used, it was necessary to soak them in water for 8 hours or overnight. Today that is no longer necessary. Quick cooking is the secret to producing quality pies from dried fruits without ending up with broken, mashed, or dark or yellow unpleasant colour.

Following are a few fruit pie filling formulas using regular starches and fillings using cold water starches (instant starches)

Apple pie filling using canned apples and cooking with regular starch

Ingredients	Pounds	Ounces
Juice or Water	2	0

Directions: bring water to boil

Ingredients	Pounds	Ounces
Starch	0	3-1/2
Water	0	6

Directions: suspend starch in water. Add to boiling water stirring constantly. Cook until thick and clear

Ingredients	Pounds	Ounces
Sugar	0	12
Sugar, Brown	0	12
Salt	0	0-1/4TH
Cinnamon	0	0-1/4TH
Butter	0	4
Lemon Juice	0	0-1/2

Directions: mix dry ingredients together, add to cooked slurry. Stir until sugar and salt are dissolved and the butter is melted. Butter is optional. Add lemon juice and then pour cooked slurry over fruit and blend carefully. Chill filling in refrigerator until ready to be used.

Apple pie filling using canned apples and using cold water starch (instant starch)

Ingredients	Pounds	Ounces
Sugar	0	12

Sugar, Brown	0	12
Salt	0	0-1/4TH
Cinnamon	0	0-1/4TH

Directions: blend all dry ingredients thoroughly in a dry bowl.

Ingredients	Pounds	Ounces
Juice or water	2	0TR>
Butter	0	4
Lemon Juice	0	0-1/2

Directions: gradually add water to the dry mixture above stirring constantly until smooth. Add lemon juice and stir into the mixture. Pour thickened mixture over apples and fold in gently. Chill filling until ready to be used.

Cherry pie filling using canned cherries and cooking with regular starch

Ingredients	Pounds	Ounces
Juice	2	0

Directions: Bring juice to Boil

Ingredients	Pounds	Ounces
Starch	0	4-1/2
Water	0	6

Directions: suspend starch in water. Add to boiling water stirring constantly. Cook until thick and original colour is obtained. Shut off heat.

Ingredients	Pounds	Ounces
Sugar	2	8
Salt	0	0-1/4TH

Butter	0	4
Lemon Juice	0	0-1/2

Directions: mix dry ingredients together, add to cooked slurry. Stir until sugar and salt are dissolved and the butter is melted. Butter and lemon juice are optional. Add lemon juice and then pour cooked slurry over fruit and blend carefully. Chill in refrigerator until ready to be used.

Cherry pie filling using no.10 canned cherries and using cold water starch (instant starch)

Ingredients	Pounds	Ounces
Sugar	2	8
Salt	0	0-1/4TH
Instant Starch	0	4-1/2

Directions: blend all dry ingredients thoroughly in a dry bowl.

Ingredients	Pounds	Ounces
Juice	2	0TR>
Butter	0	4
Lemon Juice	0	0-1/2

Directions: gradually add juice to the dry mixture above stirring constantly until smooth. Add lemon juice and stir into the mixture. Lemon juice and butter are optional. Pour thickened mixture over cherries and fold in gently. Chill filling until ready to be used.

Pecan Pie Formula

Ingredients	Pounds	Ounces
Sugar, granulated	1	0
Cake Flour	1	0

Salt	0	0-1/2
Nutmeg	0	0-1/2
Eggs	6	8
Karo sirup	10	0

Directions: Place all above ingredients into mixing bowl and beat in low speed with wire whip until well mixed.

Ingredients	Pounds	Ounces
Water	2	0
Vanilla	0	0-1/2

Directions: Add water and vanilla to above mixture and continue beating in slow speed for about 2 minutes or until well mixed. Place 0-1/2 Cup macaroon cocoanut into bottom of unbaked pie shell. Also add 1 to 1-1/2 cups of fresh pecans to the unbaked pie shell. Place unbaked pie shells containing coconut and pecan in 350 degree to 400 degree F. oven. Fill unbaked pie shell containing coconut and pecan with prepared pecan pie filling. Bake about 30 minutes or until filling has set.

Fresh Lemon Chiffon Pie

After the meringue has been folded into the hot fresh lemon filling, and the pre-baked pie shell filled, the pie can be finished off by crumbling some pre-baked pie crust into a sifter and forcing it through the sifter and sifted on top of the pie. The top of the pie can be garnished by placing a Maraschino Cherry in the centre of the pie. If desired, rather than adding pie crumbs on top of the pie, meringue can be applied and browned off in the oven Lemon Chiffon Pie filling

Ingredients	Pounds	Ounces
Water	15	0

Directions: Bring water to boil .

Ingredients	Pounds	Ounces
Cornstarch	2	2
Water	3	0

Directions: Suspend starch in water and add to boiling water stirring constantly and cook until mixture is thick and clear.

Ingredients	Pounds	Ounces
Sugar	9	0
Salt	0	0-3/4th
Butter	0	12

Directions: Add to cooked mixture and stir until sugar and salt are dissolved and the butter is melted.

Ingredients	Pounds	Ounces
Egg Yolks	3	0

Directions: Beat egg yolks slightly. Pour 1/4th of the cooked mixture over slightly beaten egg yolks and mix thoroughly with hand wire whip. Pour the egg mixture back into the steam kettle containing 3/4th of the cooked mixture stirring constantly and bring to a boil again. Cut off heat. Wash 30 lemons, and grate the rind from the lemons. Squeeze the juice from the lemons and add the grated rind to the juice. Add juice and rind to the cooked mixture. Do not heat after juice has been added or the acid in the juice will break down the starch.

The meringue for the chiffon pies should be prepared at the same time that the filling is being cooked. The meringue should be folded gently into the cooked mixture while the filling is still hot. Pour the chiffon filling in pre-baked pie shells while filling is still hot.

Cooked Meringue

Ingredients	Pounds	Ounces
Water	1	8

Directions: Bring water to boil

Ingredients	Pounds	Ounces
Starch	0	3
Water	0	6

Directions: Suspend starch in water, add to boiling water and cook until thick and clear.

Ingredients	Pounds	Ounces
Sugar	2	10

Directions: Add sugar to cooked mixture and bring to a boil again. Remove from heat

Ingredients	Pounds	Ounces
Egg Whites	1	12
Salt	0	0-1/4th

Directions: Add salt to egg whites. Beat eggs in medium speed until a peak is formed on the finger when it is dipped into the whites and withdrawn. Pour hot cooked mixture over beaten eggs in slow stream while continuing beating until the mixture forms a peak when a finger is dipped into the beaten mixture and withdrawn.

Ingredients	Pounds	Ounces
Vanilla	0	0-1/4th

Directions: Add vanilla to meringue and stir in. Apply on top of Cream Pie or chiffon pie while meringue is still warm. Bake at 375 Degrees F. until golden brown.

Recipes for Pie

Myra Hardy's Basic Piecrust

1 cup shortening or butter, or half each

2 cups flour

1/2 teaspoon salt

2 tablespoons vinegar

1/3 cup cold milk

Directions

This recipe makes enough for two crusts. In a large bowl, work shortening or butter, flour and salt together with two forks, pastry cutter or fingers. Add vinegar, then milk as needed to form a stiff dough that still is easily workable. Roll out dough on floured board to thickness of 1/8 inch. For two-crust pie, cut 2 rounds about an inch larger than pie pan. Place into pan and press gently into place. Fill and top with second crust. Seal and crimp edges. Bake according to pie recipe.

Lemon Meringue Pie with Microwave Filling

Filling:

1/3 cup cornstarch

1 1/4 cups sugar

Dash salt

1 1/2 cups hot water

3 egg yolks, beaten slightly

1 tablespoon butter

2 teaspoons grated lemon rind

1/2 cup lemon juice

9-inch pie shell, baked and cooled

Meringue:

- 2 tablespoons cornstarch
- 1/2 cup sugar
- 1/2 cup water
- 2 egg whites
- Dash salt
- 1/2 teaspoon vanilla

Directions

Preheat oven to 350 degrees. To make filling: Combine 1/3 cup cornstarch, 1 1/4 cups sugar and dash salt in a large microwave-safe bowl. Slowly stir in hot water. Microwave on high power, stirring often, until mixture is thick and clear. This should take about 3 minutes total, stirring twice. Stir several spoonfuls of this mixture into the beaten egg yolks. Mix well. Pour this tempered egg mixture back into the large microwave-safe bowl. Mix well. Cook on high an additional 2 minutes, stirring after 1 minute.

Stir again and add butter and rind. Gradually stir in lemon juice. Cool 5 minutes and then pour into cooled, baked pie shell. To make meringue: In a medium microwave-safe bowl, combine 2 tablespoons cornstarch, 1/2 cup sugar and water. Microwave on high for 1 minute; stir, then microwave another minute. Mixture should be thick and clear. Let cool while preparing egg whites.

In a separate medium bowl, beat egg whites with salt until whites are stiff. To check, run a spatula across the bottom of the bowl; stiff whites will not run back together. Pour sugar and cornstarch mixture into egg whites. Add vanilla. Beat until mixture is smooth. Spread meringue on top of pie filling, making sure it extends to edges of crust. Brown meringue in preheated oven, about 15 minutes.

Two-crust Cherry Pie

1 1/2 cups sugar, divided use

4 tablespoons cornstarch

3/4 cup juice from canned cherries

3 cups canned, pitted tart red cherries (water pack)

1 tablespoon butter

1/4 teaspoon red food colouring (optional)

Pastry for two-crust, 9-inch pie

Directions

Preheat oven to 400 degrees. Combine 3/4-cup sugar with cornstarch in a medium saucepan. Stir in cherry juice. Cook over medium heat, stirring occasionally, until mixture thickens and bubbles; cook 1 minute longer. Add remaining sugar, cherries, butter and food colouring, if using. Mixture will be very thick. Let stand while preparing pastry. Line 9-inch pie plate with pastry; fill. Adjust top crust, cutting slits for escape of steam. Bake for 55 minutes.

6

Making of Cookies

All cookies are more or less a variation of one basic formula. Because of its numerical balance of ingredients, the pound cake recipe could well serve as a basis for cookie recipes. Leavening agents, added to most varieties, have an effect on the size, colour, and eating qualities. Additional ingredients are added for flavor and texture. There are two main types of cookies, hard and soft. Soft (soft batter) cookies contain a maximum amount of moisture. They may require a greater percentage of eggs to produce the necessary structure. Characteristics of good soft cookies are moistness and softness. Hard cookies contain a minimum amount of moisture. Desired characteristics of hard cookies are crispness and brittleness.

Ingredients

Each ingredient used in cookie baking is employed for the specific characteristics it has and/or the result it has on the finished product. If these effects are understood, the ingredients may be selected with the assurance that the products produced will be good.

Flour

Flour is the principle structure builder or binding agent

in most cookie formulas. It provides the framework around which the other ingredients are grouped in varying proportions. Flour is a toughening agent because when it comes in contact with the liquids in the formula, the flour proteins (gluten) form the main structure of the cookies. Either bread flour, pastry flour or cake flour can be used to make hard and soft cookies. Hard flour may cause some types of cookies to lack spread while cake flour may cause some cookies to spread too much. It is best to use the type of flour specified in the formula. If a recipe calls for pastry flour and it is not available, use 50 percent bread flour and 50 percent cake flour or use all purpose flour.

Sugar

Sugar in some form is used in all cookie recipes. It is an important tenderising ingredient. Undesolved sugar crystals melt during baking which contributes to the flow or spread of the cookies.

Granulated sugar

The finer the granulation, the less the cookie will spread. On the hand, the larger the granulation, the more the cookie will spread.

Powdered Sugar

Powdered sugar is used when a fine-grained compact interior with little spread is desired.

Brown Sugar

Brown sugar is used to obtain a certain flavor. When brown sugar is used in place of granulated sugar, a small amount of baking soda should be used to neutralise the acid in brown sugar. By neutralising the acid in brown sugar, the colour of the crust will be improved because acids interfere with carmelisation of sugar. Because of the soft sugar crystals present in brown sugar and because all brown sugar crystals dissolve during mixing, proper spread will not be obtained without formula adjustments.

Molasses

Molasses is used to obtain a particular flavor. Formulas containing molasses have to be adjusted by the addition of baking soda to compensate for the acidity. Also, an adjustment must be made in the amount of liquid.

Honey

Honey is also used to obtain a particular flavor. Honey adds a chewy quality to cookies.

Shortening

Shortening promotes tenderness in the cookie and prevents excessive gluten development during mixing. Without shortening, cookie dough would be tough and rubbery, which would result in cookies being dry and lacking in eating qualities. Shortening contributes to the spread of the cookies. Regular hydrogenated shortening having a bland flavor is preferred and commonly used. Butter and margarine produce cookies with a more desirable taste and flavor.

Eggs

Eggs are both tenderisers and tougheners in cookie baking. Egg yolks contain a very large percentage of fat which helps to tenderise the cookie. On the other hand, the egg whites acts as structure builders because of the proteins which coagulate during baking. Eggs also contribute moisture. Some cookie recipes such as macaroon coconut cookies call for egg whites. Whole eggs contribute the combined characteristics of shortness, aeration, and tenderness.

Liquids

Liquid in some form is necessary in all cookie recipes. Water is necessary to moisten the flour proteins to form gluten so the structure can be formed. Water also contributes to dough consistency and helps to control the temperature of the dough or batter. It also dissolves the chemicals in baking powder so that carbon dioxide can be produced to leaven the cookie and to assist in the spread of the cookie during baking.

Milk solids

Dry milk solids tend to exert a slight binding action on the doughs. Milk solids are a valuable addition to cookie recipes because they provide added nutritional value. The milk sugar lactose, adds to the richness of the crust colour.

Leavening Ingredients

Leavening ingredients help to control spread or size, produces volume and promote proper crust colour through regulation of acidity or alkalinity (pH) of the dough. Cookies that are hand dropped, deposited with a pastry tube, or machine deposited, need the spread control of leavening agents. An alkali such as baking soda exerts a weakening effect on the flour proteins (gluten) and helps to promote spreading. Since most cookie ingredients are neutral to slightly acid, they are usually sufficient to neutralise any reasonable quantity of baking soda added.

Baking Soda (Bicarbonate of Soda)

Bicarbonate of soda, an alkali salt can be added alone or as a component of baking powder. The alkalinity of baking soda lowers the carmelisation point of sugar in the cookie dough or batter, causing faster and darker colouring of the crust. Soda also has a weakening effect on flour proteins. This action is more pronounced when soda is used without the counteracting food acid. Some of the alkalinity of the soda is neutralised by the natural acidity of the other ingredients in the formula. Too much soda will result in a chemical reaction between the fat in the formula and the soda. This will result in the cookie having a soapy taste.

Cream of Tartar

Cream of Tartar, a baking acid is used with soda as a leavening agent. A fast acting baking powder can be made by combining two parts of cream of tartar with one part baking soda. When cream of tartar is used without a counteracting alkali such as baking soda, it increases the acidity of the dough or batter. The higher acidity produces a whiter crumb colour and a lighter crust colour because sugar carmelises at a higher temperature in an acid media.

Baking Powder

Baking Powder is a mixed blend of food acids of which there are several types and Bicarbonate of Soda with starch added to prevent the Baking Powder from lumping during storage. During the baking process the acid ingredients and the baking soda contained in the baking powder are dissolved in the liquid forming carbon dioxide gas. None of the acid nor the the soda remain in the finished product, because they neutralise each other.

Flavoring Ingredients

Salt

Salt is used for its own flavor and to bring out the natural flavors of the other ingredients. The amount of salt must be reduced slightly when salted butter is used as the shortening ingredient in the recipe.

Cocoa

The use of cocoa requires an adjustment to the formula. Sugar is added to counteract the tendency of cocoa to toughen the dough and to sweeten and enhance the cocoa flavor. A slight adjustment in liquid may be necessary also.

Oatmeal, peanuts, spices, chocolate chips and nuts can be used to produce particular flavors and textures. They are added as prescribed by the formula.

Methods of Mixing

There are two ways of mixing dough ingredients.

One-stage Method

The One-Stage Mixing Method is preferred because its simplicity reduces the possibility of errors. Sift the dry ingredients together. Place all the ingredients, both wet and dry into the mixing bowl and mix in slow speed until they form a smooth dough or batter

Creaming Method

The main portion of the sugar (about 75 percent), shortening, salt and vanilla are creamed together. The eggs are added in three parts and mixed until smooth after each addition. Add the liquids and stir. The flour, baking powder, and remaining sugar (about 25 percent) are sifted together and added. In this method of mixing, the longer the sugar and shortening are creamed, the less spread the product will have because the sugar crystals are broken up more thoroughly.

Mixing for a long period of time after the flour has been added will develop the gluten excessively. This will result in cookies with insufficient spread. When a portion of the sugar is added in the final step with the flour, better spread will be obtained. Nut, raisins, and chocolate chips are added at the end of the mixing period and mixed only long enough to incorporate them in the dough or batter.

Cookie Recipes

Sugar cookies-Yield about 200 3-inch diameter cookies

Ingredients	Pounds	Ounces
Cake flour	4	12
Sugar, granulated	4	6
Salt	0	1
Baking powder	0	3
Shortening	1	9
Eggs	1	0
Milk non-fat dry	0	1-1/4th
Water	0	10
Vanilla	0	0-1-1/2

Ice box cookies-Yield about 250 cookies 3/4th ounce each/

Ingredients	Pounds	Ounces
Bread flour	5	0
Sugar, granulated	3	4
Salt	0	1
Shortening	3	4
Eggs	0	11
Vanilla	0	0-1-1/2

Fruit bar cookies-Yield about 200 1-1/2 inch by 3-1/2 inch bars

Ingredients	Pounds	Ounces
Bread flour	4	0
Sugar, granulated	2	0
Salt	0	1
Baking soda	0	1
Cinnamon	0	1
Ginger	0	0-1/2
Shortening	1	8
Eggs	1	8
Molasses	1	8
Raisins or Nuts *	2	10

* Note: Other fruits such as coconut, chopped dates, fruit cake mix, etc can be used also.

Demonstrations for Hand-made Cookies

Stiff-Batter Cookies

Care must be taken to keep the dough uniform during the rolling operation . Avoid the use of excessive dusting flour on the work bench. Some of the excessive dusting flour can be removed by using the bench brush. If excessive dusting flour still remains on the cookies after they are panned, they can be washed with egg wash.

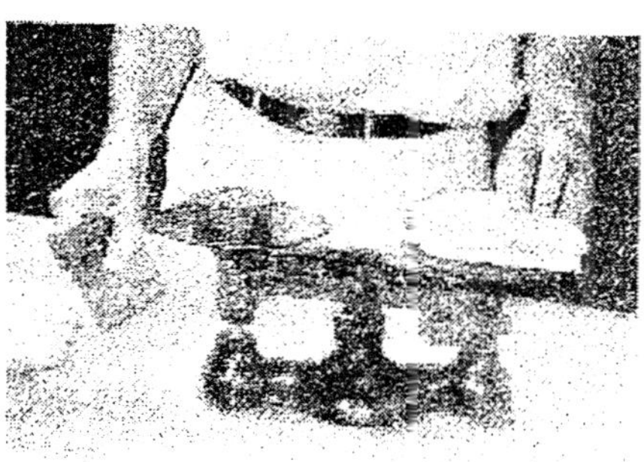

Scale the dough into one or two pound pieces depending upon the length of the work bench.

Roll the dough into a round cylinder.

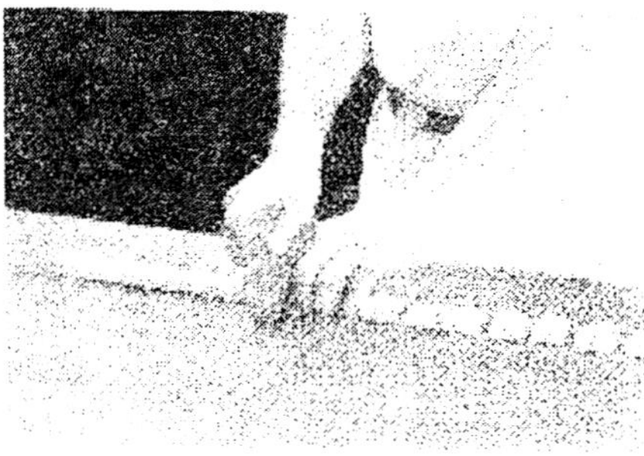

Using the dough scraper, cut the individual cookies about 1 inch long.

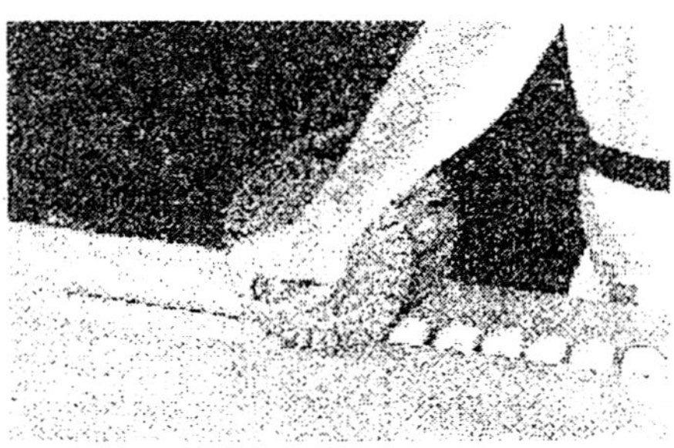

Using the squirrel cage cutter, cut individual cookies about 1 inch long. After cutting the cookies, dip each cookie into a pan of granulated sugar. Pan cookies with sugar facing up on greased pans or on pans lined with a sheet of parchment paper.

Use a can or hands to flatten cookies slightly before baking. Bake at 375 degrees F. for about 10 minutes. When cookies fall while baking, they are usually baked. Another method of determining if cookies are baked, lift cookie up with a spatula. If brown on the bottom, the cookie is baked.

Loosen cookies from pan with spatula while still warm.

After icebox cookie dough has been divided into about 1 pound pieces, the dough is rolled into round cylinders about 2 inches in diameter, wrapped in parchment or wax paper, and chilled in the refrigerator. Then they are sliced with a sharp knife and panned fairly close together on ungreased pans since icebox cookies do not spread as most other types of cookies do.

To make pinwheel design icebox cookies, roll one piece of plain dough about 8 inches wide, 12 inches long and 1/2 inch thick. Prepare one piece of dough by mixing either chocolate or cocoa into the dough. Roll the dark piece of dough about 7-1/2 wide, 12 inches long and 1/2 inch thick. Note that a food colour can be added to the smaller piece of dough if desired. Brush a very light coating of water or egg wash on the larger piece of dough. In the next step, the dark piece will be stacked on top of the plain piece.

Leave about 1/2 inch of plain dough not covered with dark dough. When the dough is rolled into a cylinder in the next step, the small portion not covered with the dark dough will be utilised to seal the cylinder. The cylinder is then rolled in parchment or wax paper and chilled in the refrigerator before slicing into individual cookies.

After the cylinder is formed, it is rolled in parchment or wax paper and chilled in the refrigerator.

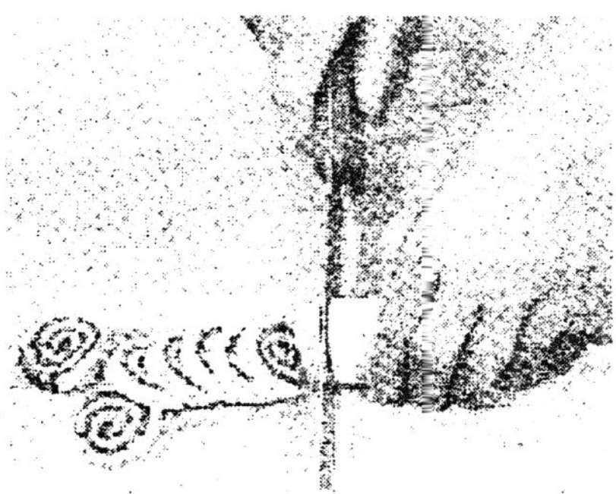

With sharp knife slice cookies 1/4th inch thick and place on ungreased pans for baking.

Roll two pieces of plain dough and two pieces of dark dough to about 3/8th inch thick. Each of the 4 pieces should be of the same weight.

Brush 3 of the 4 pieces with a light coating of water or egg wash. Then stack the pieces on top of each other alternating the colours. Place the piece that has not been washed on last.

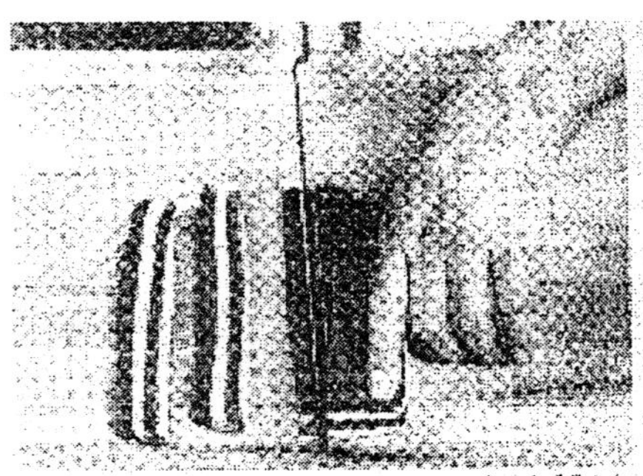

Using a sharp knife cut the four layers into strips approximately 1/2 inch thick. Then brush cut surface with a light coating of water or egg wash.

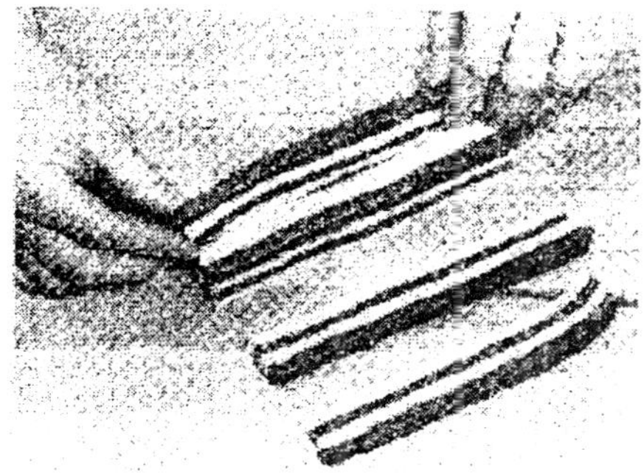

Stack the pieces on top of each other alternating the colours. Then carefully wrap the dough in parchment or wax paper and chill in the refrigerator.

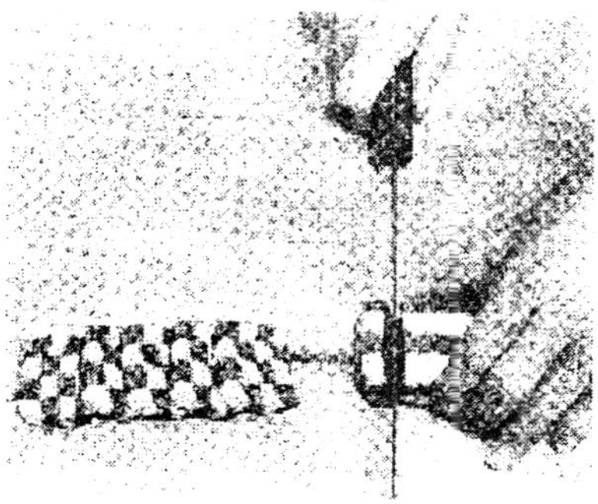

Using a sharp knife slice cookies 1/4th inch thick. Then place on ungreased pans for baking

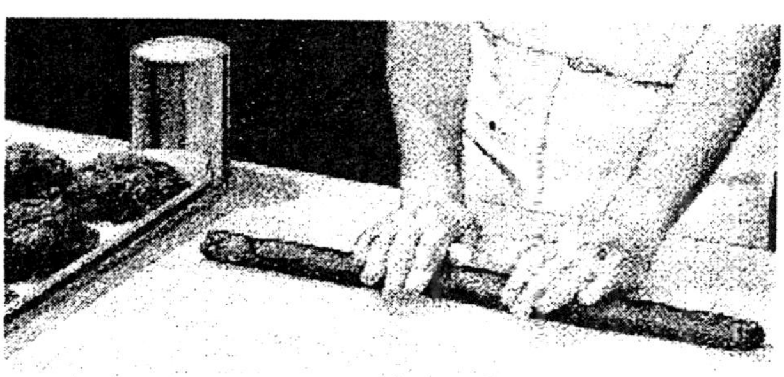

After the Fruit Bar Cookie dough has been scaled into 1 pound pieces, roll the dough into a cylinder about the length of the pan in which they will be baked.

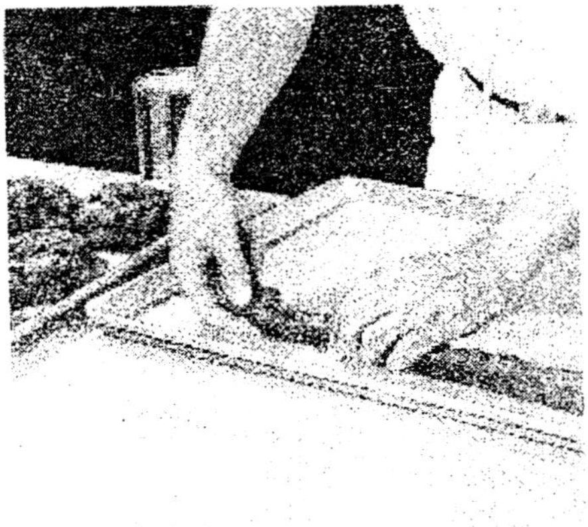

On greased sheet pan or pan lined with parchment paper, transfer the Fruit Bar Cookie rolled dough from the work bench to the pan.

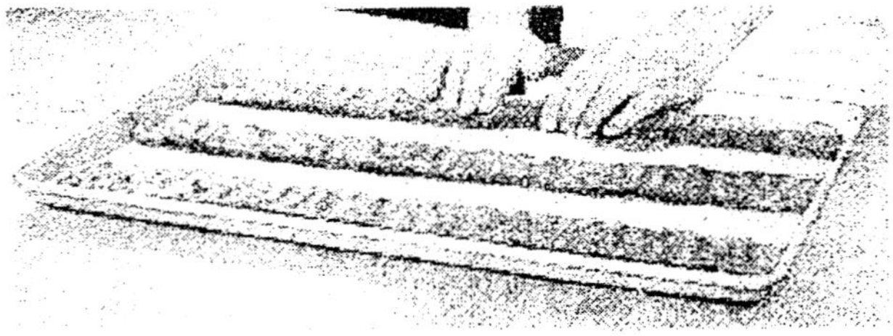

Using hands, flatten the cylinders slightly. At this point the cylinders can be washed with egg wash to produce a shine. Bake at 375 degrees F. for about 10 to 12 minutes. They can be iced with a Sweet Roll Icing while still hot or they can be left uniced.

The Fruit Bars should be cut with a cough scraper or other object while still hot.

Soft Batter Cookies

Vanila Wafers, Oatmeal cookies, etc. are classed as Soft Batter Cookies and may be dropped by hand or squeezed from a pastry tube, paper cone or cropped by machine.

Vanilla Wafers-Yield about 200 3/4 h ounce cookies

Ingredients	Pounds	Ounces
Bread flour	1	8
Cake flour	1	4
Sugar, granulated	2	4
Salt	0	1
Baking Powder	0	1
Shortening	2	4
Eggs	1	8
Vanilla	0	1

Oatmeal cookies -Yield about 200 3-inch diameter cookies

Ingredients	Pounds	Ounces
Bread flour	3	8
Sugar, granulated	2	8
Sugar, brown	1	0
Salt	0	1
Baking soda	0	1/2
Cinnamon	0	8
Shortening	1	8
Eggs	0	8

Water	1	0
Vanilla	0	1
Raisins, chopped *	1	5
Oats, rolled	1	8

Brownies

Brownies come under the classification of Soft Batter Cookies. The following recipe yields 2 sheet pans or about 200 brownies 2 inches square.

Ingredients	Pounds	Ounces
Shortening	2	11
Sugar, granulated	5	4

Mixing procedure: Cream shortening and sugar in medium speed until light (about 5 minutes

Ingredients	Pounds	Ounces
Eggs	2	11
Vanilla	0	2

Mixing procedure: Add eggs and vanilla in slow speed in about 3 parts until all ingredients are well mixed.

Ingredients	Pounds	Ounces
Gread flour	1	2
Cake flour	1	0
Baking powder	0	½
Salt	0	½
Cocoa	0	13

Mixing procedure: Sift flour, baking powder, salt, and cocoa together. Add the sifted ingredients to above and mix in slow speed until a smooth batter results.

Ingredients	Pounds	Ounces
Nuts, chopped	2	14

Mixing procedure: Add nuts to above mixture and mix in slow speed only long enough to incorporate. Do not overmix.

Baking the bownies: Divide into two equal portions and spread evenly in two greased sheet pans or preferably grease bottom of sheet pan and line the pan with parchment paper. Bake at 325 to 350 degrees F. oven for about 30 minutes. Do not overbake. Cool and cut into 2-inch squares.

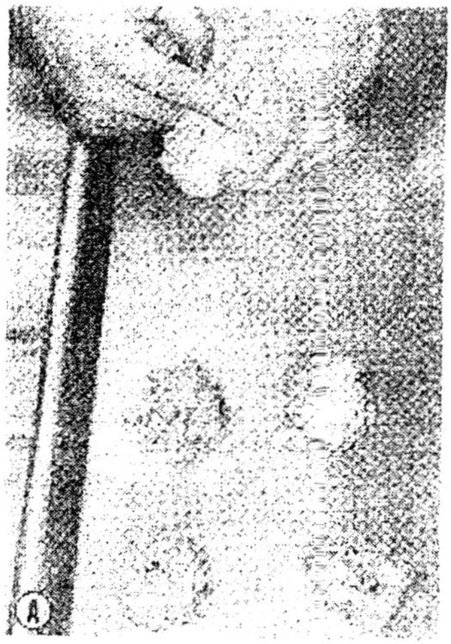

Dropping the oatmeal cookies on greased sheet pan by hand.

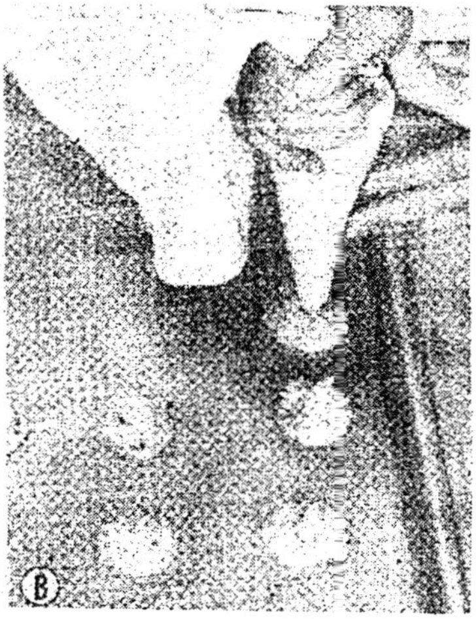

Dropping the oatmeal cookies on greased sheet pan using pastry bag.

Shown are a variety of baked icebox cookies.

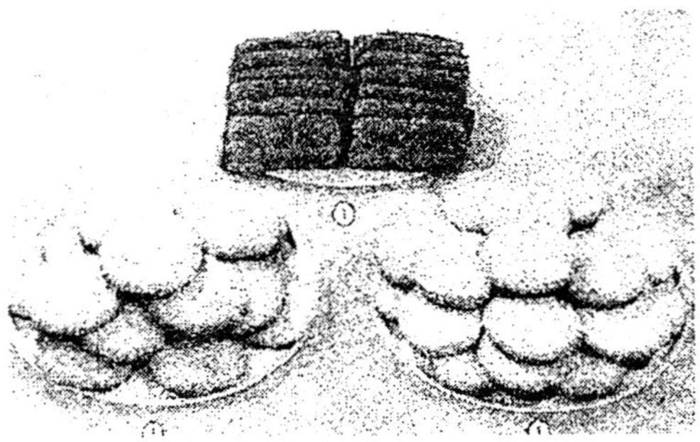

Shown are a variety of baked cookies and fruit bars.

Hand operated cookie machine

The hand operated cookie machine shown above is for use in small retail bakeries. Capable of producing up to 300 cookies per minute. Cookie weight can range between 1/4th ounce to 5 ounces. Over 60 different shapes of cookies can be made.

COOKIE MAKING TECHNIQUES

Cookies are fun, easy and quick to make. All you need are the basic ingredients and a few minutes of your time. Aone way to satisfy your sweet tooth's craving.

Golden rules applied to all types of cookies:

— Have all ingredients measured out and at room temperature; this allows for ready handling of dough

– Pre-heat your oven according to recipe specifications

Adding Flour

Avoid adding too much flour in your dough.

— Too much flour in the dough will prevent the cookies from spreading well

— Too much flour will also make hard cookies

A common mistake when mixing the dough is usually adding more flour, because the dough seems 'sticky' to the hands.

— Always add the least amount of flour as indicated in the recipe and let the dough stand in the mixing bowl, for about ten minutes, before adding more flour to it. This allows the flour to fully absorb the liquids

— If the dough still feels sticky after the stand time, then add small increments of flour and mix well after each addition

— After each flour addition, make one cookie and bake it: if you see that the cookie needs more flour, then add more; this might take two or three tries, though, it will save you from the disappointment of having spent all those ingredients, time and energy on making undesirable cookies

— When you have finally chosen the proper dough consistency, prepare and bake the rest of the dough accordingly

Mixing Ingredients

Cream the fat with the sugars vigorously well.

— The longer you mix, the more the sugar will dissolve

— This might take a few minutes, but will make for a more consistent dough

Add some butter flavour extract along with the other flavourings.

— This boosts the butter flavour

Avoid overmixing when incorporating eggs and flour into the fat and sugar mixture.

— These steps usually create tough textured cookies; they toughen the egg's protein and they activate the flour's gluten

Making Flat Cookies

Omit some of the egg whites (they make cookies puff up).

— Avoid adding to much of a leavening agent (they make cookies rise high during the baking period, they will spread too much, and they will taste bitter)

— Too much baking powder will make your cookies spread out, rise too much and collapse

— Too much baking soda will make your cookies rise too high too quickly, without spreading and they too will collapse

— Try combining different types of flour .

— Different flours will give you different results

Cookie Baking Techniques

Cookies are fun, easy and quick to make. All you need are the basic ingredients and a few minutes of your time. A-one way to satisfy your sweet tooth's craving. Golden rules applied to all types of cookies:

— Have all ingredients measured out and at room temperature; this allows for ready handling of dough

— Pre-heat your oven according to recipe specifications

Preventing Burnt Bottoms

A good indicator of a well-baked cookie is its colour (golden brown).

— Both top and bottom of cookie should have the same colour

In order to bake even-coloured cookies, when baking two sheets at a time, one on each level, only line the oven's lower baking rack with a few sheets of foil paper (dull side facing upwards).

— This prevents the cookie's bottom from over-baking and turning dark in colour

Chewy Cookies

Bake cookies for about two thirds of suggested baking time; usually about 6 - 7 minutes, depending on their size.

— Bake one cookie and time it to see how it has baked

— Calculate how much time is necessary and continue baking the rest of the cookies accordingly

As they are baking, look out for these signs:

— When cookies have stopped spreading and are beginning to rise; reaching their ultimate peak

— When tops begin to dry out and lose their wet look (glossiness)

— When they become slightly dark in colour, remove them from the oven and let them cool in their pans for three minutes; they will deflate and become flat

Dry Cookies

Bake cookies for entire suggested baking time; usually about 10 - 15 minutes, depending on their size.

Preservation of Cookies

Once cookies are baked, allow them to 'stand' (cool down) in their pan(s), on a cooling rack.

— Wait three to five minutes before transferring cookies from their pan(s) to their cooling racks

— Let them cool in their pan(s) to allow them to harden, as well as maintain their shape

Do not place a hot cookie sheet onto any solid surface other than a cooling rack.

— When a hot pan comes into contact with a cold surface, or cold air mass, condensation might develop and cookies might absorb the water droplets, causing their bottoms to become slightly soggy

Storing Cookies

When cookies have reached room temperature, store them in airtight containers. Avoid using cookie jars for cookie storage.

— Air, whether too dry or too humid, can pass through the cookie jar's lip and lid, and thus affect the cookies' texture

— Chewy cookies lose their 'bite' if exposed to dry air, while dry cookies can become pliable if exposed to humid air.

7

Baking of Hot Rolls and Buns

Soft Roll Dough Recipe

A wide variety of soft rolls can be made from the following basic soft roll dough recipe:

Ingredients	
Sugar, granulated	10 oz
Salt	2.5 oz
Shortening or margarine	12 oz
Water	4 lb

Procedure: Place all ingredients into mixing bowl and using the dough hook mix until sugar and salt are dissolved.

Ingredients	
Flour, bread	8 lb
Milk, dry nonfat	6 oz
Yeast, active dry	2.5 oz
Water, 110 degrees F.	10 oz

Procedure: Suspend yeast in water, stir and let set about 5 minutes. Add flour and dry milk to above. Then stir yeast

& water mixture, pour on top of flour. Mix until gluten is fully developed. Mixed dough should come out of mixer at 78 to 82 degrees F. Ferment in warm place for about 1-1 1/2 hours.

Buns

1 cup scalded milk	1/2 teaspoon salt
1/3 cup butter	1/2 cup raisins stoned and cut in quarters
1/3 cup sugar	
1 yeast cake dissolved in	1 teaspoon extract lemon
1/4 cup lukewarm water	Flour, cinnamon

Add one-half sugar and salt to milk; when lukewarm, add dissolved yeast cake and one and one-half cups flour; cover, and let rise until light; add butter, remaining sugar, raisins, lemon, and flour to make a dough; let rise, shape like biscuits, let rise again, and bake. If wanted glazed, brush over with beaten egg before baking.

Hot Cross Buns

1 cup scalded milk	3/4 teaspoon cinnamon
1/4 cup sugar	3 cups flour
2 tablespoons butter	1 egg
1/2 teaspoon salt	1/4 cup raisins stoned and quartered, or
1/2 yeast cake dissolved in	
1/4 cup lukewarm water	1/4 cup currants

Add butter, sugar, and salt to milk; when lukewarm, add dissolved yeast cake, cinnamon, flour, and egg (well beaten);

when thoroughly mixed, add raisins, cover, and let rise over night. In morning, shape in forms of large biscuits, place in pan one inch apart, let rise, brush over with beaten egg, and bake for twenty minutes; cool, and with ornamental frosting make a cross on top of each bun.

Hard Roll Dough Recipe

A wide variety of hard rolls can be made from the following hard roll dough recipe:

Ingredients	
Sugar, granulated	4 oz
Salt	4 oz
Shortening or margarine	4 oz
Water	5 lb 8 oz
Egg Whites	8 oz

Procedure: Place all ingredients into mixing bowl and using the dough hook mix until sugar and salt are dissolved.

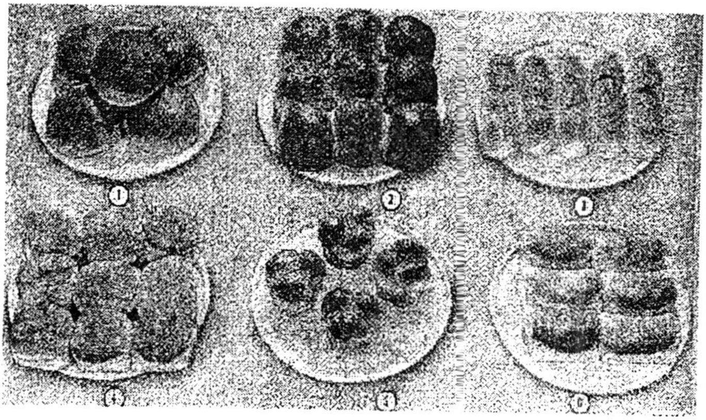

A variety of baked rolls

Ingredients	
Flour, bread	12 lb
Yeast, active dry	2 oz
Water, 110 degrees F.	1 lb

Procedure: Suspend yeast in water, stir and let set about 5 minutes. Add flour to above. Then stir yeast & water mixture, pour on top of flour. Mix until gluten is fully developed. Mixed dough should come out of mixer at 78 to 82 degrees F. Ferment in warm place for about 1½ hours. Make-up into hard rolls. Proof until double in size. Bake with lots of low pressure steam if available for a shiny and tough crust.

Brown and Serve Dinner Rolls

Brown and serve dinner rolls provide an answer to maximum table freshness, for example, freshness that is achieved by the final browning or baking of a partially baked bakery product just prior to serving time. General mills, inc developed the concept in the late 1940's. The major principle in producing brown and serve rolls involve the use of a modified formula.

The objective is to bake the rolls to a point of rigidity and full volume without browning the crust. This means that the roll will hold it's shape after being partially baked. It is fully formed, or set, with the dersired body for ease in handling and transportation. It requires controlled baking at reduced oven temperature as compared to normal baking temperature for a relatively short period of time. Following are the changes that must be made:

1. The oven temperature must be reduced to 275 to 300 degrees F. To prevent excessive ovenspring. At those temperatures, the roll should be baked as long

as possible without colour being formed. This will normally take 10 to 15 minutes.

2. Conditioning the dough to cut down on the ovenspring. This is accomplished by mixing the dough slightly stiffer than normal to produce the desired rigidity when the roll comes out of the oven. The dough should also come out of the mixer at a temperature of 90 to 95 degrees F.

3. Other recommended changes.
   - Use a richer formula. Increase the shortening and eggs. This will produce a roll with better flavour, taste, and eating quality.
   - Proof the roll at a temperature of 105 to 110 degrees F.
   - When the roll comes out of the oven, the internal temperature should not exceed 180 degree F. To prevent excessive shrinkage while cooling.

Suggested Formula

Ingredients	
Sugar, granulated	1 lb 5 oz
Salt	$4^1/_2$ oz
Shortening or margarine	1 lb 10 oz
Eggs	13 oz
Water, variable	7 lb

Procedure: place all ingredients into mixing bowl and using the dough hook mix until sugar and salt are dissolved.

Ingredients	
Flour, bread	14 lb
Milk, dry nonfat	12 oz

Yeast, active dry	$2^1/_4$ oz
Water, 110 degrees F.	12 oz

Procedure: Suspend yeast in water, stir and let set about 5 minutes. Add flour and dry milk to above. Then stir yeast & water mixture, pour on top of flour. Mix until gluten is fully developed. Mixed dough should come out of mixer at 90 to 95 degrees F. Ferment in warm place for about 1 to $1^1/_2$ hours. (One full rise). Make-up immediately into a variety of rolls, some of which can be panned in Cup Cake Pans.

Proofing: Proof in 105 -110 Degrees F. proof box to about 3/4th proof to prevent excess ovenspring.

Baking: Bake at temperature of 275-300 Degrees F. Bake as long as possible without colouring the roll. Rolls should be white in colour, fully formed, and have good body for ease in handling. This will normally take 10 to 15 minutes. After cooling, rolls can be packaged and frozen until ready to be fully baked at 350 to 375 Degrees F. until golden brown to be served hot right out of the oven.

8

Preparation of Icings

To make a cake more appealing to the eye and palate it is covered with an icing or topping. Icing is not only a sweet covering for the cake, it is the decision maker because it is the primary factor that creates the buying impulse in the customer's mind. Icings are a combination of ingredients that when mixed together for the proper length of time at the proper temperature and at the proper speed will result in an eye appealing and delicious icing or frosting.

Icings are classified into three categories, cooked, uncooked and a combination of the two. They are further classified as follows:

1. *Flat Icings*: Flat Icings do not have any or very little air beaten into it. They are usually composed of 80-85 percent powdered sugar, 15-20 percent water and optional ingredients such as glucose syrup, shortening and flavourings.
2. *Creme Icings*: Creme Icings can be whipped up into a fluffy consistency. They generally contain a large amount of shortening, butter or margarine. There are special emulsified type shortenings available which are capable of holding liquids into an emulsion so the water will not separate out. They are also capable of

retaining air beaten into the icing. Creme Icings contain 70-80 percent powdered sugar, 20-40 percent fat, and 5-10 percent liquid. Optional Ingredients such starches, gums, and driers are used to aid in moisture retention which helps to maintain a dry surface so the cake can be overwrapped by Wholesale Bakeries.

3. *Fudges and Fondants*: Fudge Type Icings and Fondants are icings similiar in appearance to each other, with a fine dense texture and a candy-like eating quality. They are used to enrobe wedding cakes, and small cakes such as petit fours that have decorations applied to them. The reduction in the size of the sugar crystals during the cooking process account for the smooth, glossy appearance of these icings.
4. *Marshmallows*: Marshmallow Icings are classed as foam type icings because of the large amount of air beaten into the mix. Gelatin and/or egg whites, and a syrup solution are used to produce an extremely light and highly aerated icing. Marshmallow Icings are not adaptable for cakes that are to be wrapped unless they are topped with coconut or a crunch. Marshmallow Icings consist of 40-45 percent powdered sugar, 25-30 percent moisture, 1-2 percent gelatin, 15 percent glucose and 15 percent Invert Sugar. Flavourings can be added to the icing.
5. *Combinations*: A Combination Icing is an icing which has been prepared by combining two or more of the above icings. A Water Icing can be combined with a Creme Icing to obtain a Flat Creme Icing, or a Marshmallow Icing can be blended with a Creme Icing to obtain a Marshmallow Creme Icing.

FLAVOURINGS

Flavoring is a complex sensation that is the source of much of the delight which a person derives from foods. There are natural flavours in all ingredients and they should only be enhansed by other complimentry flavours. Flavours are generally classed as compatible or complimentary.

1. *Compatible Flavours*: Compatible flavours are cherry-vanilla, butterscotch-rum, maple-rum, chocolate-vanilla, orange-pinepple, maple-walnut, butter-vanilla, butter-chocolate, and almond-vanilla.
2. *Complimentary Flavours*: Coconut-vanilla, coconut-maple, orange-lemon, chocolate-mint, banana-orange, cherry-almond, cherry-rum, chocolate-rum, coffee-rum, butterscotch-lemon, butterscotch-coffee and rum-butter.

Do' and Don'ts for Handling Icings

1. Cooked Icings must be applied while warm (100-125 degrees F is best).
2. Icing to be thinned must be thinned with a glucose solution or a simple syrup-never use plain water.
3. For wrapped cakes, stabalizers and driers should be used.
4. While mixing the icing, do not add liquid faster than the shortening can readily form an emulsion.
5. Do not overmix the icing. Breaking down of the icing begins when the icing is subjected to extra minutes beyond it's normal tolerance.
6. Use flavoring sparingly. Don't mask the true flavour of the cake.

Icing Stabalizers A

The majority of Bakers purchase complete Icing Stabilizers because of the difficulty of working with individual gums. Icing Stabalizers can be purchased in plastic or powder forms. They are also available in the boiling or non-boiling types. Boiling types are generally used in icings that involve boiling the sugar and liquids. Boiled Icings are usually more stable, because the stabalizer has been fully activated, and because less water is needed to produce the proper consistency and viscosity required for machine application.

Recipes for Icing

White Buttercreme Icing

Ingredients	Pounds	Ounces
Sugar, powdered	10	0
Salt	0	3/4th

Procedure: Place powdered sugar and salt into the mixing bowl.

Ingredients	Pounds	Ounces
Boiling water	1	8
Icing Stabalizer	0	12

Procedure: Dissolve icing stabalizer in boiling water. Add to mixing bowl.

Ingredients	Pounds	Ounces
Shortening, emulsified	2	8
Vanilla	0	1

Procedure: Add shortening and vanilla to the mixing bowl and creme until light and fluffy.

Hot Weather White Creme Icing

Ingredients	Pounds	Ounces
Shortening, emulsified	4	0
Flour, Cake	1	8

Procedure: Cream shortening and flour at slow speed.

Ingredients	Pounds	Ounces
Hot water	2	8
Glucose	1	8
Salt	0	2-1/2
Sugar, powdered	20	0

Procedure: Bring the water, glucose and salt to a boil and add gradually to the creamed mixture above while mixing in the powdered sugar. Do not add too rapidly or you will melt the shortening. Cream moderately.

Ingredients	Pounds	Ounces
Water	0	9
Sugar, granulated	0	12
Gelatin (variable)	0	2-1/2

Procedure: Dissolve gelatin and sugar in the water and heat to 140 degrees F. to clarify. Stir into above creamed mixture. To aid in easy spreading and quick drying, use at 90 -100 degrees F. If it is necessary to thin the icing, use simple syrup (2 parts granulated sugar dissolved in 1 part hot water). Add flavoring as desired.

Butterscotch Creme Icing

Ingredients	Pounds	Ounces
Maragarine or butter	3	0
Sugar, brown	6	0
Glucose	2	0
Milk, nonfat dry	1	0

Procedure: Cook to 246-250 degrees F. to develop flavour.

Ingredients	Pounds	Ounces
Simple Syrup	6	0

Procedure: To make simple syrup mix 4 pounds of granulated sugar to 2 pounds of hot water. Add the simple syrup to the above cooked mixture. Allow the cooked mixture to cool.

Ingredients	Pounds	Ounces
Shortening, emulsified	2	0
Milk, dry nonfat	1	0

Procedure: Add the emulsified shortening and the dry milk to the cooled mixture and cream until smooth.

Ingredients	Pounds	Ounces
Sugar, powdered	10	0
Salt	0	1

Procedure: Add the powdered sugar and salt to the above creamed mixture and cream until smooth and light.

Boiled icing

Ingredients	Pounds	Ounces
Sugar, granulated	8	0
Corn Syrup	3	0
Invert sugar	2	0
Salt	0	½
Water	2	8

Procedure: Cook to 242 degrees F.

Ingredients	Pounds	Ounces
Egg Whites	4	0

Sugar, powdered	2	0
Cream of tartar	0	½

Procedure: Whip the egg whites to a wet peak (a wet peak is when you put your finger into the whipped whites and it forms a peak that feels wet). While continuing to beat in medium speed, add the hot syrup in a thin stream and continue beating until the desired consistency has been reached.

Marshmallow Icing

Ingredients	Pounds	Ounces
Gelatin	0	7
Water, cold	0	12

Procedure: Soak gelatin in cold water then heat to140 degrees F.

Ingredients	Pounds	Ounces
Granulated sugar	8	0TR>
Water, hot	2	8TR>
Invert sugar	5	8
Corn Syrup	3	0

Procedure: Dissolve the sugar in hot water, then stir in the invert sugar and the corn syrup until the sugar is dissolved. Add the gelatin solution and beat to the fluffy consistency of marshmallow. Add flavour and food colouring as desired.

Marshmallow should be applied while warm. If some is left over, refrigerate until needed. Before it can be applied, the marshmallow must be warmed to about 100 degrees F.

French Creme Icing

Ingredients	Pounds	Ounces
Sugar, granulated	10	0
Water	3	0

Procedure: Boil to 232 degrees F.

Ingredients	Pounds	Ounces
Egg whites	5	0

Procedure: Beat egg whites to soft wet peak with wire beater.

Ingredients	Pounds	Ounces
Sugar, granulated	2	8

Procedure: Add sugar to beaten egg whites gradually in a slow stream until stiff. Then slowly add hot syrup while beating in high speed. Change to slow speed and continue beating until cold.

Ingredients	Pounds	Ounces
Margarine or butter	6	0
Shortening, emulsified	6	0
Milk, nonfat dry	1	0

Procedure: Cream the margarine or butter together in a separate bowl then add to above and mix until smooth and light.

Fondant Creme Icing

Ingredients	Pounds	Ounces
Fondant	10	0

Shortening, emulsified	4	0
Margarine	1	0
Salt	0	1

Procedure: Blend at slow speed, then cream for 2 minutes at medium speed.

Ingredients	Pounds	Ounces
Liquid milk	1	0

Procedure: The liquid milk should be 100 -120 degrees F. Add the hot milk slowly to above while creaming at slow speed until very light and smooth.

Chocolate Fudge Icing

Ingredients	Pounds	Ounces
Sugar, granulated	4	8
Water	3	12

Procedure: Heat water to 180 degrees F.

Ingredients	Pounds	Ounces
Shortening, emulsified	1	10
Salad oil	0	4
Glucose	1	0
Salt	0	1-14th
Vanilla	0	2-1/2
Cocoa (Dutch)	1	5
Milk, nonfat dry	0	8
Sugar, powdered	10	0

Procedure: Place all the ingredients of this step into the mixing bowl. Add 1/3rd of the sugar and hot water solution. Mix until smooth. Continue slow addition of sugar solution while mixing in medium speed until good spreading consistency is reached. Cocoa can be replaced with 2 pounds of melted chocolate liquor after the first sugar solution has been added. For variation, 1 pound 8 ounces of toasted pecan pieces (coarsely chopped) can be added to a portion of the icing.

Dipp Icing

Ingredients	Pounds	Ounces
Water, hot	1	8
Gelatin	0	3/4th

Procedure: Mix the gelatin and hot water together in the mixing bowl and let set for a few minutes.

Ingredients	Pounds	Ounces
Sugar, powdered	10	0
Corn syrup	1	0
Shortening, regular	0	8
Vanilla	0	1

Procedure: Add all of the above ingredients to the mixing bowl and mix to a smooth flowing icing. NOTE: The dipp icing should be warmed to about 100 degree F. when used.

White Icing

The icing for freezing is a flat boiling-type icing for hot application at 115-130 degrees F. The stabalizer used is a starch-and gum-based especially selected for its freezing stability. If this icing is to be thinned, use simple syrup, not water.

Ingredients	Pounds	Ounces
Sugar, granulated	25	0
Water	24	0
Stabalizer	1	0

Procedure: Bring water, sugar and stabalizer to a boil.

Ingredients	Pounds	Ounces
Shortening, regular	4	1

Procedure: Add shortening to hot syrup and melt.

Ingredients	Pounds	Ounces
Sugar, powdered	100	0
Vanilla	0	8

Procedure: Place powdered sugar into mixing bowl. Add the hot syrup and vanilla to the bowl and mix until smooth.

9

Measuring Ingredients

Baking is a science, and when you mix together ingredients, you're creating chemistry, albeit edible chemistry, so being precise is important. There is balance between flour, leaveners, fats, and liquids. Proper measuring is critical to baking. Extra salt or baking soda can ruin otherwise perfect cookies. Too much flour makes muffins taste dry and flavourless. No beginning cook should be nonchalant about measuring.

Proper measuring of ingredients is important to successful cooking and baking. Some foods can be greatly affect by too much or too little of certain ingredients, such as salt, baking soda, baking powder and hot or spicy ingredients, such as cayenne pepper. The following information will assist in ensuring that your ingredients are measured proper.

There are several different types of measuring utensils used for proper measuring of ingredients. The most common types of measuring equipment are follows:

Measuring Spoons

Measuring spoons come in sets of four or six, ranging from 1/4 teaspoon to 1 tablespoon. You can use the same measuring tools for both liquids and dry ingredients. For liquids, fill the spoon until it's full. For dry ingredients, pour or scoop into the spoon until it's full, levelling off the spoon with the straight edge of a spatula or knife. Never measure over the bowl of ingredients you're using for the recipe. If you overpour or level extra into the bowl, your measurements will not be accurate.

A set of 4 will include basic measures of ¼ teaspoon, ½ teaspoon, 1 teaspoon, and 1 tablespoon. Sets are also available that include 1/8 teaspoon, ¾ teaspoon and ½ tablespoon.

There is also an adjustable type measuring spoon produced that does not resemble the typical measuring spoons. It is an elongated slotted container that has a slide bar with a "stopper" to control the specific amount of measure desired.

Measuring Cups

Measuring cups are essential for every kitchen. You won't find many recipes that don't require measurements of some kind. Measuring cups come in two basic types:

— *Graded*: Graded cups range in sizes from 1/4 cup to 1 cup and can range from 4 to 6 cups in a set. Use graded cups to measure dry ingredients and solid fats, such as shortening.

— *Glass*: Glass cups are available in a wide range of sizes, the most common being 1 cup, 2 cups, and 4 cups. Use these cups for measuring liquids.

Dry Measuring Cups

Plastic or metal individual cups of various sizes used for single measures. There are also adjustable measuring cups available. The cup has a slide bar that can be adjusted so the cup can measure different amounts. Graduated and adjustable measuring cups are used to measure dry ingredients, such as flour, sugar, oats, rice and solid ingredients, such as shortening and peanut butter.

The single measure cups are generally found nested in a graduated set of ¼ cup, 1/3 cup, ½ cup, and 1 cup measures. You may also find some sets that will include 1/8 cup, 2/3 cup, and/or ¾ cup. They are used to measure dry ingredients, such as flour, sugar, oats, rice and solid ingredients, such as shortening and peanut butter.

Single measure cups

There are also adjustable measuring cups available. One type of adjustable cup has a slide bar that can be adjusted so the cup can measure different amounts. Adjustable measuring cups are used to measure dry and solid ingredients.

Adjustable measuring cups

The cylinder type measuring cup shown here is another adjustable cup that is handy for measuring solids, such as shortening and peanut butter. The tube is adjusted to the appropriate location for the amount desired and then filled with the ingredient. The ingredient is then extracted easily by pushing the tube to force the ingredients out of the cup.

Cylinder type measuring cup

Liquid measuring cups: Glass or clear plastic containers with a pour spout and handle. They are generally available in 1 cup, 2 cup, 4 cup and 8 cup sizes, which have graduated measures on the side.

Liquid Measuring Cups

The full measure on the liquid measuring cup is marked down from the rim of the cup to help prevent spilling the cup's content when it is at its full measure. The spout on the cup assists in pouring the content when adding to other ingredients.

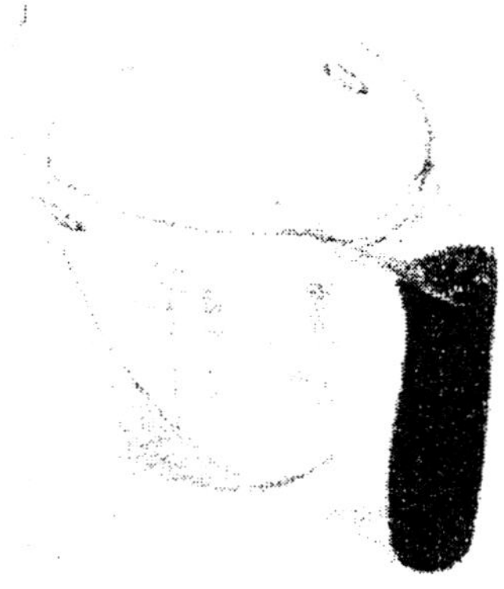

Angled measuring cups

Angled measuring cups are also available, which are designed to provide easily viewable levels so that measured amounts can be readily viewed looking either down into the cup or by viewing the measured levels at the side of the cup.

Scales

Balance or spring scales are used to measure the weight of ingredients. Balance scales will be more accurate than spring scales.

Using a scale will result in a more accurate measurement of the ingredients than by measuring it by volume.

Portioning Scoops

Scoops, available in mechanical and non-mechanical types, that are used for measuring, portioning, and forming. The measuring scoops are a popular utensil used by professional chefs.

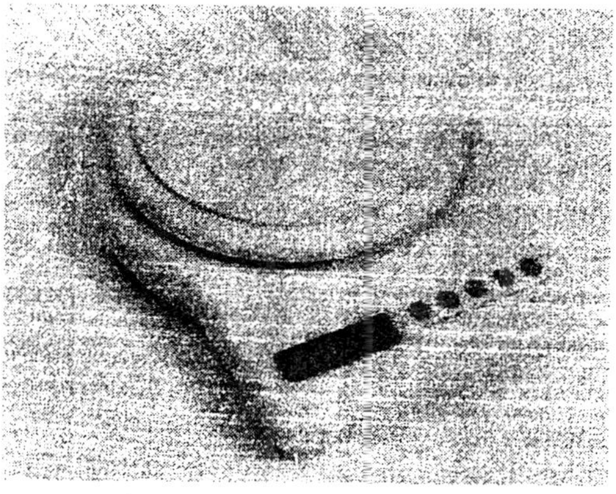

Portioning Scoops

The scoops are handy for measuring even portions when making foods such as cookies, muffins and meatballs. A #16 may be used for a medium size cookie where as a #30 or #24 would be used for a small cookie. A #16 may be used for making meatballs and a #6 for portioning meat for a patty. Using a scoop for measuring is not necessary, but if you have a size that works for a food that you make frequently, it can be a handy time saver.

Mixing Bowls: Plastic, glass, or metal containers of varying shapes into which ingredients can be placed, measured, and mixed.

There are glass and plastic measuring cups also available that are large enough to be used as mixing bowls. They range from an 8 to 12 cup capacity and are similar to liquid measuring cups in that they have a spout, handle, and measure markings down the side. They work well for large jobs and can be used for mixing and pouring batters, such as pancake or waffle batter.

Measuring Dry and Solid Ingredients

To measure flour, sugar, breadcrumbs, and other dry ingredients (with the exception of brown sugar in many cases), spoon the ingredients lightly into the measuring cup. Do not shake the cup to make level! Take the straight edge of a knife (not the cutting edge) and level off the ingredient. Levelling it off gives you one level cup. If the recipe calls for a heaping cup, do not level off the cup. Instead, leave a small mounded top of ingredients.

Sometimes ingredients, such as brown sugar, shredded cheeses, coconut, or herbs, are called for as lightly or firmly packed. Why pack? Generally, these ingredients are bulkier and can form big air pockets if you use the traditional spoon-and-level method of measuring. If you apply light or slightly firm pressure to the ingredients, you eliminate some of the air pockets and get a more accurate measurement. Never push the ingredients in so much that you actually crush them or pack them in so tightly that you have difficulty getting them out the of cup measure.

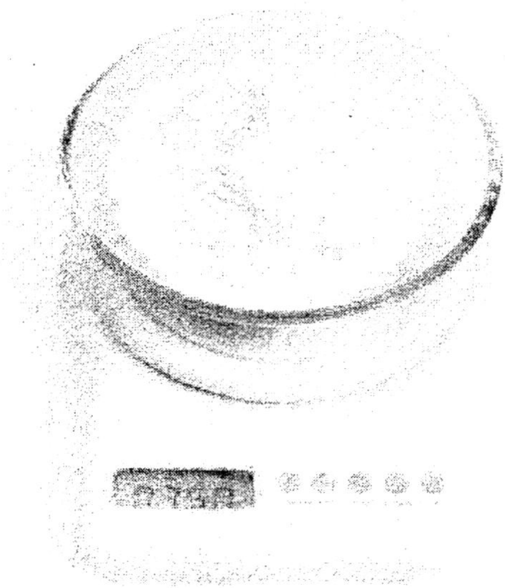

If you do so, you will overmeasure, adding too much of the ingredient. A good visual cue that you have lightly packed something is that after you pour it out of the measuring cup, it will lose the shape of the cup it was in. If it's firmly packed, it will slightly retain the shape of the measuring cup after it's dumped out into the bowl, but it will be easy to stir apart.

Measuring Chopped Nuts

To measure chopped nuts, shredded cheese, fresh herbs, and coconut, spoon the ingredients into the measuring cup and pack down lightly.

MEASURING FATS AND OTHER SOLIDS

To measure shortening, spoon the ingredients into a cup and pack down firmly with a spoon or rubber spatula to eliminate any air holes. Bakers, these days, don't often have to measure fats because butter and margarine come in conveniently measured sticks. One stick equals 8 tablespoons or 1/2 cup. Two sticks equal 1 cup. You still have to measure solid shortening, but now they make shortening sticks, so even that task has been greatly simplified.

Methods for Dry and solid Measurment

Ingredients are measured to the rim of the graduated measuring cup or spoon. Shown below are the methods that should be used for some of the common ingredients measured using graduated measuring cups and spoons.

Flour, sugar, baking soda, baking powder, and cocoa

Add enough dry ingredients so that it is heaping over the top of the measuring cup by lightly spooning the ingredients into the cup.

It may be beneficial to stir dry ingredients, such as flour, powdered sugar, baking soda and baking powder to aerate the ingredient and remove any lumps that have developed. Do not shake, tap, or pack the ingredient into the cup. If measuring with spoons, pour or scoop the ingredients into the spoon.

Using the back edge of a knife or another straight edged object, level the ingredient by running the straight edge along the rim of the cup or spoon to push the excess ingredient off.

Oats, cornmeal, breadcrumbs, and cereal

Add these types of dried ingredients to a measuring cup by pouring it into the cup until it is over filled and then level the ingredients using a straight edge, such as the back edge of a knife.

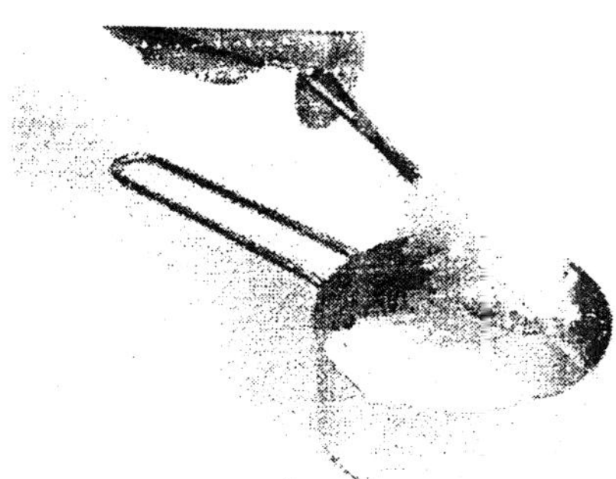

Shredded cheese, rice, coconut, chopped nuts, baking chips and other bulky dry ingredients

Bulky dry ingredients, such as shredded cheese, chopped nuts, and coconut, should be spooned into the measuring cup.

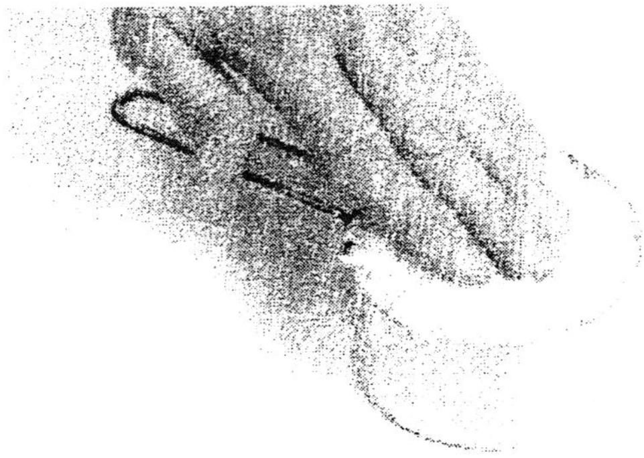

When the cup is full enough, pat the ingredients lightly and use your fingers to level the contents. Do not pack the ingredients down.

Brown sugar and shortening

Brown sugar and shortening (and other solid fats) should be spooned into the measuring device and packed firmly to eliminate any air pockets. After it is firmly packed it can be leveled with a straight edge, such as the back edge of a knife.

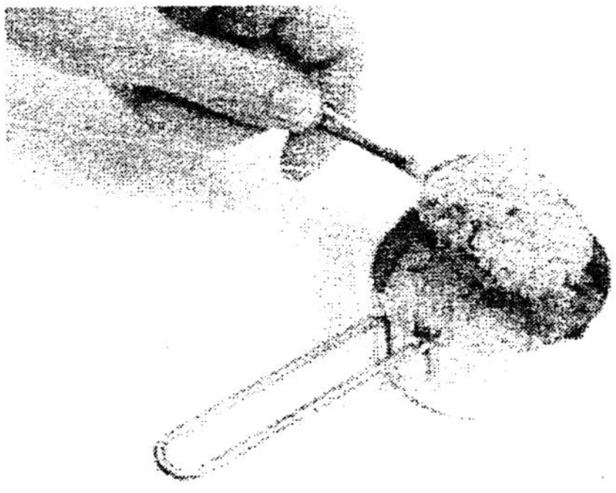

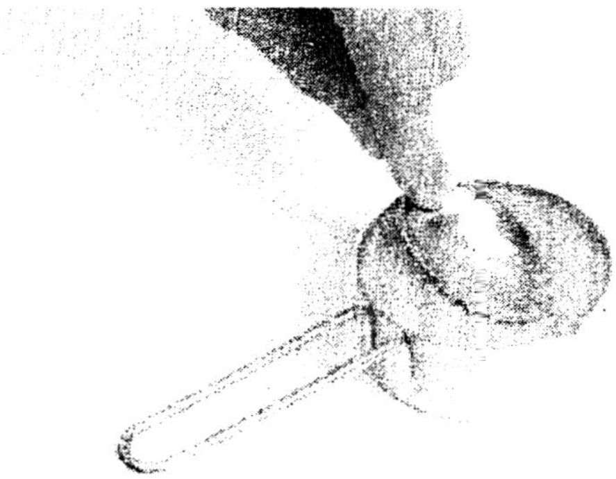

When removed from the measuring cup, the brown sugar will be molded into the shape of the cup if packed properly.

Butter or margarine in stick form

Butter and margarine in stick form have measurements marked on their wrappers, making it very convenient to measure the required amount.

You simply cut off the amount needed according to the markings on the stick. They are generally marked in tablespoons. One stick equals 8 tablespoons or ½ cup. One half of a stick equals 4 tablespoons or ¼ cup and 1 tablespoon (1/8 of a stick) equals 3 teaspoons.

Measuring Liquids

When measuring thick, sticky liquids such as honey, molasses, and corn syrup, spray the inside of the measuring glass with nonstick cooking spray or grease it a little with oil. The liquid will then be much easier to remove.

Liquid Ingredients

Pour the liquid ingredient into the measuring cup until it is at the desired measure.

When checking to see if the ingredient is at the desired level, have the measuring cup sitting on a flat, level surface and bend down to view the measurement at eye level. Do not hold the cup up to eye level because the cup may not be level when viewing and it may result in an inaccurate reading.

When using measuring spoons, hold the spoon flat and pour the ingredient into the spoon until it reaches the top rim of the spoon.

Sticky Ingredients

When measuring sticky ingredients, such as jelly, honey, molasses, and syrup, lightly coat the inside of the measuring cup with vegetable oil or spray with cooking oil. The oil will allow sticky ingredients to slide out easily. If the recipe calls for oil, you can measure the oil before the sticky ingredients and then use the same cup, without washing it, to measure the sticky ingredient.

Removal of sticky ingredients can also be assisted by using metal measuring cups or spoons and filling them with boiling water for a few minutes and then pouring the ingredients in them. The heated metal will allow the sticky ingredients to pour out easily.

Miscellaneous Measures

At times a recipe will call for a quantity that is not a standard measurement, such as pinch, dash, jigger, scant or heaping. Measurements of this type can be confusing as to the exact quantity to use. The descriptions below may help when these measurements are called for in a recipe.

Pinch

A measurement used on dry ingredients that is the amount you can pinch between your forefinger and thumb. It is less than a dash and equivalent to approximately 1/16 teaspoon.

Dash

A small amount of an ingredient equivalent to over 1/16 teaspoon but less than 1/8 teaspoon when measuring dry ingredients. A dash used to measure liquid ingredients equals approximately 3 drops.

Jigger

A measurement equivalent to 3 tablespoons or 1 ½ fluid ounces.

Scant

A scant measurement indicates that you should use slightly less than the actual measure.

Heaping

A term, used when measuring dry ingredients, indicating that enough ingredient should be added in the measure so that it heaps over the rim of the measuring cup or spoon.

Tips for Measuring

— When measuring ingredients, never measure over the mixing bowl containing the other ingredients. You may accidentally tip the measuring device or over pour the ingredients and excess ingredients would fall into the mixture. This could ruin the whole batch, depending on the ingredient and how much was spilled. Measure over the sink, another bowl, or a sheet of wax paper to catch any excess spillage. Spillage caught on wax paper can be returned to that ingredient's container.

— If you do not have two sets of measuring cups or spoons that can be used separately for the dry and liquid ingredients, measure the dry ingredients first and then use the measuring cups and spoons for the liquid ingredients.

— Store loose dry ingredients, such as salt, in a lidded container. The ingredient can then be spooned out and leveled, rather than trying to pour it into a measuring spoon and having it spill over the edges.

— To help you keep track of which ingredients you have measured and added in the mixing bowl, place all the ingredients on one side of the mixing bowl and once you have measure and add an ingredient, move its container to the opposite side of the bowl.

— When cooking, learn to estimate the small measurements of ingredient to save time. Pour the measured amount of an ingredient into the palm of your hand. Observe the look and feel of the quantity and then try to pour that same amount into your palm without measuring first. Measure the amount you poured out to see how close you are to the actual measure. Practice doing this and soon you will be able to measure the ingredients by look and feel. This will greatly reduce your prep time when cooking. Do not use this technique when measuring ingredients for baking. When baking, it is more critical to that all ingredients are measured accurately.

— Do not mistake fluid ounces for ounces. Ounces measure weight and fluid ounces measures volume.

10

Properties and Functions of Baking Powder

Baking powder is a solid mixture that is used as a chemical leavening agent in baked goods. It can be composed of a number of materials, but usually contains baking soda (sodium bicarbonate, $NaHCO_3$), cream of tartar (potassium bitartrate, $C_4H_5KO_6$), and cornstarch. (A base, an acid, and a filler respectively.) Baking powder is made by generating these solids, combining them in unique proportions, and then transferring them to packaging. First developed in the mid 1800s, baking powder formulations have changed little since.

To modify the final characteristics of baked goods, leavening agents such as baking powder or yeast are added to recipes. A leavening agent is a material that releases carbon dioxide (CO_2) under certain conditions. This creates gas bubbles in the dough making it expand. When the product is baked, air pockets are created resulting in food that is light and crispy. Baking powder is generally preferred to yeast because it produces bubbles much faster. Yeast leavened dough takes anywhere from two to three hours to rise. Baking powder dough takes about 15 minutes.

Baking powder is a white solid that typically has three components, including an acid, a base, and a filler. When water is added to the baking powder, the dry base and acid dissolve into a solution. In this form, the compounds react to produce carbon dioxide bubbles, however, the amount of carbon dioxide produced by this reaction varies. Baking powder determines the final texture of the food and can affect the flavour, moisture, and overall palatability.

History of Baking Powder

The development of baking powders began with the discovery of carbonate materials. One of the first carbonates was potash (potassium carbonate, K_2CO_3), a material that was extracted from wood ashes. During the eighteenth century, potash production had become a major commercial industry. American colonies exported huge amounts to England where it was used by glass factories and soap manufacturers.

Potash's usefulness to the baking industry was discovered during the 1760s. Prior to this time bakers had to hand knead dough for long periods to get the proper amount of air mixed throughout. For recipes which called for sourdough, pearlash (concentrated potash) was added to offset the sour taste. By chance, bakers found that these types of dough rose quickly. Evidently, the pearlash reacted with the natural acids in the sour-dough to produce carbon dioxide gas. This discovery revolutionised the baking industry.

Over time, wood sources became scarce in England and other sources of carbonates were sought. In 1783 the French Academy of Sciences ran a contest for inventors who could develop a process for converting salt (sodium chloride, NaCI) to soda ash (sodium carbonate, Na_2CO_3). This contest was won by Nicolas LeBlanc in 1791. In his process, salt was reacted with sulfuric acid, coal, and limestone to produce soda ash. The soda ash was tried by bakeries as a leavening agent and found to be equivalent to potash. Baking soda was soon after extracted from soda ash and used to sooth stomach acids. The superior leavening properties of this material were discovered by American bakeries by the 1830s. It released gas quicker and the aftertaste was not as bitter as soda ash.

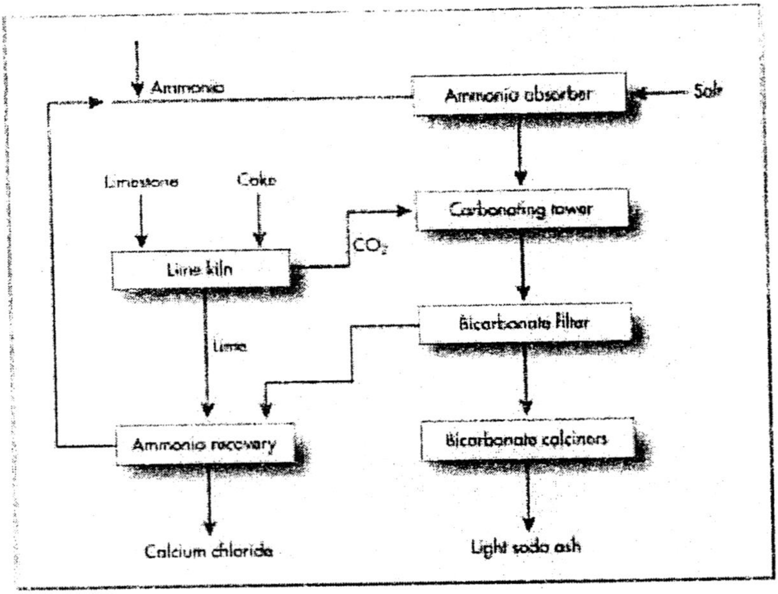

Another important development in America was the development of potassium bicarbonate ($CHKO_3$) by Natha Read in 1788. He suspended lumps of pearlash over fermenting molasses. This converted the potassium carbonate into potassium bicarbonate. Unfortunately, this process resulted in a less dependable leavening agent when compared to that manufactured in Europe. In 1834, Dr. Austin Church developed a different process for making baking soda from soda ash.

During the 1860s, various companies introduced other ingredients in their baking soda formulas and sold them as baking powders. These ingredients behaved in a more controlled way in recipes. Over time, different carbonate and acid mixtures have been sold as baking powders. Today, sodium bicarbonate and tartaric acid mixtures remain the most popular.

Primary Components of Baking Powder

As suggested, the primary components of a baking powder are a dry acid, base, and filler. Each of these materials can have a significant impact on the texture and taste of the finished product.

The most common dry base used in baking powders is baking soda, also called sodium bicarbonate. It is a water soluble white crystalline material, and produces carbon dioxide gas by itself when heated above 122°F (50°C). In addition to its use in baking, it is also used in the production of effervescent salts in medicine to prevent excess stomach acidity and in various types of fire extinguishers.

The type of acid used in a baking powder formula is more varied. The first baking powders used cream of tartar, a powdered acid. It was quick reacting and had to be put in the oven quickly or the gas would be spent. This material was perfect for products like pancakes or muffins. Today, there are four major acids used in commercial baking powders including monocalcium phosphate ($CaHO_4P$), sodium acid pyrophosphate ($H_2Na_{207}P_2$) sodium aluminium phosphate ($H_{304}P$), and sodium aluminium sulfate ($NaA_{108}S_2$).

Monocalcium phosphate is a fast reacting acid which produces a large amount of gas within three minutes of its addition to baking soda. This is about twice the speed of other acids. Sodium acid pyrophosphate is a slower reacting acid and is used in refrigerated biscuit dough recipes. Sodium aluminium phosphate and sodium aluminium sulfate are also slow reacting acids which generate gas when heated. While these compounds are used, most bakers prefer aluminium-free baking powders due to the unpleasant flavour the aluminium can cause in the baked good.

The third major component of baking powders is an inert filler. The most common of these is cornstarch. The cornstarch has three purposes. First, it helps keep the product dry and easily flowing. Without it, containers of baking powder could bind up and form one large mass. Second, it keeps the acids and bases separated and prevents them from reacting during storage. Finally, it adds bulk to the powder to make it easier to measure and standardise.

While a variety of baking powders are available, all of them meet basic standards and generate almost identical amounts of carbon dioxide. The basic difference between all types is the reaction time. There are two categories of baking powders: single acting and double acting. Single-acting baking powders immediately produce most of their gas when mixed with a liquid. They are classified by the type of acid they utilise. Those that contain cream of tartar and tartaric acid (C_4H_{606}) create gas rapidly when mixed with baking soda and a liquid. These batters must be cooked quickly or they will go flat. Slower single-acting baking powders are phosphate baking powders that contain either calcium phosphate (Ca_3O_8 P_2) or disodium pyrophosphate ($H_2Na_2O_7P_2$). Aluminium sulfate ($AI_{2012}S_3$) powders react more slowly at room temperature but give a bitter taste to the batter.

Most commercial baking powders are double-acting. These means that initially a small amount of gas is released when it is mixed with a liquid. The primary generation of gas occurs when the batter is heated during cooking. These types of powders allow a batter to be left in an unbaked condition for long periods of time. Often double-acting baking powders have two acids, one which reacts immediately and one that reacts when heated.

A less often used third type of baking powder is baker's ammonia. It results in a light, airy product but can impart an ammonia flavour if not used properly. It is best used in the production of flat cookies, helping to dissipate the ammonia odor during cooking.

Manufacturing Process

Baking powder is made in a batch process and involves production of the component raw materials, blending, and packaging.

Production of raw materials

1 The manufacture of baking powder begins with the production of sodium carbonate. Known as the Solvay ammonia process, it was first developed in 1861. In this process ammonia and carbon dioxide are passed through a saltwater (NaC_1H_2) solution in an absorption tower. This results in a compound called ammonium bicarbonate (CH_5NO_3) which reacts with the salt to produce crude sodium bicarbonate crystals and ammonium chloride (C_1H_4N).

2 The bicarbonate crystals are filtered out using vacuum filters or centrifuges. They are then washed with water to remove any residual chloride. The resulting solid is then conveyed to the calcining operation. Here, the material is heated and reacted with carbon dioxide to produce sodium carbonate, or soda ash.

3 The soda ash is dissolved, carbonated, and cooled which results in crystallised sodium bicarbonate. This solid bicarbonate material is of a purer concentrate

than the intermediate bicarbonate formed earlier in the process. It is then laid out on driers to remove most of the moisture. The product is passed through metal screens to produce the desired particle size and filled into drums for storage.

4 The solid acid for many baking powders is tartaric acid. This material is made using potassium hydrogen tartrate, which is a waste product from wine making. The potassium hydrogen tartrate is first purified and converted to calcium tartrate. Using sulfuric acid, the calcium tartrate is hydrolysed to produce calcium sulfate and tartaric acid. These materials are then separated and the resulting tartaric acid is purified and dried.

Blending the powders

5 The sodium bicarbonate, tartaric acid, and cornstarch are transferred to a blending area. Compounders pour the appropriate amount of each solid into mixing containers. These mixers have large, stainless steel blades that thoroughly combine the powders into a single, homogeneous blend. This material is then transferred through vacuum tubing to the filling machine.

Filling and packing

6 The baking powder is placed in a covered hopper and dispensed into the desired package. Baking powders are packaged in a variety of ways depending on the manufacturer. For home use it is typically sold in a 4 or 10 oz (113 or 264g) can. Restaurants can get baking powder in 5 or 10 lb (2.3 or 4.5 kg) metal cans. Industrial bakeries buy it in 50 or 100 lb (23 or 45 kg) fiber cartons. Filling is typically performed by a rapid, carousel filler which forces a specific amount of baking

powder into the package which is then sealed. The sealed containers are placed into cardboard boxes and stacked on pallets. The pallets are transferred to trucks or railroad cars and shipped to local grocery stores or commercial bakeries.

Quality Control

To ensure the quality of each batch of baking powder manufacturers monitor the product at each stage of production. The starting raw materials are subjected to various physical and chemical tests to determine if they meet previously determined specifications. Some of the characteristics that are tested include pH, appearance, and density. The finished product is also tested. Typically, the particle size is checked as are the micro-biological characteristics of the powder.

While baking powders have changed little over the last 100 years, manufacturers are always looking for new ways to make a greater profit. The baking powders of the future may be blended with different ingredients to enhance flavour. They may also be specially formulated for specific types of batter to accentuate characteristics such as gas evolution speed, residual flavour, or blending ease. Certainly, in the future manufacturers will find less expensive production methods.

Types of Baking Powder

There are different types of baking powders.

Single-Acting Baking Powders

Single-acting baking powders are characterised by the type of acid they include. Tartrate baking powders contain both cream of tartar (potassium acid tartrate) and tartaric acid. These create gas quickly when combined with baking soda in the presence of liquid, so the batter must be cooked quickly or it will go flat. Phosphate baking powders contain either calcium phosphate or disodium pyrophosphate (source of sodium pyrophosphate). They work a little slower than the tartrate baking powders, but most of the gas is still created outside of the oven and therefore can be lost. S.A.S. baking powders have sodium aluminium sulfate (alum) as the acid. S.A.S. baking powders react slowly at room temperature and release more of the gas when heated.

The phosphate and tartrate baking powders react rapidly at room temperature to release the leavening gas, which means that the batter has to be cooked quickly after the liquid ingredients have been added. On the other hand, the S.A.S. baking powders are better for products that will sit a while before being cooked. The problem with S.A.S. powders is that they have a bitter taste. They are used in combination with other leavening agents so not as much is needed. S.A.S. is often used in D.A. powders.

Double-Acting (D.A.) Baking Powders

Double-acting (D.A.) baking powders are the most common type of baking powder in US supermarkets. The first "action" refers to the release of gas when the baking soda in the powder reacts with an acidic liquid.

D.A. baking powders contain a dry acid which does not react with the baking soda in the powder until water is added; at that point the baking soda dissolves, the acid dissolves, and the two can now mix and the reaction shown above occurs. The second "action" refers to the release of gas when the batter is heated in the oven or on a griddle. This relies on the presence of the slower acting acid, S.A.S. which only combines with soda when the temperature increases.

Most baking powders are double-acting, which means they the baking powder reacts twice; contain one acid that dissolves when it comes in contact with water and another acid that does not dissolve until it reaches a higher temperature in a hot oven. This type of double action ensures that the finished product is light and fluffy. Single-acting baking powders are mainly used by manufacturers and are usually not available for retail sale.

Tips for Purchasing Baking Powder

When buying and stocking up on baking powder, try to find the most current date available by looking for a manufacturing or expiration date on the product. Keep in mind when it was manufactured and how long it has been sitting on the store's shelf, because time weakens its potency. Just because you bought it last week, it doesn't mean it was made last week and is as fresh as possible. Once a can is opened, fresh baking powder should be good for 3 to 6 months.

Food Chemistry Rule in Using- When creating a recipes, the food chemistry rule is 1 to 11/4 teaspoons of baking powder per 1 cup flour. If too much baking powder is used in your recipe, this produces big bubbles that run into each other, rise to the surface and pop. There goes the leavening, and the result is that the muffins, cakes or quick breads become heavy or sunken.

Storing Your Baking Powder

Baking powder should be stored at room temperature in a dry place. A cabinet or pantry away from the sink or heat source (such as the stove, direct sunlight), is a perfect place. Do not store baking powder in the refrigerator as it may shorten the shelf life due to condensation that occurs on the can.

Testing Your Baking Powder:

— First stir the contents of the can to see if there are any lumps. Lumps are an indication that the baking powder has picked up moisture, and has started a reaction in the can.

— Check the code dates on the bottom of the can to make sure it is still in code. Baking powder has a usual useful life of 24 months from the date of manufacture.

— Stir 1 teaspoon baking powder into 1/3 cup of hot water. If it bubbles gently, it is fine to use.

— If you have a can of baking powder that has been sitting in your cupboard for a year or more - toss it out!

Making of Baking Powder

If you have run out of baking powder you may be able to make a substitution by using the following:

— For one teaspoon baking powder = mix 1/2 teaspoon cream of tartar plus 1/4 teaspoon baking soda.

— If you are not using immediately, add 1/4 teaspoon cornstarch to absorb any moisture in the air and to prevent a premature chemical reaction between the acid and alkali.

Remember that a recipe for baked goods is like a formula. The ingredients work together to create an acceptable finished product. Substitutions don't always work as well as the original ingredients called for in the recipe. Any substitutions that you make come with the risk that the recipe will not turn out as intended.

11

YEAST-FREE BAKING

Baking yeast-free is possible and can produce excellent results—once you adapt to the rules

RULE 1. KNOW THE ENEMY

For most people, baking yeast-free requires making significant dietary changes. Depending on your particular medical or dietary needs, some or all of the following may need to be avoided:

— Refined or starchy flours and grains

— Yeast or yeast-leavened products

— Fermented products such as vinegar, cream of tartar, cheese and many condiments milk or milk products (except unsweetened yogurt)

— Cocoa or chocolate

— Mold containing foods such as vanilla bean, other spices and many nuts

— Refined or high sugar content foods

— Fruits

— Highly processed products

Rule 2. Become Friendly with New Foods

So what's left to eat? Quite a bit in fact. No need to feel lonely. Get acquainted with these delicious tastes:

- Unprocessed, minimally or unrefined whole grains (amaranth, brown rice, buckwheat, corn, kamut, millet, oats, quinoa, teff, and wild rice)
- Minimal starches
- Many vegetables and legumes
- Plain, unsweetened yogurt
- Unsalted butter
- Alcohol-free, sugar-free flavorings or extracts
- Baking powders that don't contain cream of tartar
- Freshly squeezed lemon juice
- Many seeds and nuts
- Seed and nut butters
- Seed and nut oils
- Eggs
- Cold processed, unrefined fats

Some yeast-free diets also allow certain starchy vegetables in moderation such as red skin potatoes, sweet corn, water chestnuts, winter squash, artichokes, Jerusalem artichokes, lima beans and English peas.

Rule 3. Learn Substitution Solutions

Baking powder is a great leavening agent and an excellent substitution for yeast in most recipes. Unfortunately, commercial baking powders typically contain cream of tartar. The fermentation used to produce cream of tartar makes this ingredient unacceptable for use in yeast-free diets. You can easily make your own baking powder using small amounts of unbuffered vitamin C (ascorbic acid) crystals or freshly squeezed lemon juice mixed with baking soda.

Many bakers use baking powder instead of yeast because the leavening reaction occurs so much faster. Yeast takes 2 to 3 hours to produce its bubbles, whereas baking powder is instantaneous. For this reason, baked goods leavened with baking powder are called quick breads.

Rule 4. Have Fun in the Kitchen

For starters, try these great recipes and start saying "yes" to yeast-free baking.

Yeast-free Baking Recipes

Onion & Herb Focaccia Bread

Toppings

1	small onion, sliced paper thin
1	tablespoon olive oil
2	cloves garlic, sliced thinly
8	olives, cut in half (optional)
3	sun-dried tomatoes, cut into small pieces
2	tablespoons fresh chopped herbs (e.g., thyme, marjoram, parsley, rosemary, chives) or 1 teaspoon dried mixed herbs

Dough mix

1	level teaspoon gluten-free mustard
1 ⅓	cups of warm water or 1 cup of warm water plus 1 egg
⅔	cup rice flour

⅓	cup corn meal or polenta flour
⅔	cup soy flour
1	teaspoon dried parsley
¼	teaspoon dried thyme
½	teaspoon dried sage
1	heaped tablespoon baking powder
1	tablespoon olive oil
2	scallions, chopped or 1 tablespoon minced onion

1. Preheat oven to 400°F.
2. Mix the onion with the olive oil and set aside with other topping ingredients.
3. Grease two baking trays.Make the dough. If using a food processor, place all the ingredients in the goblet and process until smooth and well mixed. If mixing by hand, place the mustard in a bowl and beat with a little water until smooth. Add the remaining ingredients and beat well with a wooden spoon.
4. Spread dough into two 9-inch rounds on the greased trays.
5. Spread the onion, garlic, olives, tomatoes and herbs over the surface and bake near the top of the oven for 25-30 minutes until the bread is quite brown and crisp.
6. Serve while still hot.

Caribbean Cobbler

2	large mangoes
2	large firm bananas
3-4	wrinkled passion fruit grated rind of ½ an orange juice of ½ an orange

Scone topping

¾ cup plus 2 tablespoons rice flour
¾ cup plus 2 tablespoons potato flour
2 teaspoons baking powder
2 tablespoons margarine
1 egg mixed with soy or another milk alternative to make a scant ¾ cup

1. Preheat oven to 475°F.
2. Peel and pit the mangoes and cut the flesh into large chunks.
3. Slice the bananas into 1-inch pieces.
4. Halve the passion fruit and scoop out the seeds.
5. Combine the mango flesh, sliced banana, passion fruit seeds and grated orange rind. Place mixture in a deep-sided baking dish and pour orange juice over it.
6. Place the flours, baking powder and margarine in a food processor and process. Beat the egg and the soy milk and add to the processor. Process until the mixture forms quite a soft dough. If you do not have a food processor, rub the margarine into the flours and baking powder, then mix in the egg and milk.
7. Place even-sized spoonfuls of the scone topping on to the fruit base. Do not level the surface but leave the topping in mounds.
8. Bake for 10-12 minutes or until the fruit is bubbling and the topping is firm and beginning to brown. Do not overcook or the fruit will begin to break up.
9. Allow to cool slightly before serving.

12

The Physics of Baking

Cooking methods can be characterised into two main types: dry heat cooking and moist heat cooking. Dry heat cooking heats foods in the absence of water, and includes methods such as baking and roasting, broiling, and pan-frying. Moist heat cooking uses water to heat food, and includes boiling, simmering, braising, and steaming.

Dry Heat Cooking

Ovens are used in many dry heat methods. Electric ovens use two heating coils, located at the top and bottom of the oven. The bottom coil is used for baking and roasting; the top is used for broiling. The heating coils are simply resistive elements which are heated by passing an electric current through them. In traditional ovens, when bottom coil is heated, the air inside the oven is heated primarily by conduction and natural convection. The heat is then transferred to the food, which is heated by the natural convection current.

In a convection oven, the heat transfer is enhanced by the use of a fan. The fan creates forced convection within the oven, which not only heats the food faster but also encourages even distribution of heat. Heat transfer inside an oven is actually more complicated than simply conduction and convection. The heating element emits a considerable amount of radiation which also contributes to the heating of the food. In addition, the walls of the oven become heated as well, emitting their own radiation. Broiling takes advantage of the radiation from the coil to rapidly heat the top of the food.

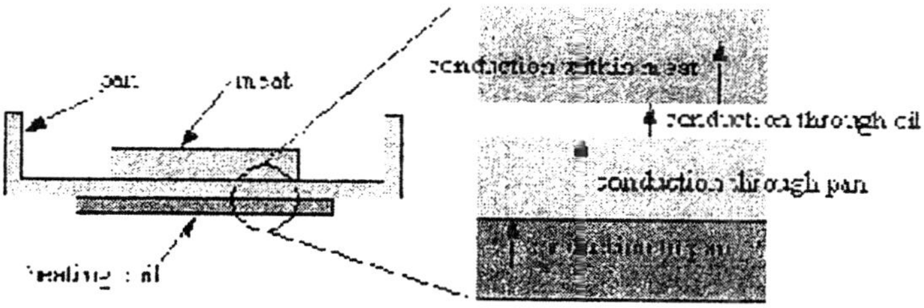

Figure 1. Dry heat cooking on the stovetop

Broiling uses only the top coil, which suppresses the natural convection current since the hot air is blocked by the ceiling of the oven. The radiation from the top coil heats the surface of the food to high temperatures, promoting browning. Cooking on the stovetop utilises conduction through a pan to heat the contents. In dry heat cooking, heat is conducted directly from the pan to the food, as shown in Figure 1. The heat is generated by resistive elements on electric stovetops.

Moist Heat Cooking

Moist heat methods use water in various states to heat food. The most common state is boiling, where water is heated by conduction through a pot on stovetop, and the heat is transferred to the food through convection (Figure 2). Water boils at 100 °C at sea level; therefore, boiling allows the food to cook at a constant temperature of 100 °C. Although this provides a convenient, consistent cooking medium, one major disadvantage is the lack of browning and the flavours that accompany the browning reaction. Browning of food occurs at temperatures above ~150 °C, which cannot be reached by moist cooking methods. The boiling point of water can be raised slightly to reduce the cooking time. Impurities, such as salt, increase the boiling temperature by a few degrees. Pressure cookers also effectively raise the temperature of boiling water by increasing the pressure inside the pot.

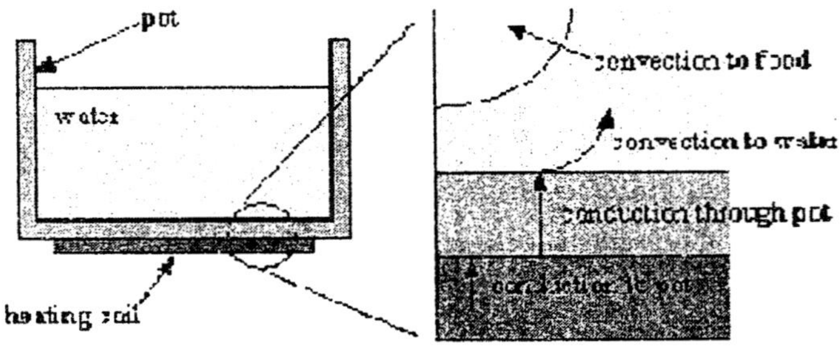

Figure 2. Moist heat cooking

Various states of water are used for cooking, and there are special names for some of them. Poaching uses warm water just before boiling. Simmering refers to cooking in water when it has just begun to boil. Boiling uses vigorously boiling water. Steaming uses the water in vapour phase to heat the food. The temperature of steam is typically a few degrees hotter than the temperature of boiling water.

Other Methods

Deep frying is classified as a dry heat method, although the food is heated indirectly by convection through the oil. Oils can reach a higher temperature than water; this is why the food can be browned during frying but not in

boiling. Typical frying temperatures range between 180 and 200 °C. Microwave has become a popular cooking appliance, mainly because of its ability to heat foods quickly. It uses electromagnetic radiation to agitate water molecules, which then produce heat due to friction. You will study more about microwave ovens later on in the course.

Cooking Materials

Most cooking containers are metal or ceramic, due to their ability to withstand high temperatures. Common metals are aluminium, copper, tin, stainless steel, and cast iron. Ceramics include glass, porcelain, earthenware, and stoneware. In addition, non-stick coatings and enamel coatings may be used on the surfaces to improve the properties.

Thermal conductivity, density, and specific heat are three very important factors in determining what cooking material is best suited for the kitchen. A high conductivity material tends to have a more even temperature distribution than one with a low thermal conductivity. A material with low specific heat requires less thermal energy to heat; therefore, it heats faster than one with a high specific heat.

Thermal diffusivity (α) is the combination of the three properties:

$$\alpha = \frac{k}{\rho \cdot c_p} \tag{1}$$

where k is the thermal conductivity, ρ is the density, and c_p is the specific heat. Thermal diffusivity measures the effectiveness by which a material conducts thermal energy with respect to its ability to store thermal energy.

A material with high α is characterised by a quick response to the changes in surrounding temperatures. A material with low a takes longer to reach a steady state condition, but is excellent at retaining heat once heated. In addition to the effect of material properties, the thickness (mass) of pot also changes the heating characteristics. Thicker pots tend to allow more uniform heat distribution; however, they take longer to heat. The size and placement of the heating element with respect to the pot is also of concern.

Table 1. Properties of common cooking materials

	ρ(kg/m^3)	k(W/mK)	c_p(J/kgK)	α(10^{-6}m^2/s)
Aluminium	2780	170	880	70
Cast iron	7870	70	450	21
Copper	8900	400	385	117
Stainless steel	8000	15	480	3.7
Glass	2600	4	800	1.9

Metals

Copper has the highest conductivity among all common cooking materials. Copper pans are highly treasured due to their ability to distribute heat evenly. However, copper tends to oxidise, leaving a black tarnish on its surface which also reduces its conductivity. To prevent oxidation, copper pots need constant cleaning and polishing. In addition, copper can be toxic if it diffuses into food in large amounts. Copper pans are often lined with tin to overcome these difficulties. The high cost of copper pans keep them from being used in majority of households.

Aluminum pans have the second highest thermal conductivity, next to copper, and are available at much lower cost. Aluminum is non-toxic and non-reactive, and also lightweight. Anodised aluminum has a thin coating

of aluminum oxide (Al_2O_3) on the surface. Aluminum oxide forms as a result of aluminum reacting with oxygen at high temperature, and is a very hard ceramic suitable for protection from scratches. Stainless steel is an alloy of iron, chromium, and nickel, and is desired due to its strength and resistance to corrosion. However, it is a relatively poor conductor, and often must be lined with copper or aluminum to improve the heat distribution.

Cast iron is typically iron alloyed with a small amount of carbon to increase strength. Cast iron pans are very prone to corrosion, and must be "seasoned" to prevent rusting. Seasoning a pan involves coating it with oil, then heating at moderate heat for a few hours. This fills up the small pores on the surface of the pan. Cast iron pans are generally heavy duty, allowing even heating despite the low conductivity and an ability to retain heat well.

Ceramics

Ceramic is usually a compound of metallic and non-metallic elements. They are poor conductors of heat, but are excellent at retaining heat once heated. They are very resistant to corrosion and are non-toxic. Due to their ability to retain heat, dishes made of thick ceramic materials will keep foods warm longer than metallic serving dishes of comparable shape and size.

A common problem with most ceramics is their tendency to crack due to thermal stress. Since they are such poor conductors of heat, there may be a large temperature difference between one side of the pan and the other. In such case, the warmer side expands more than the other, causing the pan to crack. Ceramic pans are seldom used on stovetops for this reason; ovens allow even heating from all sides, preventing large thermal gradients.

Coating Materials

Two common materials used for coating are Teflon (non-stick coating) and glass enamel. Teflon is a polymer which covers small roughness on pan surfaces and forms a very smooth finish, preventing food from sticking. Teflon coated pans are popular for low fat cooking; however, the coating is easily scratched by metal utensils. Glass enamel is formed by fusing glass powders to pan surfaces. The enamel improves the chemical resistance of the pan and prevents corrosion; however, they may crack when exposed to sudden changes in temperature.

THERMODYNAMICS OF COOKING

Cooking is based upon the three elements of thermodynamics: (a) conduction; (b) convection and (c) radiation. Any energy exchange between bodies occurs through one of these modes or a combination of them.

Conduction

Conduction is the transfer of heat through solids or stationery fluids. Convection uses the movement of fluids to transfer heat. Radiation does not require a medium for transferring heat; this mode uses the electromagnetic radiation emitted by an object for exchanging heat.

When there is a temperature difference in the solid, the hot side of the solid experiences more vigorous atomic movements. The vibrations are transmitted through the springs to the cooler side of the solid. Eventually, they reach an equilibrium, where all the atoms are vibrating with the same energy. Solids, especially metals, have free electrons, which are not bound to any particular atom and can freely move about the solid. The electrons in the hot side of the solid move faster than those on the cooler side. This scenario is shown in Figure 3.

As the electrons undergo a series of collisions, the faster electrons give off some of their energy to the slower electrons. Eventually, through a series of random collisions, an equilibrium is reached, where the electrons are moving at the same average velocity. Conduction through electron collision is more effective than through lattice vibration; this is why metals generally are better heat conductors than ceramic materials, which do not have many free electrons.

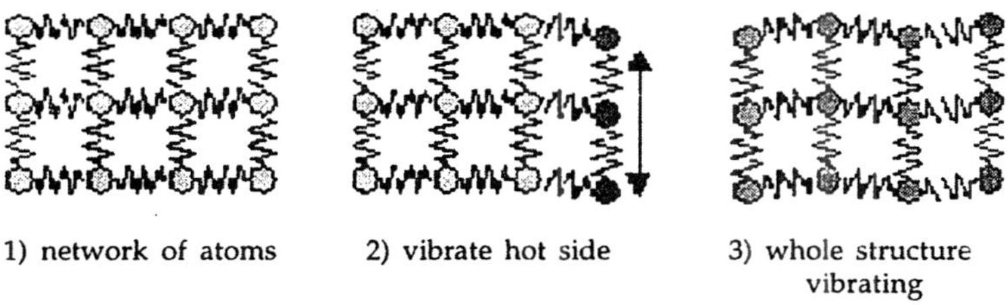

Figure 3. Conduction by lattice vibration

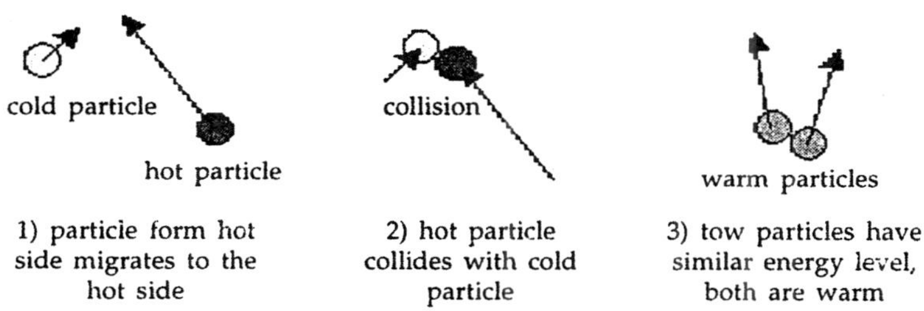

Figure 4. Conduction by particle collision

In fluids, conduction occurs through collisions between freely moving molecules. The mechanism is identical to the electron collisions in metals. The effectiveness by which heat is transferred through a material is measured by the thermal conductivity, *k*. A good conductor, such as copper, has a high conductivity; a poor conductor, or an insulator, has a low conductivity. Conductivity is measured in watts per meter per Kelvin (W/mK).

Steady-State Conduction

If you heat a pan on a stove, it takes a while for the pan to heat up to cooking temperature, after which the temperature of the pan remains relatively constant. The latter state is called the steady state, where there is no temporal change in temperatures. When the system is still changing with time, it is in transient state. The rate of conduction through an object at steady-state is given by:

$$q = -kA\frac{\Delta T}{\Delta x} \tag{2}$$

where k is the conductivity of the material, A is the cross-sectional area through which the heat is conducting, and ΔT is the temperature difference between two surfaces separated by a distance Δx.

One-Dimensional Conduction

One-dimensional heat transfer refers to special cases where there is only one spatial variable - the temperature varies in one direction only. A model used often to calculate the heat transfer through a 1-D system is called the thermal circuit model. This model simplifies the analysis of heat conduction through composite materials. In this model, each layer is replaced by an equivalent resistor called the thermal resistance. An analysis much like a circuit analysis follows. For conduction, the thermal resistance is expressed as:

$$R_{conduction} = \frac{L}{kA} \tag{3}$$

where L is the thickness of the layer, k is the thermal conductivity of the layer, and A is the cross-sectional area. When there is more than one layer in the composite, the total resistance of the circuit must be calculated. The total resistance for layers in series is simply the sum of the resistances:

$$R_{series} = \Sigma R \tag{4}$$

For resistors in parallel, the total resistance is given by:

$$\frac{1}{R_{parallel}} = \sum \frac{1}{R} \tag{5}$$

The convection at the surface must also be expressed as a resistor:

$$R_{convection} = \frac{1}{hA} \tag{6}$$

Once the total resistance of a structure is found, the heat flow through the layers can be found by:

$$q = \frac{T_{initial} - T_{final}}{R_{total}} \tag{7}$$

where $T_{initial}$ and T_{final} refers to the temperatures at the two ends of the thermal circuit (analogous to voltage difference in an electrical circuit) and q is the heat flow through the circuit (current).

Transient Conduction

Transient state is a state of non-equilibrium, when the temperatures are still changing with time. Some simple examples are the pre-heating time of ovens, when you are waiting for the temperature inside the oven to reach a specified temperature. However, even in a heated oven, cooking occurs by transient conduction because the heat is conducted through the food as it cooks. The internal temperature of the food, therefore, is usually increasing during cooking.

Lumped Body Model

The simplest case of transient conduction is the lumped body model. This model considers the heat transfer between a solid and a surrounding fluid. The main assumption in the model is that the solid can be approximated to have a uniform temperature; that is, there are no temperature gradients within the solid. Furthermore, the surrounding fluid must be large enough so that its temperature remains a constant throughout the process. This approximation can be made when the following condition is satisfied:

$$B_i = \frac{hL_c}{k} \leq 0.1 \tag{8}$$

where h is the convection coefficient, k is the conductivity of the solid, and L_c is the characteristic length of the solid, defined as the volume/area ratio. For a wall of thickness $2L$, $L_c = L$; for a cylinder, $L_c = r/2$; for a sphere, it is $r/3$. The number Bi is a dimensionless parameter called the *Biot* number. For a lumped body, the temperature of the solid which is introduced into the fluid at time $t = 0$ is given by:

$$\frac{\theta}{\theta_i} = \frac{T(t) - T_\infty}{T_i - T_\infty} = \exp\left(\frac{hA}{mc_p}t\right) \tag{9}$$

where A is the surface area of the solid, m is the mass of the solid, c_p is the specific heat of the solid, T_i is the initial temperature of the solid, and T_∞ is the temperature of the fluid. This equation assumes that T_∞ remains constant through the process. The newly introduced variable θ/θ_i is the "normalised temperature," which measures the temperature in relative terms. That is, the normalised temperature measures the percent change in temperature rather than exact change in temperature in °C. Once the normalised temperature is known, the actual temperature of the solid can be calculated based on initial temperature and the surrounding temperature.

Semi-Infinite Solid

The second approximation often applied to analyse transient conduction is the semi-infinite approximation. Semi-infinite solids can be visualised as very thick walls with one side exposed to some fluid. The other side, since the wall is very thick, remains unaffected by the fluid temperature. This condition is expressed as $T(\infty, t) = T_i$, where T_i is the initial wall temperature. This is illustrated in Figure 5.

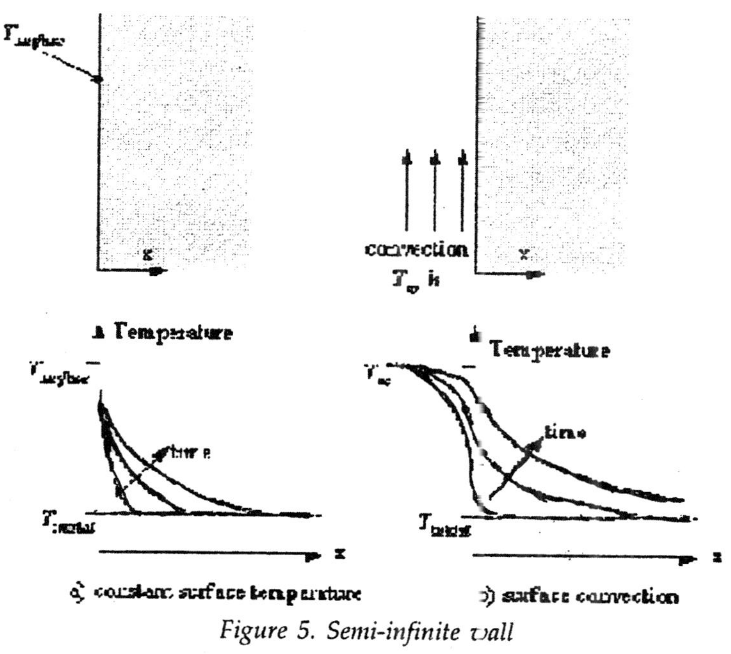

Figure 5. Semi-infinite wall

The condition at the exposed side of the wall is called the boundary condition.

Convection

Convection uses the motion of fluids to transfer heat. In a typical convective heat transfer, a hot surface heats the surrounding fluid, which is then carried away by fluid movement such as wind. The warm fluid is replaced by cooler fluid, which can draw more heat away from the surface. Since the heated fluid is constantly replaced by cooler fluid, the rate of heat transfer is enhanced.

The heat transfer rate for convection is given by the following equation:

$$q = -hA(T_{surface} - T_{\infty}) \tag{10}$$

where h is the convection coefficient, A is the surface area, and $T_{surface}$ and T_{∞} are the surface and ambient temperatures, respectively. The convection coefficient is a measure of how effective a fluid is at carrying heat to and away from the surface. It is dependent on factors such as the fluid density, velocity, and viscosity. Generally, fluids with higher velocity and/or higher density have greater h.

Natural Convection

Natural convection (or free convection) refers to a case where the fluid movement is created by the warm fluid itself. The density of fluid decrease as it is heated; thus, hot fluids are lighter than cool fluids. Warm fluid surrounding a hot object rises, and is replaced by cooler fluid. The result is a circulation of air above the warm surface, as shown in Figure 6. In general, fluids expand as the temperature rises, and thus the density decreases (density is the mass per unit volume).

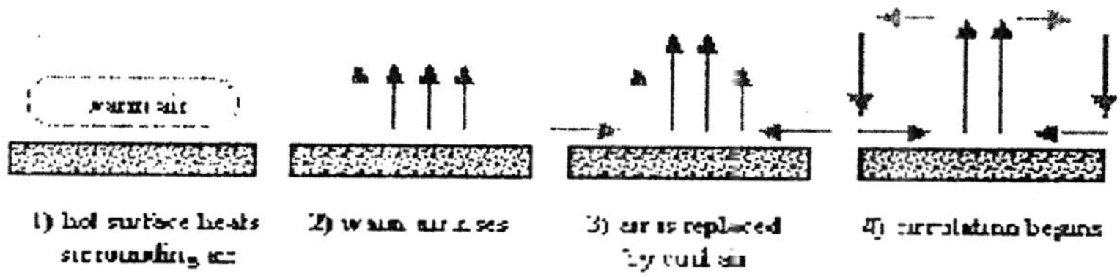

Figure 6. Natural convection

Warm fluids therefore are more buoyant than cooler fluids. A hot object will heat the surrounding fluid, which rises due to the buoyancy force. The warm fluid is then replaced by cool (unheated) fluids. Similarly, cool objects will draw heat away from the surrounding fluid which then fall due to the increased density. The cool fluid is then replaced by the warm fluid, initiating convective currents. For a hot horizontal plate surrounded by air, convection currents form when the air above the plate start to rise, as shown in Figure 7.

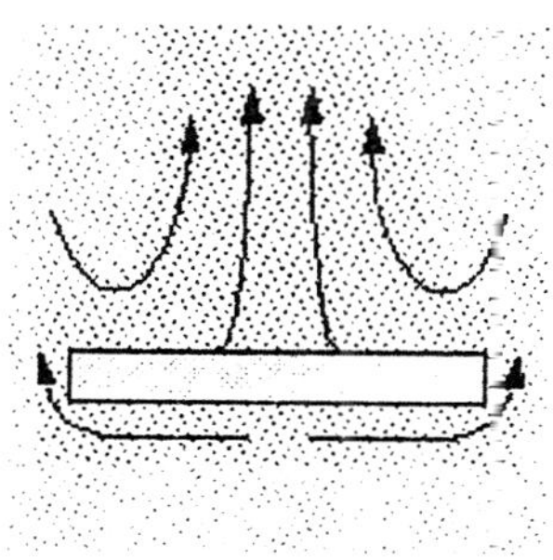

Figure 7. Natural convection around a horizontal hot plate

The air below the plate, however, cannot rise because the plate is blocking the flow. The heated fluid under the plate will eventually escape through the sides of the plate; however, the convective flow below the plate is very small compared to the flow on top. In general, natural convection is more pronounced (has a higher *h*) on the topside of a hot plate or the bottom side of a cold plate. The convection coefficient for natural convection in gas is generally between 1 W/m^{2}K and 20 W/m^{2}K; typical values for liquids fall between 100 W/m^{2}K and 1000 W/m^{2}K.

Forced Convection

Forced convection uses external means of producing fluid movement. Forced convection is what makes a windy, winter day feel much colder than a calm day with same temperature. The heat loss from your body is increased due to the constant replenishment of cold air by the wind. Natural wind and fans are the two most common sources of forced convection. Convection coefficient, *h*, is the measure of how effectively a fluid transfers heat by convection. It is measured in W/m^{2}K, and is determined by factors such as the fluid density, viscosity, and velocity. Wind blowing at 5 mph has a lower h than wind at the same temperature blowing at 30 mph.

Fluid flow caused by a fan or any other external forces create forced convection. Forced convection is generally more efficient than natural convection due to the faster velocity of the currents. In forced convection, buoyancy has little effect on the direction of flow. The relation between flow velocity, direction, and temperature is illustrated in Figure 8 for a hot, horizontal surface under forced convection.

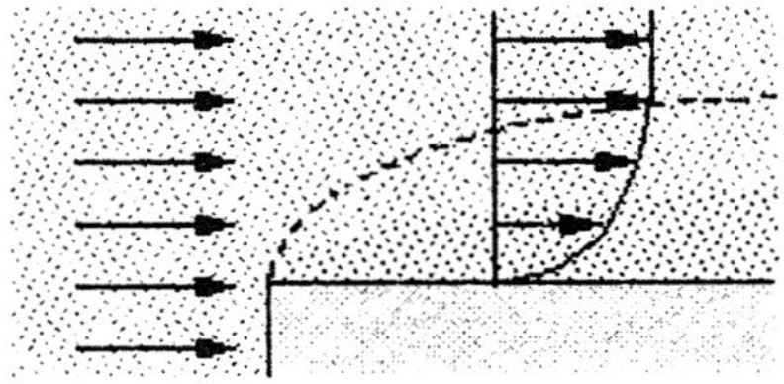

Figure 8. Forced convection currents

Since the buoyancy does not affect the flow, the bottom side of the plate will have the same patterns. Close to the surface, the flow velocity is inevitably slowed down due to friction. Right at the surface, the velocity is actually zero. This region of retarded flow is called the boundary layer. The region of warm airflow is generally well within the velocity boundary layer, and is called the thermal boundary layer.

The convection coefficient for forced convection in gasses generally range between 50 W/m^2K and 250 W/m^2K. For liquids, values start around 100 W/m^2K, and can be as high as 10,000 W/m^2K.

Convection in Ovens

Conventional ovens use natural convection to heat foods while baking. Ovens typically contain two heating elements, on top and bottom of the oven. During baking, the bottom element heats up, which heats the air inside the oven. The hot air rises and creates a current, which helps to distribute heat throughout the oven.

Natural convection currents are easily blocked by large pans, creating non-uniform temperatures within the oven. Convection ovens improve temperature distribution by using a fan, located inside the oven, to create forced convection currents. The forced convection currents efficiently mix the air inside an oven, creating uniform temperatures even in the presence of large pans. Furthermore, the increased airflow results in a higher convection coefficient, which reduces cooking time.

Radiation

Radiative heat transfer does not require a medium to pass through; thus, it is the only form of heat transfer present in vacuum. It uses electromagnetic radiation (photons), which travels at the speed of light and is emitted by any matter with temperature above 0 degrees Kelvin (-273 °C). Radiative heat transfer occurs when the emitted radiation strikes another body and is absorbed. We all experience radiative heat transfer everyday; solar radiation, absorbed by our skin, is why we feel warmer in the sun than in the shade.

The electromagnetic spectrum classifies radiation according to wavelengths of the radiation. Main types of radiation are (from short to long wavelengths): gamma rays, x-rays, ultraviolet (UV), visible light, infrared (IR), microwaves, and radio waves.

Radiation with shorter wavelengths are more energetic and contains more heat. X-rays, having wavelengths ~10-9 m, are very energetic and can be harmful to humans, while visible light with wavelengths ~10-7 m contain less energy and therefore have little effect on life.

A second characteristic which will become important later is that radiation with longer wavelengths generally can penetrate through thicker solids.Visible light, as we all know, is blocked by a wall. However, radio waves, having wavelengths on the order of meters, can readily pass through concrete walls. Any body with temperature above 0 Kelvin emits radiation.

The type of radiation emitted is determined largely by the temperature of the body. Most "hot" objects, from a cooking standpoint, emit infrared radiation. Hotter objects, such as the sun at ~5800 K, emits more energetic radiation including visible and UV. The visible portion is evident from the bright glare of the sun; the UV radiation causes tans and burns.

The incident radiation is determined by the amount of radiation emitted by the object and how much of the emitted radiation actually strikes the surface.

Microwaves

The electromagnetic spectrum consists of various types of radiation, characterised by wavelength (λ) and frequency (v). The frequency of the wave can be visualised as the number of wave crests that move by an observer in a second. Microwave radiation refers to the region of the spectrum with frequencies between ~10^9 Hz to ~10^{11} Hz. This type of radiation lies between infrared radiation and radio waves.

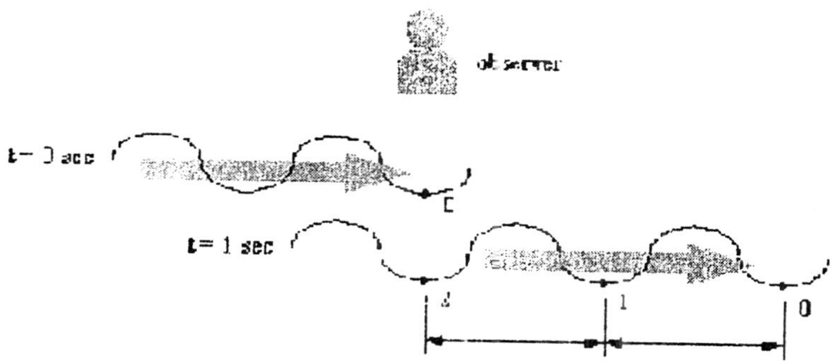

Figure 9. Wave representation of electromagnetic radiation . In the case shown here, the frequency of the wave is 2 Hz, or 2 cycles per second.

Recall that waves with shorter wavelength (lower frequency) have higher energy. The frequency of visible (green) light at λ = 500 nm is approximately 6×10^{14} Hz. The frequency of radiation used in microwave ovens is approximately 2.5×10^9 Hz, with a wavelength of 12 cm. This means that the microwave radiation used to heat food is actually less energetic than visible light.

Microwaves and Water Molecules

Water molecules (H_2O) are polar; that is, the electric charges on the molecules are not symmetric (Figure 10). The alignment and the charges on the atoms are such that the hydrogen side of the molecule has a positive (+) charge, and the oxygen side has a negative (-) charge. Electromagnetic radiation have electric charge as well; the "wave" representation shown in Figure 9 is actually the electric charge on the wave as it flips between positive and negative.

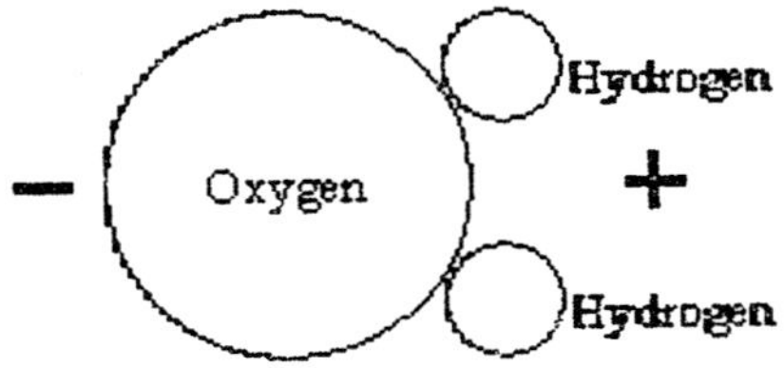

Figure 10. Water molecule

For a microwave oscillating at 2.5×10^9 Hz, the charge changes signs nearly 5 billion times a second (2 times 2.5×10^9 Hz). Electric charges are similar to magnetic charges: opposites attract. When oscillating electric charge from radiation interacts with a water molecule, it causes the molecule to flip (Figure 11).

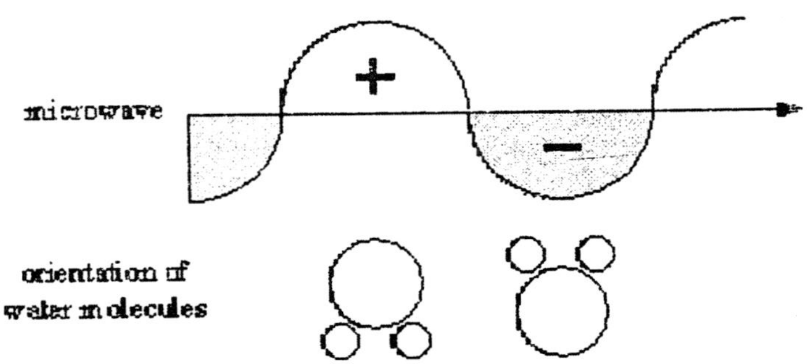

Figure 11. Interaction between water molecule and microwave

Microwave radiation used in ovens are specially tuned to the natural frequency of water molecules to maximise this interaction. Therefore, as a result of the radiation hitting the molecules, the water molecule flips back and forth 5 billion times a second. Because temperature measures how fast molecules move in a substance, the vigorous movement of the water molecules raises the temperature of water.

Reactions Inside a Working Microwave Oven

Microwave radiation is produced by a device called the magnetron. The magnetron consists of four major components: an anode block, a cathode filament, a pair of permanent magnets, and an antenna. The anode block is a hollow cylinder with fins coming out to the inside. The top view is shown in Figure 12.

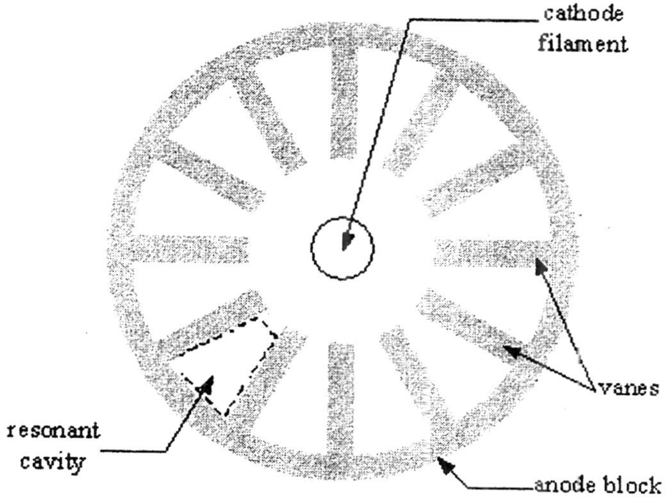

Figure 12. Anode block inside a magnetron

The fins are called the anode vanes, and the spaces between them are the resonant cavities. The cathode filament is a cylindrical rod located at the center of the anode block. It serves as the source for electrons during the emission of microwave radiation. Two permanent magnets are located at the top and bottom of the anode block, as shown in the side view in Figure 13. These magnets create a magnetic field inside the anode block that are parallel to the cathode filament.

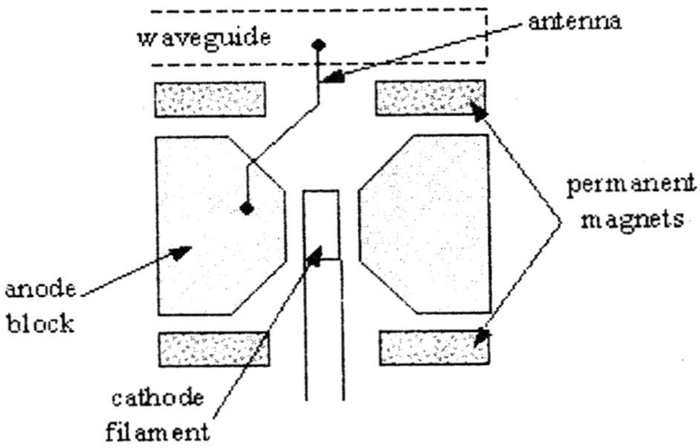

Figure 14. Side view of the magnetron

An antenna is positioned so that one end goes into one of the resonant cavities in the magnetron. The other end is located in the waveguide, which transfers the microwave radiation to the cooking chamber, much like how fiber optics is used to transfer light to remote locations. The events which lead to the production of microwave radiation begins when an electron is emitted by the cathode filament. Since the filament and the electron both have negative charge, the electron accelerates outward, toward the anode vanes.

The magnetic field due to the permanent magnets around the anode block are pointing out from the paper. Moving electric field from the electrons induce a magnetic field around itself. If you were travelling with the electron, facing forward, the induced magnetic field appears clockwise. The positive and negative magnetic field below the electron cancels each other, while the positive magnetic fields above the electron builds up to create a net force downward, perpendicular to the electric force. This net force pushes the electrons clockwise around the cathode filament. When electrons are continuously emitted all around the cathode tube, they collectively move clockwise around the cathode filament, forming a cloud (Figure 15).

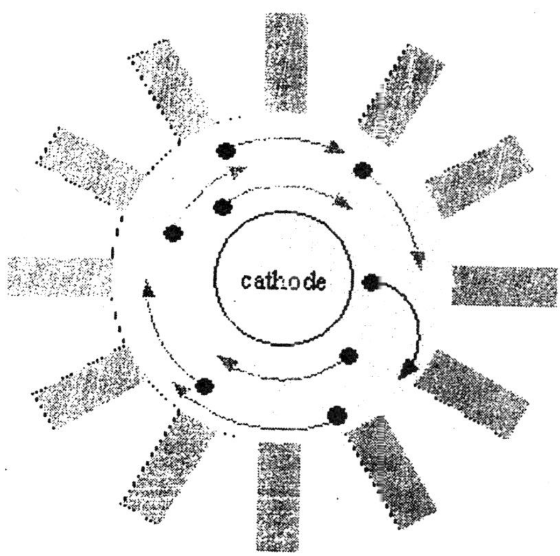

Figure 15. Clockwise motion of electrons in the anode block

When an electron approaches an anode vane, a positive charge is induced within the vane as electrons in the vane are repelled by the approaching free electron. Meanwhile, negative charge is induced in the neighbouring anode vanes due to the accumulation of the repelled electrons. The negatively charged vanes then repel electrons rotating within the electron cloud. The effect of this is that the electron cloud forms into a pinwheel shape, as shown in Figure 16.

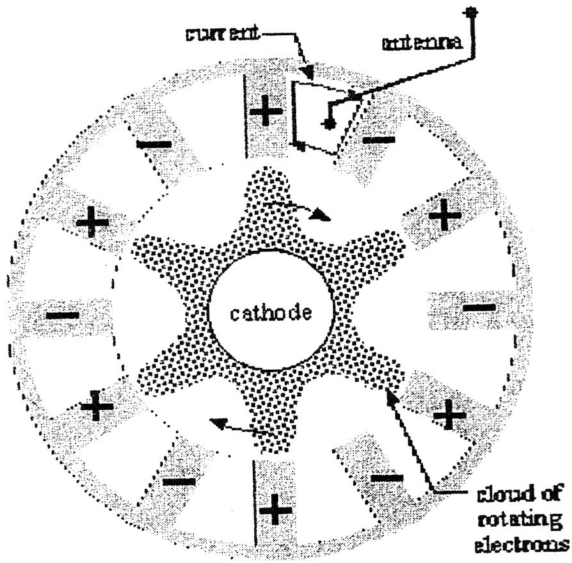

Figure 16. Cloud of electrons and the induced current

As the cloud rotates, the anode vanes closest to the "spokes" of the pinwheel has an induced positive charge, and the neighbouring vanes are negatively charged. As the cloud rotates, the positive and negative charges in the vanes rapidly oscillate. The oscillation of charges create an alternating current within the resonant cavities of the anode block. This current is carried by the antenna, and released as microwave radiation inside the waveguide.

The final step in a microwave oven is to release the radiation within the cooking chamber. When the radiation exits the waveguide, it is often reflected by a rotating fan blade to evenly distribute the radiation throughout the cooking chamber (Figure 17).

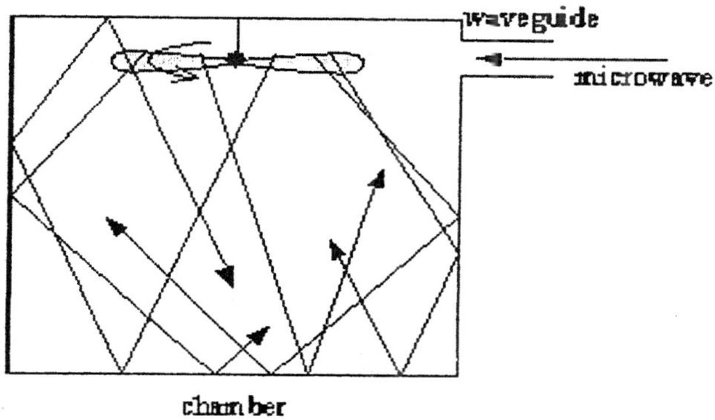

Figure 17. Microwave chamber

Once entering the chamber, the radiation is reflected by the chamber walls until it is absorbed by the food.

Properties of Microwaves

Microwave heats food by agitating the water molecules. Most foods we consume contain over 70% water by weight, making this an effective way for heating foods. However, the negative side is that food with low water contents take longer to heat in a microwave. Furthermore, frozen foods take longer to heat because the water molecules are not moved as much as in liquid water. One must also remember, when using a microwave, that foods heated in a microwave cannot become hotter than the boiling point of water, or 100°C. This is why foods cannot be browned in a microwave, and pies reheated in a microwave do not have a crisp crust like a freshly baked pie would have.

One often hears the statement "microwaves cook foods from inside." This is because foods heated in a microwave tend to heat most rapidly in the center, and cooks much faster than what we intuitively believe from conduction of heat from outside. The statement, however, is not entirely true; microwave radiation simply has the ability to penetrate several inches into the food. Radiation with longer wavelengths tend to penetrate deep into matter, evident by the ease by which radio waves go through concrete walls.

Since microwaves penetrate into food several inches down, it can heat foods quicker than if it were being heated through conduction by boiling water on the outer surface. If the food were larger than several inches, however, the middle of the food still needs to be heated by conduction. One of the major drawbacks of microwave is the hot spots. Microwave radiation, as it reflects around the cooking chamber, interact with other reflected radiation in such way that hot and cold spots are formed inside the microwave. The phenomenon responsible for this is the interference of waves.

Waves overlapping in such way that the crests match one another interfere constructively, forming a hot spot. Waves which interfere destructively result in a cold spot. The uneven heating caused by these effects can be reduced significantly by using a rotary platform, included in many of the newer models of microwave ovens.

13

BAKING APPLIANCES

Baking appliances or bakeware are types of food preparation containers commonly found in the kitchen. Bakeware comprises cooking vessels intended for use inside an oven.

History of cooking vessels before the development of pottery is minimal due to the limited archaeological evidence. It has been possible to extrapolate likely developments based on methods used by latter peoples. Among the first of the techniques believed to be used by stone age civilisations were improvements to basic roasting. In addition to exposing food to direct heat from either an open fire or hot embers it is possible to cover the food with clay or large leaves before roasting to preserve moisture in the cooked result. Examples of similar techniques are still in use in many modern cuisines.

Of greater difficulty was finding a method to boil water. For people without access to natural heated water sources, such as hot springs, it was possible to prepare a small pit lined with stones and filled with water. Heated stones could then be placed in the water to raise its temperature. In many locations the shells of turtles or large mollusks provided a source for waterproof cooking vessels. Bamboo tubes sealed at the end with clay would have provided a usable container in Asia, while the inhabitants of the Tehuacan Valley began carving large stone bowls that were permanently set into a hearth as early as 7000 BC. A final cooking vessel available to early civilisations were the stomachs from animals killed by hunters.

The development of earthenware pottery allowed for the creation of fireproof cooking vessels in a variety of shapes and sizes. Coating the earthenware with some type of plant gum, and latter pottery glazes, converted the porous container into a waterproof vessel. The earthenware cookware could then be suspended over a fire through use of a tripod or other apparatus, or even designed to be placed directly into a fire or coal bed. The development of bronze and iron metalworking skills allowed for cookware made from metal to be manufactured although adoption of the new cookware was slow due to the much higher cost. After the development of metal cookware there was little new development in cookware, with the standard Medieval kitchen utilising a cauldron and a shallow earthenware pan for most cooking tasks with a spit employed for roasting.

By the 17th Century, it was common for a western kitchen to contain a number of skillets, baking pans, a kettle, and several pots along with a variety of pot hooks, and trivets. In the American colonies, these items would commonly be produced by a local blacksmith from iron while brass or copper vessels were common in Europe and Asia. Improvements in metallurgy during the 19th and 20th centuries allowed for pots and pans from metals such as steel, stainless steel and aluminum to be economically produced.

Bakeware Pan Sizes

When you are preparing to create a recipe that requires the use of bakeware, you should first determine whether you have the proper size pan. There are many different types of bakeware, such as, round pans, square pans, rectangle pans, tube pans, bundt pans, jelly roll pans, loaf pans, springform pans, roasting pans, and casseroles. Your recipe will suggest that you use a certain size and type of pan. If you do not have the pan requested, you can generally substitute a similar size pan without any problem. The information below will help you make the appropriate substitution

Select a pan that is as close to the same size as the pan the recipe indicated to use. The pan should have approximately the same volume capacity and when filled, the ingredients should be close to the same depth as they would be in the original size pan. To find the surface of a pan use the equations below:

Square or Rectangular Pans:		Width x Length *9" x 13"*	 =	Total Square Inches *117 Square Inches*
Round Pans:	Step 1.	Diameter ÷ 2 *9" ÷ 2*	 =	Radius *4.5"*
	Step 2.	Radius x Radius x (3.14) *4.5" x 4.5" x 3.14*	 =	Total Square Inches *64 Square Inches*

— If the pan is close in size and volume to the original, the cooking temperatures and times should not have to be adjusted.

— When measuring a pan, measure from inside edge to inside edge so that you are not including the thickness of the sides in your measurement. Measure the depth by placing the ruler inside the pan and measuring from the bottom up to within ¼ inch of the top edge. If the pan has slanted sides, be sure to keep the ruler straight and do not slant when measuring.

— Measure volume by filling the pan with water and then pouring the water into a measuring cup to measure how much water is in the pan, or you can also measure the water as you pour it into the pan.

— Be careful substituting a pan when the recipe indicates that a tube pan should be used, such as an angel food cake pan or a bundt pan. Generally the recipe requires this type of pan so the heat is distributed properly. This is necessary for the food to get to its proper doneness.

— If you select a glass bakeware dish or a dark, non-stick pan, reduce the oven temperature by 25°F.

— If selecting a pan that results in the depth of the ingredients to be thinner, reduce the cooking time by 10 to 15 minutes. If the depth of the ingredients is thicker, increase the cooking time by 10 to 15 minutes. Be sure to use the visual indicators to test for the proper doneness.

— If a casserole dish has to be substituted for a different size when making a savoury dish, use the same guidelines as for baked goods. Use a dish that holds close to the same volume and if you substitute one that will affect the depth of the ingredients, adjust the cooking times to use less time for thinner depths and more time for thicker depths.

The table below shows the sizes and volume of some of the common bakeware pans. The table will help to assist you in selecting a successful substitution.

Table 1. Bakeware Pan Sizes and Volume

Bakeware Pan Size	Approximate Volume	Bakeware Pan Size	Approximate Volume
Round		Loaf	
6″ x 2″	4 cups	8″ x 4″ x 2 1/2″	4 cups
8″ x 1 1/2″	4 cups	8 1/2″ x 4 1/2″ x 2 1/2″	6 cups
8″ x 2″	6 cups	9″ x 5″ x 3″	8 cups
9″ x 1 1/2″	6 cups	Bundt	
9″ x 2″	8 cups	7 1/2″ x 3″	6 cups

9" x 3"	12 cups
10" x 2"	11 cups
Pie	
8" x 1 1/2"	4 cups
9" x 1 1/2"	5 cups
9" x 2"	8 cups
10" x 1 1/2"	6 cups
Square	
8" x 8" x 1 1/2"	6 cups
8" x 8" x 2"	8 cups
9" x 9" x 1 1/2"	8 cups
9" x 9" x 2"	10 cups
10" x 10" x 2"	12 cups
Rectangular	
11" x 7" x 2"	6 cups
13" x 9" x 2"	14 cups
Jelly Roll	
10 1/2"x 151/2" x 1"	10 cups
12 1/2" x 17" x 1"	12 cups
Springform	
8" x 3"	11 cups
9" x 2 1/2"	10 cups
9" x 3"	12 cups
10" x 2 1/2"	12 cups
9" x 3"	9 cups
10" x 3 1/2"	12 cups
Tube	
8" x 3"	9 cups
9" x 3"	12 cups
9 1/2" x 4"	16 cups
10" x 4"	16 cups
Tart	
11" x 1"	4 cups
Muffin Cup	
1 3/4" x 3/4"	1/8 cups
2 3/4" x 1 1/8"	1/4 cups
2 3/4" x 1 1/2"	1/2 cups
3" x 1 1/4"	5/8 cups
Casserole	
1 quart	4 cups
1 1/2 quart	6 cups
2 quarts	8 cups
2 1/2 quarts	10 cups
3 quarts	12 cups
4 quarts	16 cups

When selecting a substitution according to the pan's volume , that you need to allow room for expansion of certain foods as they cook. Certain types of food need more room for expansion than others. Fill the pan as directed by the instructions on the recipe. If you select a pan that has a comparable volume to the original size suggested by the recipe, it should allow the necessary room for expansion.

Bakeware Materials

Bakeware can be found made from several different materials. Each can have a different effect on the outcome of your baked goods. A dark colored surface on the bakeware will cause the food to brown more easily because it absorbs the heat of the oven. An aluminum pan conducts heat evenly but its shiny surface reflects the heat, resulting in food that does not brown well. Some of the common materials are shown below.

Aluminum

Aluminum is a good material for bakeware because it is a good conductor of heat, which results in even baking, but the shiny surface of the aluminum may prevent proper browning. Aluminum bakeware is durable and will not rust. For best results, look for bakeware made of a heavy-gauge aluminum rather than thinner, flimsy aluminum. The thinner bakeware may be less expensive but may warp when exposed to high temperatures, causing uneven baking. Aluminum bakeware should not be used for preparing acidic foods, such as tomato based dishes, because the acid in them and the aluminum may cause a reaction and result in food that has an off taste. It may also cause the aluminum pans to have to be replaced more often.

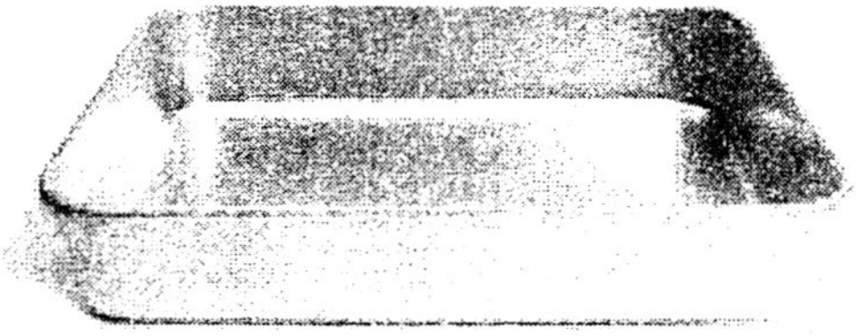

Aluminum bakeware

Aluminum bakeware can be cleaned with hot soapy water and if extra cleaning is required, a nylon scouring pad or a special powder cleanser made for aluminum and stainless steel can be used.

Aluminum foil

Constructed of a heavy gauge aluminum foil, this type of pan is used for a variety of foods and food chores such as pizza baking, bread baking, meat roasting, and numerous other cooking activities. They are generally considered to be a disposable pan, but often they can be reused if they are thoroughly cleaned after use. However, this type of pan is typically a good way to prepare baked goods and deliver them still in the pan to friends and family without worrying about the bakeware being returned. It is also a good alternative to regular bakeware or cookware when you need several pans for serving large groups of people. They work well for preparing egg, potato and casserole dishes.

Aluminum foil bakeware

Aluminum foil bakeware can be cleaned with hot soapy water and if extra cleaning is required, a nylon scouring pad can be used. The foil pans require careful handling to prevent them from being dented or bent out of shape. They can be cleaned in a dishwasher but to prevent denting or bending, care should be taken to not stack heavier dishes on top of them.

Anodized aluminum

There are two types of anodized bakeware available, silver anodized and hard anodized bakeware. The difference between the two is explained below.

Silver anodized aluminum

Silver anodized: Silver anodized bakeware is made of aluminum that has gone through an electro-chemical process that hardens its outer layer. This process gives the aluminum a scratch resistant surface that is also stick resistant and it will not rust, peel or blister. The aluminum maintains its ability to conduct heat quickly and evenly, and becomes more durable. Cooking times may need to be reduced when using anodized bakeware. It is recommended to check for doneness about two-thirds through the suggested baking time. Anodized aluminum is scratch resistant but you should be careful not to use sharp knives to cut food that is baked in these pans to prevent damage to the surface.

Hard anodized: Hard anodized bakeware is very similar to silver anodized only the aluminum is processed for a much longer period of time. This produces a hardened surface finish that is about 10 times thicker than the silver anodized.

Metal utensils, other than knives, can be used on hard anodized bakeware without causing any damage to the surface. Sharp knives may cause some scoring. The hard anodized bakeware has a darkened surface, which may affect the baking time required when using this type of bakeware. Check for doneness about two-thirds through the suggested baking time. Hard anodized bakeware can be cared for in the same manner as the silver anodized.

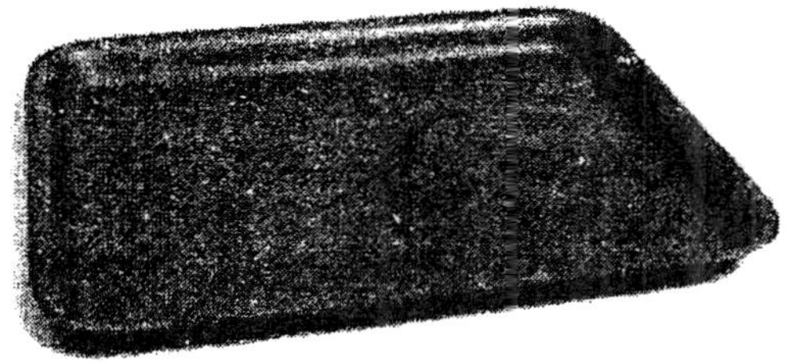

Hard anodized aluminum

Silver anodized and hard anodized bakeware should be care for in the same manner. They are easy to clean and do not rust. The bakeware should be washed by hand in hot soapy water and any stuck on food should be soaked in water to loosen. Do not wash in the dishwasher.

Glass

Glass ovenproof bakeware is made from tempered glass and conducts heat well. Food browns well in the glass ovenware. It is fairly easy to clean, does not stain, and does not react to acidic foods. Ovenproof glass can be taken from the refrigerator or freezer and placed in a hot oven, but when it is removed from a hot oven, it should be allowed to cool to room temperature before being placed in a refrigerator or freezer, and before being exposed to water.

Because glass bakeware heats up faster and holds the heat longer than metal bakeware, you may want to reduce the baking temperature by approximately 25°F from what is suggested on the baking instructions for the item you are baking. Experimenting with reducing the temperature will be the best way to decide if it gives you the results you desire.

Glass bakeware

To care for glass bakeware, wash with hot soapy water. Stuck on food may be loosened by soaking in warm water or a nylon scouring pad can be used. Avoid using metal scrapers or abrasive cleaners.

Insulated Bakeware

Insulated bakeware is made of two layers of metal that are cushioned by a layer of air. The layer of air provides for even baking and browning without burning. Look for insulated bakeware with a good non-stick coating, otherwise food has a tendency to stick to the bakeware.

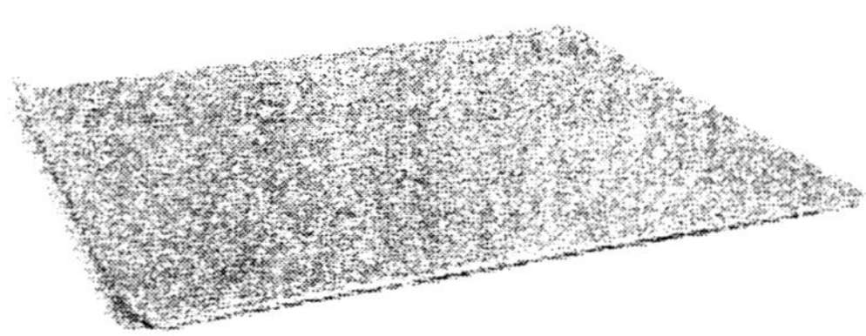

Insulated - Top

The layers of the insulated bakeware provide more protection from the heat source than other bakeware so it heats up slower and the food does not brown as nicely.

This may cause additional baking time to be necessary for properly baked food.

Insulated - Bottom

To care for insulated bakeware, use care methods according to the type of material it is made from and if it has a non-stick coating, care for it in the same manner as other non-stick bakeware.

Non-Stick Bakware

Non-stick bakeware consists of steel or aluminum that has had a non-stick coating applied to it. The coating allows the food to easily release from the surface of the bakeware and makes for easy clean up. Non-stick bakeware works well when you are baking items that need to be removed from the bakeware once they are done, such as muffins, cupcakes, and layered cakes.

Non-stick bakeware

The non-stick coating gives the bakeware a darker surface, which absorbs heat more quickly than bright or lighter colored bakeware. This can cause a problem with food becoming overly browned or dried out. Baking times and temperatures may have to be adjusted to compensate for this.

To care for the non-stick bakeware, wash in hot soapy water and for stuck on food, soak in warm water to loosen. To prevent the coating from becoming scratched, do not use metal utensils or knives on the non-stick surface. Scratches in the coating may cause the coating to peel.

Silicone Bakeware

Silicone bakeware is a fairly new product that is used for baking molds. It is made of a flexible and bendable silicone material that can be used in the oven, microwave, and freezer. The silicone bakeware does not absorb the heat like other bakeware but allows the heat to transfer evenly to the food.

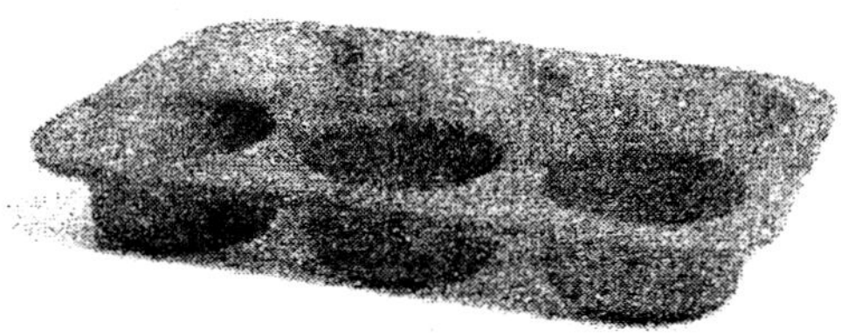

Silicone bakeware

The cooking process stops immediately when food is removed from the oven, preventing additional browning of the bottom and edges of the food. It can withstand temperatures ranging from minus 40 to 580 degrees F. The bakeware can be folded for convenience in storing without damaging the material. The silicone bakeware is also dishwasher safe.

Stainless Steel

Stainless steel bakeware is available in stainless steel but is not recommended because it is not a good conductor of heat. It distributes heat unevenly and causes inconsistency in baked goods.

Stainless steel bakeware

To care for stainless steel bakeware, wash with hot soapy water. A nylon scouring pad or nylon scraper can be used to remove stuck on food. There are stainless steel cleaners that can also be used to remove stains and bring a shine back to the stainless steel.

Stone Bakeware

Stone bakeware is made from stone fired at a very high temperature. The firing of the stone produces a nonporous surface on the bakeware, which does not need glazing. The stone bakeware absorbs the heat from the oven and distributes it evenly. It retains the heat well and keeps food warm long after it is removed from the oven. Because the stone bakeware distributes heat so evenly there is less chance of burning and overcooking the food.

Stone bakeware

It does take a little longer to initial heat up so if you are baking something that only requires a short baking time, you may have to extend the time by a couple of minutes. Most stone bakeware must be seasoned before using for the first time. To season, rub or spray the surface with oil. To care for stone bakeware, do not use soap to clean. Scrap off excess food and wash with warm water.

Tinned Steel

Tinned steel bakeware is made from sheets of steel that have been covered with a thin coat of tin, making it a good conductor of heat. The tin coating allows the bakeware to heat slowly and evenly. Tinned steel bakeware will darken after extended use, which may have an effect on baking times.

Tinned steel

Special care must be taken to prevent tinned steel bakeware from rusting. It should be washed in hot soapy water and then dried thoroughly. It should then be placed in a warm oven and allowed to dry completely. Do not use metal scouring pads or abrasive materials to clean tinned steel.

Types of Bakeware

There are many different types of bakeware available for baking a variety of baked goods. Some types are essential in producing the desired end product and some bakeware can be used for several purposes. Bakeware can be found made of many types of material, such as tinned steel, stainless steel, glass, silicone and stone. The type of material the pan is made from can have an affect on the cooking times or temperatures, requiring that they be adjusted.

A flat pan with straight sides that are ¾" or taller, which is used for baking food in the oven. There are many sizes and depths available, but the most common sizes is 9 x 13 x 2 inches. Other common sizes include: square pans 8 x 8 inches in size or rectangular pans that are 7 x 11 inches, 10 x 15 inches, and 12 x 18 inches by 1 or 2 inches in depth. Baking pans are used to hold runny batters, such as cake batter and thick solid masses of food, such as savory hot dish recipes. The foods are placed in the pan and the baked in the oven. Baking pans are available in a variety of sizes, materials, surface textures, and colors. Typical foods that are prepared in baking pans include: cakes, brownies, bars, cornbreads, lasagna, egg dishes, cobblers, potato dishes, meats, poultry, and fish.

Baking Sheet

A flat pan or sheet of metal that is used to bake products that are thick or stiff enough to stand on their own, such as cookies, freestanding breads, biscuits, pastries, and meringues.

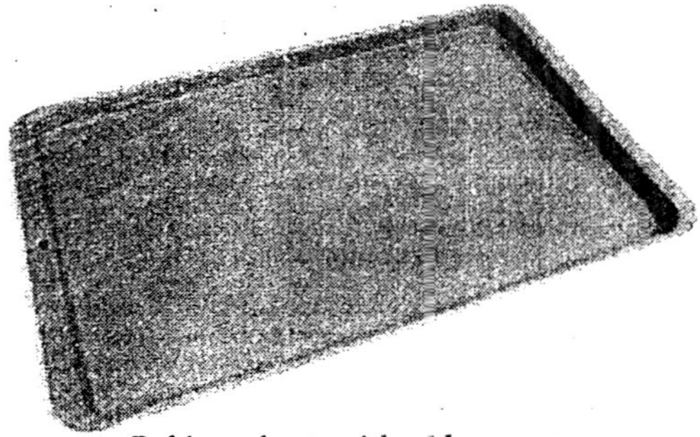

Baking sheet with sides

Baking sheets, which are also referred to as cookie sheets, consist of a metal sheet that is flat with one or both of the short ends containing a slightly turned up lip that is used to aid in handling the sheets. Baking pans with ¾ inch sides all around are often referred to as baking sheets or cookie sheets also. They can be used in the same manner as a flat baking sheet but they can also be used for baking runny batters.

Bread Pan

A baking pan, rectangular in shape and deep walled, that is used to bake a single loaf of bread. Bread pans are typically made from aluminum, steel, glass, pottery, or stoneware. They range in size from 5 to 14 inches in length and 3 to 5 inches in width, with a 3 to 4 inch sidewall.

Standard bread pans

Lo-Fat Loaf Pans

Bread pans are produced to make a semi-crisp crust with an evenly textured crumb, however not all pans are effective in doing so. Small sized pans work well for making gift or dessert loaves that can be cooked in less time. The larger loaf pans are also often used to bake meat loaf. Bread pans are also referred to as a loaf pan.

A loaf pan made as two separate pans, one fitting within and above the bottom or base pan. The insert piece allows grease to drip to the lower pan and away from the food. This pan is typically used for a dish such as meat loaf but the bottom pan can be used on its own for baking bread.

Brioche Pan

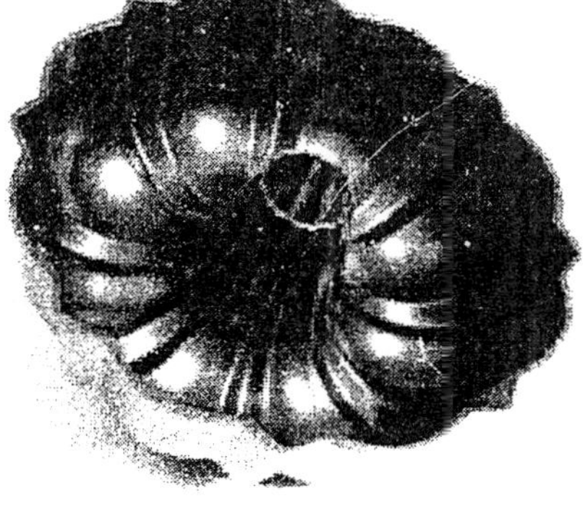

Bundt Pan

A type of pan or mold with fluted sides that is used to form the base of the traditional French bread known as Brioche. When making Brioche, a French dough is prepared with a light yeast and portionally larger amounts of butter and eggs. It is baked using a Brioche pan or mold to form the alternating columns on the base of the bread.

A heavy-walled baking pan formed with a decorative indented curvature and a hollow tube in the center. The heavier walled construction and hollow center tube allows cake batters to rise and bake more uniformly, creating a golden crust on the outside of the cake. After the ingredients are baked, the pan is turned over so the firm cake falls out onto a plate or counter, producing a decoratively shaped cake. This pan is generally used for baking coffee cakes and sweet cakes, referred to as bundt cakes.

Cake Pan

A baking pan commonly found in every kitchen. Cake pans can be round, square, or rectangular and are available in several sizes. Probably the most popular is the 9 x 13 x 2 inch rectangular cake pan that is used to bake cakes, bars, and savory dishes, such as lasagna. Standard square cake pans are 8 or 9 inches wide and 1 ½ inches in depth. Round cake pans are used to make layer cakes and are generally found in 8 or 9-inch diameters and are 1 ½ inches in depth.

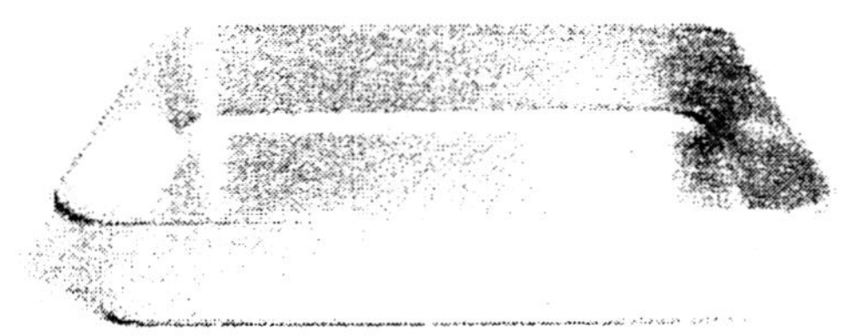

Rectangular Cake Pan

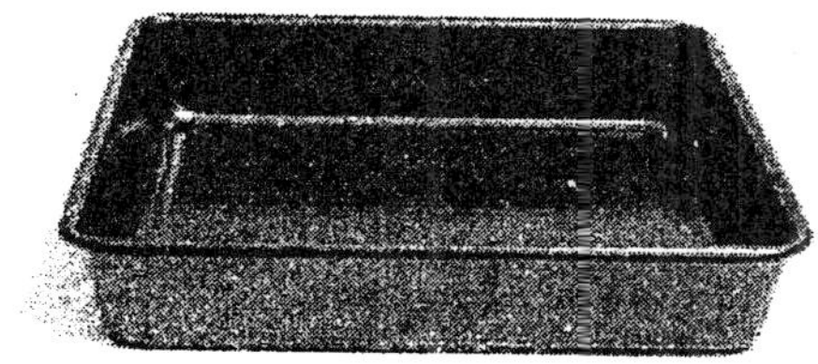

Square Cake Pan

Round Cake Pan

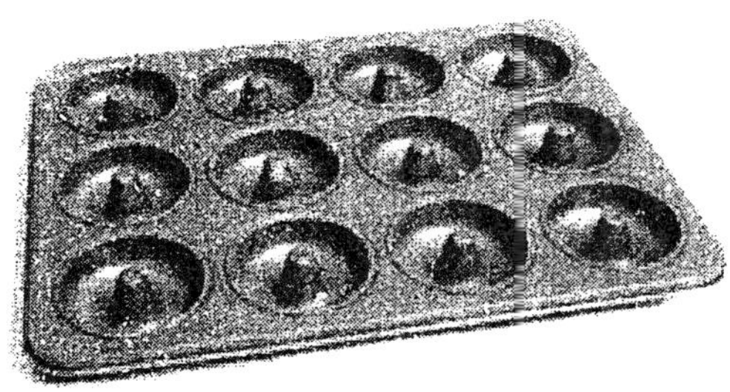

Doughnut Pan

A pan that is constructed similar to a muffin tin except that the cups have a rounded bottom and a stem that comes up in the center of each cup to create the hole in the doughnut. The pan helps produce a lower calorie doughnut because they are baked rather than fried in oil. The doughnuts can be eaten plain or glazed, frosted, or rolled in powdered sugar. The pan is typically constructed of metal or silicone non-stick materials and is available in 6 or 12 cup sizes. It is also found spelled as Donut Pan.

Flan Pan

A shallow pan that has shallow sides that are generally fluted to provide a decorative edge to the item being baked. They are available in different materials, such as tinned steel, steel with non-stick coating and ceramic. The fluted sides of the flan pans are slightly slanted and some varieties have a loose bottom.

Flan Pan

They are available in a range of sizes between 6 and 12 inches in diameter, and they are also available in a single serving size, which range in size from 3 3/4 to 4 3/4 inches in diameter. The flan pans are used to make flans and other dishes, such as tarts, cobblers, and quiches. The small individual flan pans are used to make single serving cakes, cobblers, breads, and tarts. the flan pan is also referred to as a Maryann Pan.

French Bread Pan

A type of metal or aluminum pan that is 16 to 20 inches in length and formed with a curved base so that long lengths of dough can be laid horizontally in them to be baked into baguettes. Some pans are double or triple-formed for several loaves to be baked at once.

French Bread Pan

If made of metal, the pan may have numerous small holes that are perforated throughout the metal to allow steam to escape while the bread is baking in order to assist in making the crust a deeper brown in color and crispier in texture that is familiar to traditional French crust. The pan is available constructed with non-sticking metal surfaces as well as porus clay stoneware.

Jelly Roll Pan

A large rectangular shaped baking pan with shallow sides, which generally have a rolled rim. The most common size is 10 x 15 inches but larger sizes are available, such as 12 x 17 inches and 13 x 18 inches. They have shallow sides that are 3/4 to 1 inch deep. They are found made from a variety of materials, such as aluminum, stainless steel, or steel. Many have a non-stick surface. It is designed to bake a thin cake that is coated with a layer of jelly and then rolled up into what is known as a jelly roll. It is commonly used for baking sheet cakes, cookies and pastries.

Muffin Tin

A type of pan used for baking muffins. The pan typically will have 6 or 12 individual round pockets or holders connected to the tin and formed in the shape of a muffin. The muffin batter is poured into the individual pockets and as the muffin bakes, a top forms over the pocket while the base of the muffin is baked in the shape of the cup.

Silicone Muffin Cups

Muffin tins are also often used to bake cupcakes. The tins are commonly available in 3 sizes ranging from those that bake a small or miniature muffin that is 1 to 2 inches in diameter, standard muffins that are 2 3/4 inches and large muffins approximately 4 inches in diameter. Individual muffin cups made from silicone are also available for baking individual cupcakes and muffins. Muffin tins may also be referred to as muffin pans.

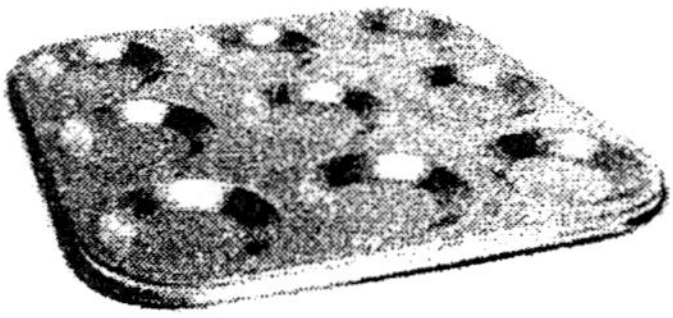

Muffin Top Pan

A muffin tin that has shallow cups that are approximately 1/2" deep. If you prefer the crusty top of the muffin, this pan will produce basically just the top of the muffin. Its shallow cups allow the top to bake up brown and crusty while eliminating the finer crumb bottom. It is available in 6 cup or 12 cup pans. This pan can also be used to bake single serving toaster cakes and individual buns. It is sometimes referred to as a bun pan.

Pie Plate

A round baking dish with shallow slanted sides that are 1 to 1 1/2 inches deep. The pie plate is available in several sizes. The most common sizes being 8, 9 and 10 inch diameters. They can be found made from many different materials, such as aluminum, glass, stoneware, ceramic, and tinned steel. Pie plates are used to bake single crust and double crust pies. They are also available in a deep dish variety that has slanted sides that are 2 to 2 1/2 inches deep and range from 9 to 11 inches in diameter.

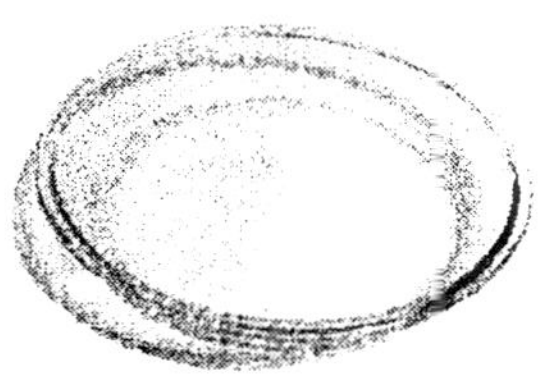

Glass Pie Plate

The deep dish pie plate is most often used to bake a savory dish. A pie plate is also referred to as a pie tin.

Popover Pan

A baking pan or baking utensil as it may also be referred, specifically designed to hold popover batter so the popover, as it bakes, can rise up into a large, airy pastry-like shape. Larger than traditional muffin tins, the popover pan will typically be made with individual cups joined by wire racks that are constructed to hold the batter and keep the baked contents away from touching the adjacent popovers as they enlarge when baked.

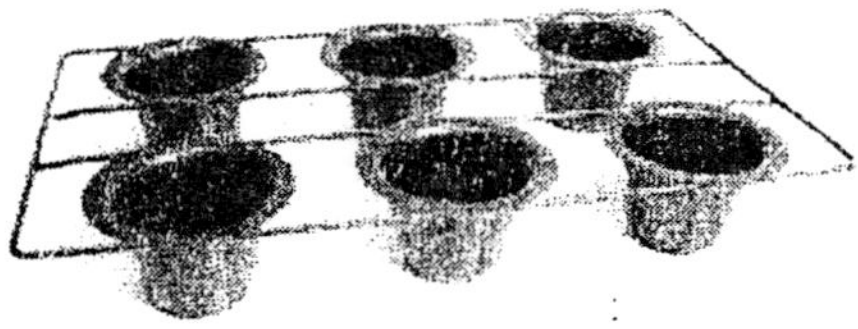

Popover Pan

The cup expands outward from the base as it moves up to the top of each cup where a lip is formed. The lip seperates the base and the head of the baked popover so the base is uniform in shape and the head is allowed to bake into an irregular, puffy shaped popover, as the baked batter "pops" up and over the cup. Popover pans generally are available in several sizes, both giant and small. The number of cups included in each pan may be 4 or 6. Smaller pans may hold as many as 8 or 10 cups for individual popovers.

Shortbread Pan

Constructed of a heavy cast iron or a coated aluminum surface, this type of pan is made for baking shortbread cookies. Typically made with a non-stick surface, the Shortbread pan is filled with the rich, buttery flavored dough that is baked into a tender but crumbly texture for cookies and dessert crusts.

Shortbread Pan

Made to produce the traditional shape that is pressed into decorative molds and baked until solid, a Shortbread Pan is contoured with square or round pattern designs that include ridges so the biscuit-like cookies can be broken apart evenly after cooking. This type of pan can bake not only shortbread cookies and biscuits, but also cakes, dessert bars, and various pan breads such as cornbread.

Springform Pan

A round pan, with tall, straight sides that are removable. They can be found made of a variety of materials, such as tinned steel, steel, stainless steel, and carbon steel. Many have non-stick surfaces.

Springform Pan

They range in size from 8 to 12 inches in diameter with high sides of 2 1/2 to 3 inches. The most common size is a 9 or 9 1/2 inch diameter. The removable sides aid in removing cakes that would otherwise be difficult to remove, such as cheesecakes and tortes. When purchasing a springform pan, be sure the sides and base fit accurately together for a tight fit.

Tart Pan

A round or oblong shallow pan with smooth or fluted sides and a removable bottom.

Tart Pan

The pan is constructed with a removable bottom so the baked tart can be pushed up out of the pan, allowing the contents to be removed easily and cleanly, while retaining the shape created by the fluted sides.

Tube Pan

Tart pans, which are manfactured with non-stick surfaces, are available in various sizes so tarts can be served as individual round servings or as pie-shaped servings.

Single Piece Construction

Pan with Removable Bottom

The small pans for individual round servings range in size from 3 to 5 inches in diameter while the larger pans for pie-shape servings commonly range in size from 9 and 11 inches in diameter.

14

Functions of Baking Ovens

An oven is an enclosed compartment for heating, baking or drying. It is most commonly used in cooking and pottery. Two common kinds of modern ovens are gas ovens and electric ovens. Ovens used in pottery are also known as kilns. In cooking, the conventional oven is a kitchen appliance and is used for roasting and heating. Food normally cooked in this manner includes meat, casseroles and baked goods such as bread, cake and other desserts.

History of Oven

Culinary historians credit the Greeks for developing bread baking into an art. Proper front-loaded bread ovens originated in Ancient Greece. The Greeks created a wide variety of doughs, loaf shapes and styles of serving bread with other foods. Baking developed as a trade and profession as bread increasingly was prepared outside of the family home by specially trained workers to be sold to the public.

The Greeks also pioneered sweetbreads, fritters, puddings, cheesecakes, pastries, and even wedding cakes. Often prepared in symbolic shapes, these products were originally served during special occasions and ceremonies.

By 300 CE, the Greeks had developed over seventy different kinds of bread.

Types of Ovens

Gas and electric ovens are available for installation in walls, under benches or as part of a conventional stove. The type of installation has no effect on energy efficiency or cooking performance.

All ovens are available in various sizes, so its important the oven you choose is sufficiently sized for your requirements. There are various types of ovens currently available on the market. Each cooks food in slightly different ways.

Conventional Ovens

Have a gas burner or an electric element(s) for cooking. As hot air rises, the highest temperature is found towards the top of the oven. This can limit the amount of food which can be cooked at the one time. These types of ovens are no longer commonly available.

Conventional Ovens

Usage

- Conventional ovens, which can operate on electricity or gas, incorporate two heating elements -- one for baking and roasting and another for broiling.
- To compensate for uneven heating patterns, some ovens cycle the top element on during baking for more even browning.
- The self-cleaning feature probably is the most-requested option for both wall ovens and ranges.
- Combination ovens are available that pair conventional oven cooking with microwave cooking or convection cooking.

Fan-forced (or Convection) Ovens

Have an inbuilt fan which circulates heated air around the oven. This results in an even temperature throughout the entire oven, allowing all shelves to be used simultaneously. Fan-forced ovens heat more quickly than conventional ovens, can cook food at lower temperatures, and use up to 35% less energy than conventional ovens. Multi-function ovens provide the option of either fan-forced or conventional cooking. This greater flexibility means, for example, a pie can be baked using the fan-forced function, and then the top can be browned using the oven conventionally.

Usage

- Convection ovens use thermal heat much like conventional ovens but circulate the heated air with an internal fan for faster, more even cooking.
- With circulated air, more of the oven space can be used and preheating time is shortened.

— The downside of forced-air circulation is that some foods can dry out or overbrown during cooking if not covered; fan noise also is a consideration.

— Combination ovens are available that pair convection oven cooking with conventional cooking or microwave cooking.

Microwave Ovens

Can be extremely energy efficient as a result of the short cooking times involved. Energy only heats the food, with little wasted in heating cooking utensils or the oven itself. This results in shorter cooking times and significantly less energy consumption than conventional ovens.

Combination Cookers

Combination cookers are also available. These combine convection and microwave cooking in the same oven. The advantage of these ovens is that food can be browned/ crisped on the outside using convection cooking, while the microwave energy reduces the actual cooking time.

Infrared Annealing Ovens

IR annealing relieves stresses by raising the temperature of the molded part to a target level, much the same as with conventional thermal annealing, but in considerably less time and with significantly less energy consumption. It is a cost-effective process that can result in parts with greater resistance to stress-cracking failure or other problems caused by high internal stresses.

Generally, IR annealing is used to help reduce part stress and can provide an extra measure of safety in applications such as complex medical devices, where stress-cracking failures can be critical. It can also help to improve certain secondary operations, such as plating or coating, by minimising part stress as a variable in the process. Conventional oven annealing can accelerate physical aging, which can lead to embrittlement and loss of impact toughness. By contrast, IR annealing may be less detrimental to physical properties such as impact strength, particularly for polycarbonate.

Infrared annealing ovens

There are many advantages over the conventional convection annealing process. The molded part is placed on the annealing oven's conveyor belt directly from the molding machine. Heat transfer takes place not through convection but by absorption of infrared light by the plastic. The process takes advantage of the residual heat in the molded part eliminating the need to heat up the part to the proper annealing temperature.

The long heating times required in the conventional oven annealing process may require large ovens with higher energy costs, or additional handling and storage facilities, all of which may be avoided with the IR annealing process. Typical infrared annealing times are 2-

3 minutes compared to 4 hours in a conventional convection oven.

Wood Fired Ovens

The dome-shaped wood burning oven in one form or another has been in use for thousands of years. Scores of ancient civilisations dating back to Mesopotamia, ancient Egypt and later Rome, used some variation of this time-proven design to bake breads, meats, seafood and more. These small wonders are considered extremely efficient miniature bakeries. Additionally, in cold climates the presence of these cooking devices provided the occupants of small dwellings with an efficient source of heat. Most ancient civilisations used the clay and straw method of construction. This was time consuming and did not always result in a usable oven. As with pottery, many of their castings cracked during the initial firing.

In modern times, refractory castings have been developed to reproduce the great heating qualities of ovens from long ago. Employing modern methods of casting, and through the use of better materials, these assemblies create ovens that are that have all the desired attributes of wood fired ovens without many of the pitfalls which burdened their predecessors. There is no cooking appliance that will give more pleasure to someone who appreciates cooking than a wood fired oven. Pizza is not the only thing you can do in a wood fired oven. Nothing is better for baking bread, and there's nothing like meats grilled or roasted by a wood fire.

A wood-fired oven will dramatically alter the way you cook, while providing you genuine enjoyment. A wood fired oven quickly becomes the center all gatherings for friends and family in your home. The sight of flames dancing in an oven provides a sense of well-being and warmth in any kitchen even when not in use for cooking.

Incidentally, there's no way that a regular oven will replicate what can be done in a wood fired oven. A pizza cooks in less than three minutes right on the oven floor. The crust will be crisp on the bottom, and the toppings always come out beautifully melted, and perfectly cooked.

Dutch Ovens

The Dutch oven, as we know it today, was developed in the early eighteenth century in England and Holland. It is characterised by three legs designed to straddle live coals, a flat-bottomed bowl with flared sides, a rimmed lid to cradle coals on top, and a bail for lifting. This basic design has remained unchanged for centuries, due, no doubt, to the delectable, tender food it produces. There is no need to alter the perfect pot.

The name, however, has had many variations. The functional titles "bake oven," "bake kettle," and "camp oven" all describe how or when the pot is used; baking and camping. The origin of the more common term "Dutch oven" is more elusive. Some writers have argued that the name originated with German and /or Dutch peddlers who sold the cast iron pots from their wagons. Others have credited the Pilgrims with introducing both the pot and the name to this country as a tribute to their former hosts in Holland. A more likely scenario attributes the origin of the name to cast iron cookware made in Holland and imported in to England in the early eighteenth century, or to a Dutch casting technique patented in England in 1708.

Beech Ovens

Beech Ovens are world leaders in stone· hearth ovens, supplying more than $2 million worth of spectacular ovens to luxury hotels, restaurants and cafes around the world every year. These ovens are entirely Australian, designed and manufactured. Typically, imported ovens are based on a modular design with a limited lifespan, conditional on how carefully the oven has been installed. A modular style oven requires complex installation, which takes days and significantly adds to the cost of the oven.

Beech Oven

The oven comes as a finished unit with an attractive stainless steel outer. It is ready to use immediately or can be built into any architectural finish.

A Beech Domestic Oven has an optional steel stand or can be easily placed onto a specially built bench, or your existing conventional oven benchtop. A Beech Domestic oven can be installed easily on a wooden deck or into a small space. The location of a modular oven is limited by its requirement for a large bench, which is capable of supporting very large weights. The modular oven sits on a concrete slab, alone weighing almost as much as the entire Beech Oven. Allowance then must be made for the weight of the oven and the render, making the completed unit exceptionally heavy. Unique design makes cooking so much easier. The location of the fire below the refractory tile cooking surface gives an even heat distribution, making cooking far simpler. A fire on the cooking surface will always give intense hot spots. Even cooking will then depend upon rotating the food. A fire on the cooking surface also occupies valuable cooking space.

All Beech Ovens can have a gas burner installed for greater temperature control and ease of use. Gas fired ovens cook as efficiently as the wood fired ovens. You can even put wood on top of the burner for flavour only! Thermometer for temperature control. All domestic Beech Ovens have a thermometer installed so you can easily see when your oven has reached the required temperature.

Beech ovens have all been designed by refractory engineers with a view to being easily installed in any location in any country. As a result the designs are simple and strong with no fragile pieces to fit together.

The oven is heavy, however 2 men are able to move the oven should you wish to remodel your outdoor living area or lend it to a friend for a party. You can even put it on the back of your ute or trailer and do outside catering. Due to the efficient design of the oven, it reaches temperature in 1hr rather than the 4- 10 hours required by traditional modular ovens. A stainless steel grill and tray allows you to cook steaks or anything you would cook on a regular conventional oven, while you are waiting for the oven to reach optimum temperature for pizzas. The oven can be used to cook anything you can cook in a conventional oven, with the additional of a wonderful wood fired flavour.

Hotplates in Ovens

Hotplates have several advantages over convection type ovens: decreased bake time, increased reproducibility and better film quality.

Stratification, the formation of different temperature zones, is a problem associated with convection ovens and can severely affect film quality and reproducibility. The heating rate of a substrate depends not only on the heated air flow past a substrate but also on its proximity to other cold substrates.

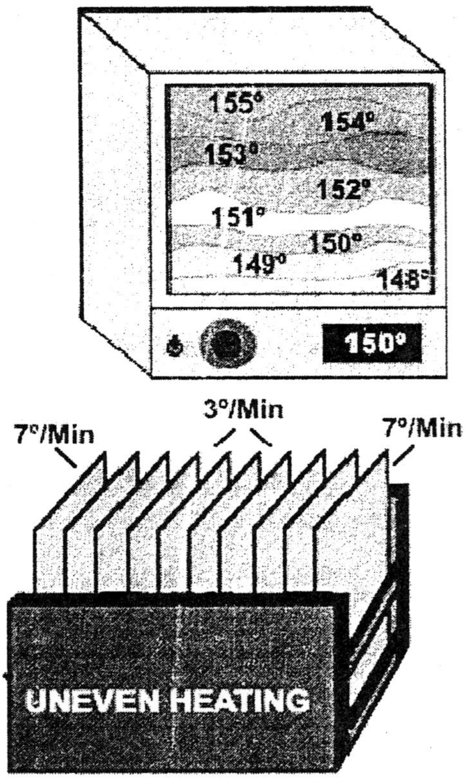

Interior of Conventional Ovens

Thus the heating rate for each substrate in a cassette of substrates being baked, will be less than if each substrate is baked alone. In addition, substrates near the ends of a cassette heat faster than the substrates in the middle, thus producing a non-uniform heating. Particle generation also occurs within a standard oven. In a forced-air, convection oven, substrates are commonly exposed to a flow of particle laden air for at least 30 minutes. During resin film cures, substrates will be exposed to considerable particulate contamination. The substrates are vulnerable since the film may still contain solvents and during this soft' state, the film is very susceptible to having particles adhere to it.

Skin Effect

Another disadvantage in normal oven baking results from baking substrates from the "outside in". Since heat is applied to the outer surface of the film first, a skin forms on the surface of the film thus trapping solvents.

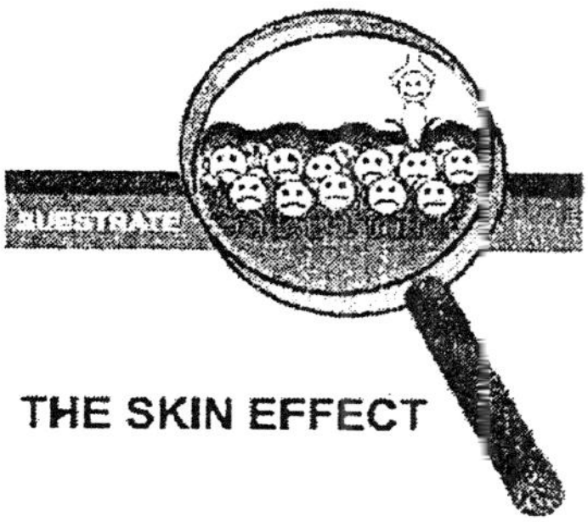

Skin effect

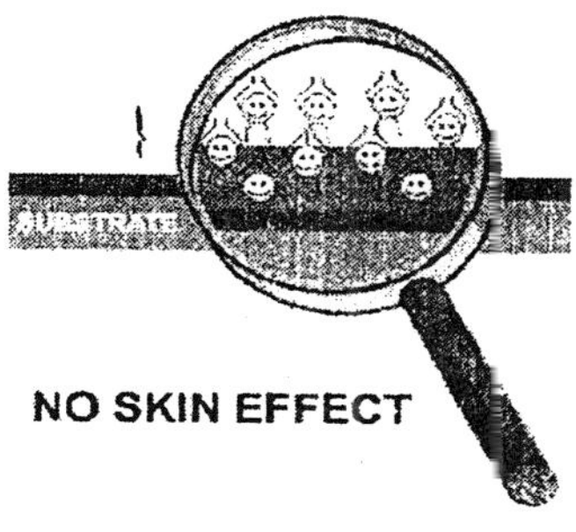

No skin effect

Upon vaporising, these solvents form blisters or bubbles which results in adhesion loss or even bulk film failure. This problem prevails in processes involving thick film resins, e.g. polyimides. No skin effect occurs on a hotplate since hotplate baking heats the substrate from the bottom up. This "inside out" approach offers advantages for thick films since solvents in the film nearest the substrate are baked off before the in surface seals over.

Uniform Baking

A well designed hotplate insures uniform baking across the substrate. Since the substrate intimately contacts a surface of a known constant temperature, it heats at a rate dependent only on the bake style selected and the thermal properties of the substrate. Increased throughput results from a faster warmup of the substrate. Bake times will be measured in seconds, rather than minutes or hours, as in conventional ovens. Reduced vulnerability to particulate contamination is a major advantage of hotplate baking. Only conditioned ambient clean room air passes over the substrate.

Silicon Wafers

Hot Plates virtually eliminate the skin effect with thick films and substantially increase throughput. There are two commonly used resins: positive photoresist and polyimide. These figures should not be use as a rigid guideline, since the best method with a particular baking application can only be achieved through experimentation.

Galium Arsinide

Segments and pieces of GaAs wafers are commonly used in research and pilot lines for economic reasons. The Hot Plates are ideally suited for these circumstances as all Hot Plates offer the proximity bake mode. This is most useful for prewarming GaAs wafers before going to a hard-contact bake and insures uniform heating without thermal shock.

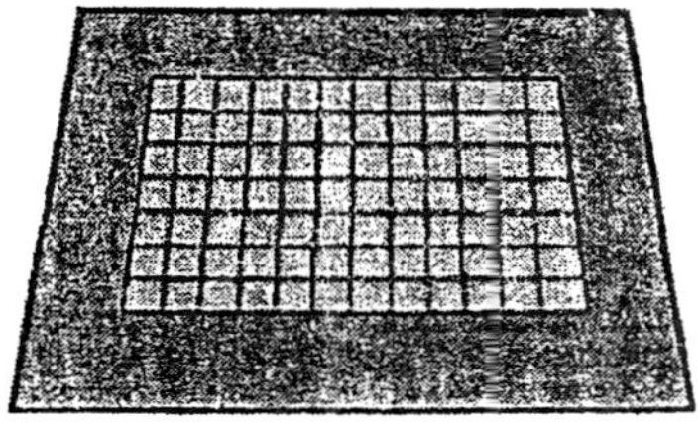

Typical bake processes are identical to those provided above, for silicon substrates.

Photomasks and Displays

Reproducibility and throughput are key issues both for photomasks and for display makers. Because of the large thermal mass of these substrates, oven baking is slow and non-uniform. A proximity bake eliminates back side defects. The Hot Plate typically reduces bake times by 90%. Reproducibility is greatly improved since the rate of heating is not dependent on batch size. All substrates are baked individually.

Exhaust Cover

The design of the exhaust cover promotes the dissipation of vapors removed from a substrate placed on the chuck, without actually drawing air across the chuck surface.

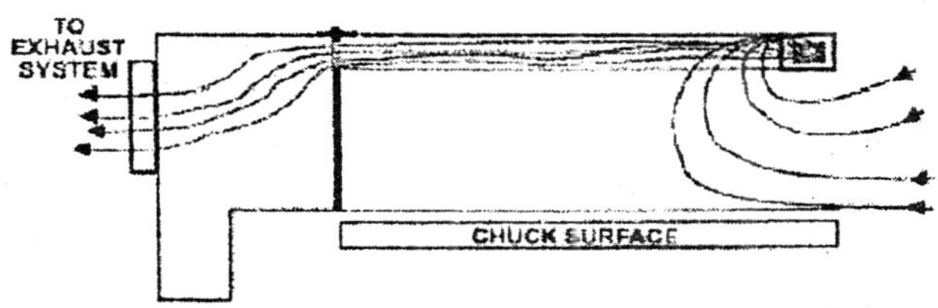

Reflow Soldering

Hotplate baking heats the substrate and the solder without applying heat directly to the devices on the board.

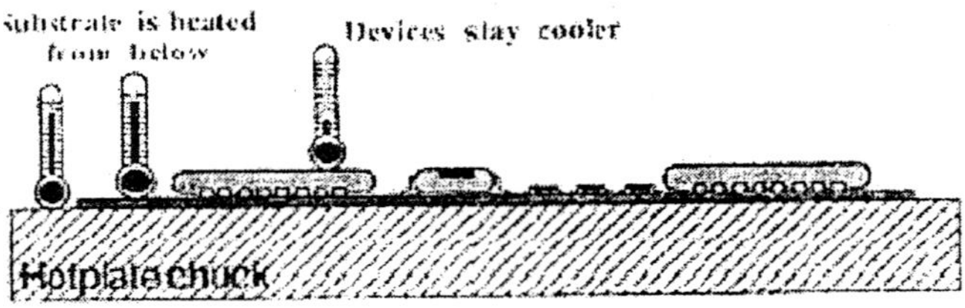

Using a combination of proximity and hard-contact bakes, the bake profile can be adjusted to suit any process.

Spin Coating Process Theory

Spin coating has been used for several decades for the application of thin films. A typical process involves depositing a small puddle of a fluid resin onto the center of a substrate and then spinning the substrate at high speed (typically around 3000 rpm). Centripetal acceleration will cause the resin to spread to, and eventually off, the edge of the substrate leaving a thin film of resin on the surface. Final film thickness and other properties will depend on the nature of the resin (viscosity, drying rate, percent solids, surface tension, etc.) and the parameters chosen for the spin process. Factors such as final rotational speed, acceleration, and fume exhaust contribute to how the properties of coated films are defined.

One of the most important factors in spin coating is repeatability. Subtle variations in the parameters that define the spin process can result in drastic variations in the coated film. The following is an explanation of some of the effects of these variations.

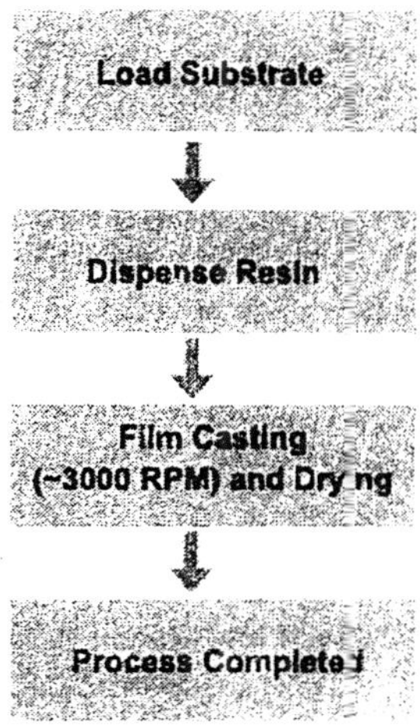

Simple spin process

A typical spin process consists of a dispense step in which the resin fluid is deposited onto the substrate surface, a high speed spin step to thin the fluid, and a drying step to eliminate excess solvents from the resulting film. Two common methods of dispense are Static dispense, and Dynamic dispense. Static dispense is simply depositing a small puddle of fluid on or near the center of the substrate. This can range from 1 to 10 cc depending on the viscosity of the fluid and the size of the substrate to be coated.

Higher viscosity and or larger substrates typically require a larger puddle to ensure full coverage of the substrate during the high speed spin step. Dynamic dispense is the process of dispensing while the substrate is turning at low speed. A speed of about 500 rpm is commonly used during this step of the process. This serves to spread the fluid over the substrate and can result in less waste of resin material since it is usually not necessary to deposit as much to wet the entire surface of the substrate. This is a particularly advantageous method when the fluid or substrate itself has poor wetting abilities and can eliminate voids that may otherwise form.

After the dispense step it is common to accelerate to a relatively high speed to thin the fluid to near its final desired thickness. Typical spin speeds for this step range from 1500-6000 rpm, again depending on the properties of the fluid as well as the substrate. This step can take from 10 seconds to several minutes. The combination of spin speed and time selected for this step will generally define the final film thickness. In general, higher spin speeds and longer spin times create thinner films.

The spin coating process involves a large number of variables that tend to cancel and average out during the spin process and it is best to allow sufficient time for this to occur. A separate drying step is sometimes added after the high speed spin step to further dry the film without substantially thinning it. This can be advantageous for thick films since long drying times may be necessary to increase the physical stability of the film before handling. Without the drying step problems can occur during handling, such as pouring off the side of the substrate when removing it from the spin bowl.

In this case a moderate spin speed of about 25% of the high speed spin will generally suffice to aid in drying the film without significantly changing the film thickness. Each program on spin coater may contain up to ten separate process steps. While most spin processes require only two or three, this allows the maximum amount of flexibility for complex spin coating requirements.

Spin Speed

Spin speed is one of the most important factors in spin coating. The speed of the substrate (rpm) affects the degree of radial (centrifugal) force applied to the liquid resin as well as the velocity and characteristic turbulence of the air immediately above it. In particular, the high speed spin step generally defines the final film thickness. Relatively minor variations of ±50 rpm at this stage can cause a resulting thickness change of 10%. Film thickness is largely a balance between the force applied to shear the fiuid resin towards the edge of the substrate and the drying rate which affects the viscosity of the resin. As the resin dries, the viscosity increases until the radial force of the spin process can no longer appreciably move the resin over the surface. At this point, the film thickness will not decrease significantly with increased spin time. All spin coating systems are specified to be repeatable to within ±5 rpm at all speeds. Typical performance is ±1 rpm. Also, all programming and display of spin speed is given with a resolution of 1 rpm.

Acceleration

The acceleration of the substrate towards the final spin speed can also affect the coated film properties. Since the resin begins to dry during the first part of the spin cycle, it is important to accurately control acceleration. In some processes, 50% of the solvents in the resin will be lost to evaporation in the first few seconds of the process. Acceleration also plays a large role in the coat properties of patterned substrates. In many cases the substrate will retain topographical features from previous processes; it is therefore important to uniformly coat the resin over and through these features. While the spin process in general provides a radial (outward) force to the resin, it is the acceleration that provides a twisting force to the resin. This twisting aids in the dispersal of the resin around topography that might otherwise shadow portions of the substrate from the fluid. Acceleration of spinners is programmable with a resolution of 1 rpm/second. In operation the spin motor accelerates (or decelerates) in a linear ramp to the final spin speed.

Fume exhaust

The drying rate of the resin fluid during the spin process

is defined by the nature of the fluid itself as well as by the air surrounding the substrate during the spin process. Just as a damp cloth will dry faster on a breezy dry day than during damp weather, the resin will dry depending on the ambient conditions around it. It is well known that such factors as air temperature and humidity play a large role in determining coated film properties. It is also very important that the airflow and associated turbulence above the substrate itself be minimised, or at least held constant, during the spin process.

All spin coaters employ a "closed bowl" design. While not actually an airtight environment, the exhaust lid allows only minimal exhaust during the spin process. Combined with the bottom exhaust port located beneath the spin chuck, the exhaust lid becomes part of a system to minimise unwanted random turbulence. There are two distinct advantages to this system: slowed drying of the fluid resin and minimised susceptibility to ambient humidity variations.

The slower rate of drying offers the advantage of increased film thickness uniformity across the substrates. The fluid dries out as it moves toward the edge of the substrate during the spin process. This can lead to radial thickness nonuniformities since the fluid viscosity changes with distance from the center of the substrate. By slowing the rate of drying, it is possible for the viscosity to remain more constant across the substrate. Drying rate and hence final film thickness is also affected by ambient humidity.

Variations of only a few percent relative humidity can result in large changes in film thickness. By spinning in a closed bowl the vapors of the solvents in the resin itself are retained in the bowl environment and tend to overshadow the affects of minor humidity variations. At the end of the spin process, when the lid is lifted to remove the substrate, full exhaust is maintained to contain and remove solvent vapors.

Another advantage to this "closed bowl" design is the reduced susceptibility to variations in air flow around the spinning substrate. In a typical clean room, for instance, there is a constant downward flow of air at about 100 feet per minute (30m/min). Various factors affect the local properties of this air flow. Turbulence and eddy currents are common results of this high degree of air flow. Minor changes in the nature of the environment can create drastic alteration in the downward flow of air. By closing the bowl with a smooth lid surface, variations and turbulence caused by the presence of operators and other equipment are eliminated from the spin process.

Process trend charts

The process trend charts are given below:

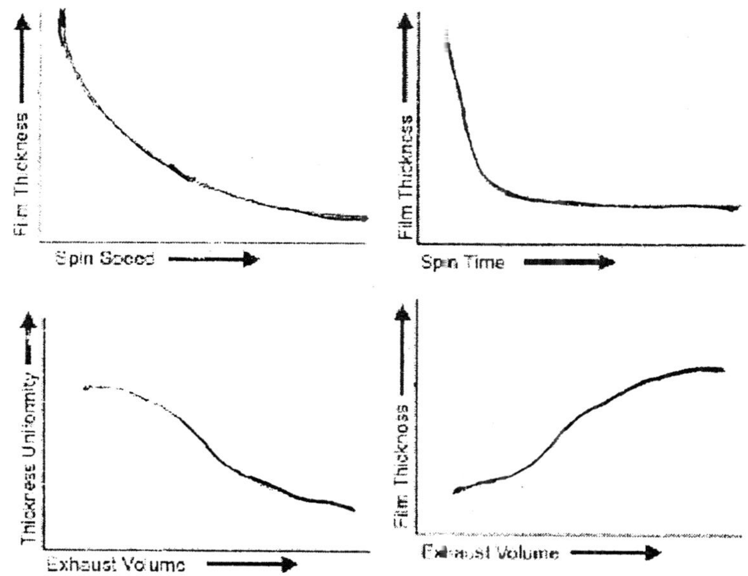

These charts represent general trends for the various process parameters. For most resin materials the final film thickness will be inversely proportional to the spin speed and spin time. Final thickness will be also be somewhat proportional to the exhaust volume although uniformity will suffer if the exhaust flow is too high since turbulence will cause non uniform drying of the film during the spin process.

Spin coater

There are several major factors affecting the coating process. Among these are spin speed, acceleration, spin time and exhaust. Process parameters vary greatly for different resin materials and substrates so there are no fixed rules for spin coat processing, only general guidelines.

Hotplate Bake Process Theory

Hotplate bake processing has increased in popularity since the early 1980s. Previously the most common technique for film drying and curing was the convection oven. Hotplates offer several advantages in the form of increased throughput, increased uniformity and reproducibility and decreased particle contamination. In a typical bake process the substrate is placed into contact with a heated surface of known temperature. The substrate quickly rises to a peak temperature slightly lower than the hotplate surface temperature. Drying and curing steps generally take about 1 minute. This is in contrast to traditional oven processes taking 30 minutes or more.

Variables and methods

A typical bake process consists of preheating the surface to a known temperature, loading the substrate onto the surface for a specific length of time and removing it promptly at the end of the cycle. The selection of the temperature and time values used as well as the bake method employed all affect the overall performance of the process.

Bake temperature

The bake temperature used is dependent on several factors. The material and substrate being baked as well as the results desired are key factors to be considered in developing a bake process. In general hotplate baking will be performed at temperatures slightly higher than those used in oven bake processes. The film being baked will reach a temperature somewhere between the temperature of the hotplate and the ambient air above the film. As an example, with a hotplate surface temperature of 115°C, a layer of photoresist on a silicon wafer will reach a final temperature of about 105°C after a few seconds. Thicker substrates and / or substrates with lower coefficients of thermal conductivity will require even higher temperatures to compensate for this phenomenon.

Another reason for using higher temperatures is to increase process throughput. In oven processes there is a problem commonly known as the "skin effect". This is a result of the outer exposed layer of the film drying and forming a skin before all of the solvents in lower layers have evaporated.

Most oven processes are adjusted to use lower temperatures and bake times measured in minutes and hours to prevent this. During a hotplate bake process the film is baked from the bottom up thus preventing the formation of a skin over the surface. Because of this it is possible to increase temperatures and adjust bake times to be measured in seconds without danger of blistering or cracking in the film.

Bake time

The selection of the bake time parameter plays an important role in the reproducibility of the bake process. Substrate thermal properties and the choice of bake method greatly affect the amount of time necessary for the substrate and therefore film temperature to stabilise during the bake. Thicker substrates and the use of proximity bake methods will increase the time necessary for the film to reach its final temperature. It is important that most of the baking action in the film takes place after this temperature is reached.

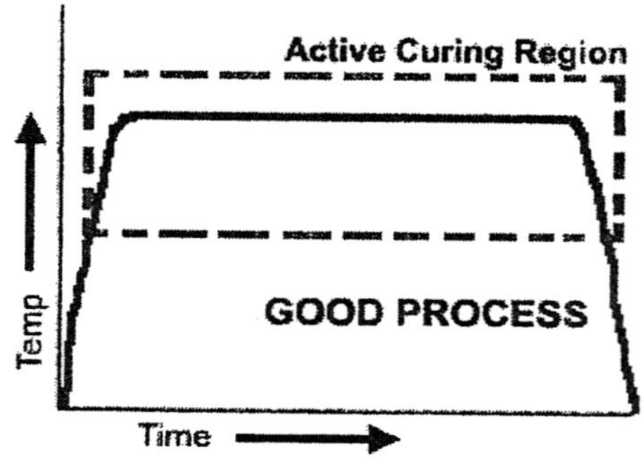

A silicon wafer will reach a stable temperature within a few seconds and so it is traditional to adjust a photoresist bake processes to be completed in 60-90 seconds with an appropriate bake temperature. For thicker substrates such as photomasks and ceramic modules the increased time necessary to heat the larger mass of the substrate results in bakes times approaching five minutes. It should be noted that these substrates can be processed with higher temperature and much shorter bake times but reproducibility may suffer.

If the bake time is too short then a significant amount of the actual bake process will take place during the loading and unloading steps as well as while the substrate is cooling after removal from the hotplate. This is an unstable condition since it is very difficult to exactly reproduce conditions during these steps.

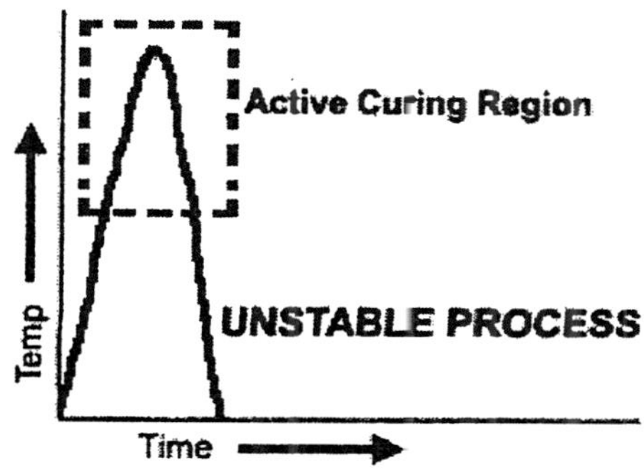

In general the temperature-time relationship in a bake process can be taken as a "dose" of the (temperature) x (time) product. Increasing the bake temperature results in a need for decreasing bake time. The limits for both of these parameters can be considered to be reached when the process is no longer reproducible or when the physical temperature limitations of the resin or substrate have been reached.

Bake method

Another important factor is the method of bake. Hotplates allow for three distinct bake methods. These are Proximity, Soft contact and Hard contact. The choice of bake method is programmable and up to three consecutive bake steps may be programmed into the onboard microcomputer.

— proximity bake,

— soft contact bake and

— hard contact bake.

Hard contact bake: In a hard contact bake the substrate is clamped onto the hotplate surface by the application of vacuum to the underside of the substrate. Small holes are machined into the hotplate surface in a pattern which optimises vacuum distribution without the formation of cold spots or warping of the substrate.

This method is usually preferred for silicon and other .at substrates where back side contact is not a problem. Soft contact baking uses gravity alone to hold the substrate onto the hotplate. This method generally offers less uniformity since the substrate-hotplate thermal interface is not as efficient as in hard contact baking and can be somewhat random in variation.

Proximity baking: Proximity baking is accomplished by forcing heated gas (usually nitrogen) through the same ports in the hotplate surface that are used for vacuum in the hard contact method. This forces the substrate to .oat at a distance of one to four mils (25-100 μm) above the hotplate surface. Proximity baking allows a slower warm-up than contact bake methods and can be advantageous when baking thick films where blistering would otherwise be a problem.

Another advantage of proximity baking in this manner is that in many cases cambered or warped substrates can be baked with a high degree of uniformity. This is usually not possible with the contact methods since it is not possible to achieve a vacuum under a substrate that is not .at to start with.

Soft contact baking: Processing cambered substrates with the soft contact method creates hot spots where the substrate touches the hotplate and cold spots where it does not. It should be noted as well that this type of proximity process is "self-leveling" in that the substrate will tend to form a uniform gap to the hotplate surface. This is a significant advantage not found in "pin lift" type systems.

Proximity baking also offers the unique advantage of allowing hotplate processing without touching the bottom side of the substrate. An example of this application is photomask processing. In processing these relatively thick glass plates it is important that the back side of the glass not directly touch the hotplate since this causes microfractures in the glass itself from rapid heating. By performing the entire bake process in the proximity mode the integrity of the substrate is not endangered and the uniformity is excellent.

Advantages of Ovens

Their advantages are many: In addition to radically reduced cooking time, (enhanced by the fact that there's no need to preheat), the absence of radiant heat in microwave cooking results in a cooler, more comfortable kitchen environment. Basic microwave operation is simple and safe enough that even children can prepare their own meals and snacks without assistance or supervision ¿ within reason, of course. Microwave cooking uses far less energy than conventional oven cooking. And, surprisingly, microwaves' increased cooking speed allows the resulting preparations to retain more nutrients than foods prepared by more traditional cooking methods.

All in all, microwaves are a natural fit for our time- and convenience-obsessed culture. And while the first home microwaves were considered a luxury or indulgent curiosity, the ensuing decades have produced a steady stream of refinements. Today's microwaves are sleeker, quieter, more powerful and efficient, and much less expensive than their forebears. In addition, they're much more versatile, with new high-tech features that enhance their ability to prepare more than just popcorn and boxed frozen entrees.

Culinary Convenience

At the heart of every microwave oven is a magnetron. It sounds like a doomsday device from a bad '50s sci-fi movie, but it's really just a specialised vacuum tube that converts electricity to radio-frequency (RF) energy. When this energy (called microwave radiation) comes in contact with food, it excites the molecules within, causing them to vibrate rapidly. The resulting friction generates heat, which cooks the food.

Contrary to a popular myth, microwaves do not cook "from the inside out," but from the surface inward just like the radiant heat from a conventional oven. The actual microwaves cannot penetrate further than about an inch from the surface of food; the inner parts of the food are cooked (as with conventional ovens) by radiant transmission of the heat from the outer layers. Metal reflects microwave energy, and since microwaves move in a straight line and cannot bend around objects, they are typically dispersed by metallic "stirrer fans" and bounced off metal surfaces within the oven cavity for more even cooking. Rotating trays, called turntables or carousels, are commonly used to help promote more even absorption of the microwaves into the food.

When selecting a microwave, one prime consideration is capacity (expressed in terms of cubic feet). Microwaves typically range in size from compact, countertop models of about 0.5-0.7 cubic feet, to monster over-the-range ovens of 2.2 cubic feet or more. The former are suitable for small reheating tasks and the old standby, microwave popcorn preparation, while the behemoths are intended for serious microwave chefs who intend to "nuke" complete family meals. When comparing capacities, bear in mind that not all ovens of a given cubic footage are created equal. Pay attention to specific interior dimensions; for example, a narrow, tall microwave will accommodate a roast, while a squat, broad cavity is better suited for a large casserole or multiple dinner plates. The shape of the cavity is worth considering as well—some ovens feature rounded cooking areas and convex doors that will allow turntable-mounted foods to rotate freely without colliding with the interior surfaces. These are also easier to clean due to the absence of recessed corners.

Absolute Power

The other primary differentiator among microwaves is power (expressed in wattage). Small countertop models may offer as little as 400-500 watts of cooking power, whereas higher-end and larger ovens can deliver as much as 1400 watts or more. The higher the wattage, the faster the oven will cook—although all but the least expensive models allow you to employ reduced wattages for specific tasks like reheating and defrosting.

Bells and Whistles

Once you've narrowed down your selection in terms of capacity and maximum wattage, it's time to consider the extra touches that can make your microwaving life more convenient and, ultimately, more satisfying. Here are a few common features to look for:

Turntable

A rotating platform helps ensure even cooking and eliminates the need for hands-on stop, rotate and restart rituals. Many models feature built-in turntables, but these often affect "headroom" within the oven, so consider a removable plate or a recessed turntable mechanism that maximises usable cooking space. Accessory turntables are available for microwaves that don't include them as standard equipment.

Sensor Cooking

This convenient feature makes it easier to reliably cook food to the desired degree of doneness on the first try, without the risk of overcooking. Sensing methods vary, from humidity sensors that gauge cooking progress by the amount of moisture that escapes during the heating process, to probe-based systems that actually monitor internal temperatures.

Integrated Browning Element

Although microwaves cook very rap[illegible] and efficiently, they're not particularly good at browning or crisping. This feature solves the problem by incorporating an actual radiant heating element inside the oven to impart the desired pleasing color and crispness to foods that require it. Ovens with a browning element typically also feature enameled interiors to facilitate easier cleanup of baked-on spills and splatters.

Exhaust Fan and Range Light

Over-the-range microwaves usually include these features, since they are generally installed just under the hood of a conventional oven and thus pre-empt built-in range fans and lights. Deluxe models may include multiple fan speeds and convertible venting systems for use in kitchens with or without exterior ventilation.

Adjustable Power Levels

Most microwaves now include this feature, designed to facilitate proper preparation in cases where a more delicate touch is warranted. Less expensive ovens may offer as few as 3 distinct power levels; more deluxe models offer greater precision, in the form of 10 or more levels. Most achieve the variations by cycling full power on and off in timed increments; more sophisticated systems actually provide constant microwave bombardment at a reduced intensity.

Preset Controls

Also common on today's microwaves, this feature allows for one-touch cooking of commonly prepared foods. Typical presets include defrost and/or reheat settings for various foods, as well as ready-to-eat preparation of popular meals like whole chickens, ground beef and fresh or frozen vegetables in various quantities. Ritzier models may also provide for quick execution of preliminary steps in more elaborate culinary undertakings, such as softening refrigerated cream cheese, or melting chocolate or butter.

Quick-start Settings

This feature provides a simple way to engage the oven at full power for short time periods (usually, 30 seconds to one minute). Convenient for quick heating of small dishes or beverages, this option also allows you to extend cooking times in short increments to "touch up" entrees to taste.

Keep-warm or Simmer Functions

Ovens with these controls allow the oven to continue running at reduced power after cooking is complete, so no last-minute reheating is necessary if you're interrupted just prior to mealtime.

Child Lock

Some ovens provide a passcode-based electronic lockout system that deactivates the panel controls to prevent innocent accidents—or thwart outright mischievous experimentation by budding young scientists.

Oven Related Energy Saving

Opportunities of improving the energy efficiency of oven operations include:

- *Fill the oven*—Whenever possible, try to schedule oven use so that it will be full.
- *Limit oven preheat times*—Preheat ovens no longer than the manufacture's recommendation—usually 10 to 15 minutes. Letting the oven preheat longer than needed increases energy cost. A high preheat setting will not heat the oven any faster but it may overheat the oven leading to wasted energy.
- *Maintain oven seals*—Make sure the seals around oven doors are in good shape, and seal snuggly with the face of the stove. Worn or loose-fitting seals allow heat from the oven to escape. Clean out debris at the stove bottom, especially in the front where the door seal meets the stove. The accumulated material can prevent the door from sealing well.
- *Use ceramic or glass pans*—Using ceramic or glass pans will allow food to cook at an oven temperature 25°F lower than would be needed for metal pans.
- *Choose an efficient oven*—The efficiency of ovens varies widely with how they're operated, but there are general comparisons that can be made. The following tables provide a comparison of the efficiency and relative operating cost of the different types. You may find that the type of oven you're using isn't as economical as you thought.

— *Maximize use of microwave ovens*—They are the most efficient, fastest and lowest cost. Use them where ever possible. They heat only the food and save significant energy compared to standard ovens. Microwave ovens operate even more efficiently if the inside is kept clean.

— *Convection ovens*—Convection ovens use fans to circulate hot air around food as it cooks. The motion of the air increases the rate at which heat is transferred to the food. As a result, convection ovens cook more quickly and at lower temperatures to achieve the same result as conventional ovens. According to one manufacturer, convection ovens cook in 10 percent less time at 25°F lower temperature. Convection ovens can also be loaded more fully than conventional ovens due to the increased circulation leading to additional productivity improvements.

Oven Repair and Care

Do not store plastic items or other utensils in oven as they may melt or burn if the oven is accidentally turned on with them inside. Use large enough cooking pans to avoid boil-overs. If spills occur, wipe them up promptly to avoid baking on.

Do not put large pieces of foil on oven floor or racks unless appliance manual recommends it. Then follow manual instructions as to size and placement of foil exactly. Foil in the oven, especially on the racks, may slow cooking and reduce browning. Do not line broiler pan with foil, as it concentrates heat and may damage the pan.

Fill a small glass bowl with 1/2 cup full strength ammonia, place in oven and close. Let stand overnight, then wipe loosened dirt with paper towels or newspapers. If then necessary, rub surfaces with a suitable abrasive, such as fine steel wool or soap-filled steel wool pad, wiping off "suds" with paper towels. Then wash with warm soapy water and rinse. Repeat the process if necessary.

Commercial oven cleaners are helpful if ovens are very soiled, but they should be used with caution. These cleaners can damage surfaces outside and around the oven. Be sure to protect these areas with layers of newspaper or other materials and cover your hands with protective gloves. If using most commercial oven cleaners, never spray in a hot oven (over 200 F.) which will make it even more caustic and can corrode surfaces. Never spray on oven light, electric elements, or pilot light in older gas ranges. Turn off the pilot light when using spray oven cleaners.

Removable parts, such as broiler pans and racks, can be cleaned more easily if allowed to soak in a sink or pan of sudsy water to which a little ammonia has been added. A lot of soaking is better for the surface, and easier, than a lot of scrubbing. Never soak aluminum in ammonia or other alkalis.

If necessary, you can scour oven racks or porcelain enamel with steel wool or a scouring pad to remove baked- on grease or food spills which have not come off in regular cleaning. Occasional use of such abrasive pads, or scouring powder on badly baked-on soil is OK, but don't use these too often or scrub too hard for you can scratch the smooth surface and make it harder to clean from then on. Do not scour mirror-finished metals, glass, or baked-on enamel.

Bibliography

Allen Bob, "A Guide to Collecting Cookbooks", *Antiques & Collectibles,* 1990

Braker Flo, "Simple Art of Perfect Baking", *Cooking,* 2003.

Braker Flo, "Sweet Miniatures: The Art of Making Bite-Size Desserts Completely Revised and Expanded", *Cooking,* 2000

David Elizabeth, "French Provincial Cooking", *Cooking,* 1999

Edwards Linda, "Baking for Health", *Health & Fitness,* 1988.

Galli Franco, "The Il Fornaio Baking Book: Sweet and Savory Recipes from the Italian Kitchen", *Cooking,* 2001

Hamelman Jeffrey, *Bread: A Baker's Book of Techniques and Recipes,* Wiley 2004.

Kleinman Kathryn, "Birthday Cakes. Recipes and Memories from Celebrated Bakers", *Cooking.* 2004

Klivans, Elinor, "Big Fat Cookies", *Cooking,* 2004

Lansky Vicki, "Baking Soda: Over 500 Fabulous, Fun, and Frugal Uses You've Probably Never Thought of", 2004

Miller Marcianne, "The Artful Cupcake: Baking & Decorating Delicious Indulgences", *Cooking,* 2004.

Morgan, Aaron, "The Artful Cookie Baking & Decorating Delectable Confections", *Cooking,* 2004

Ojakangas Beatrice A., "The Great Scandinavian Baking Book", *Cooking,*1999.

Patent Greg, "Baking in America: Traditional and Contemporary Favorites from the Past 200 Years", *Cooking,* 2002.

Reinhart, Peter, "Crust & Crumb: Master Formulas for Serious Bread Bakers", *Cooking,* 1998

_______________, "The Bread Baker's Apprentice: Mastering the Art of Extraordinary Bread", *Cooking,* 2001.

Waters Alice, "The Cheese Board: Collective Works: Bread, Pastry, Cheese, Pizza", *Cooking,* 2003

Wood Ed, "Classic Sourdoughs: A Home Baker's Handbook", *Cooking,* 2001

INDEX

Adjustable measuring cups 181
Aluminum foil 241
Angled measuring cups 183

Baked goods 1
Baking powder 55
Biscuits 8
Boston brown bread 61
Bread baking 43
Bread Pan 250
Brownies 8

Cake pan 252
Casserole dishes 241
Chain stores 3
Clear flour 42
Coffee cakes 67
Coffee rolls 68
Conventional ovens 262
Cornbread 8
Cream pies 126
Cylinder type measuring cup 181

Date bread 61
Desserts 1
Dinner rolls 64
Disintegrator 39
Dough 53
Dought 53
Dry toast 73
Dutch apple cake 70
Entire wheat bread 58

Flan pan 254
French bread pan 254
French rusks 66
Fresh lemon chiffon pie 132

German caraway bread 59
German coffee bread 67
Glass cups 180
Glass ovenproof bakeware 243
Graded cups 179
Graham bread 59

Hard anodized bakeware 242
High milling process 38
Hot cross buns 162
Hungarian system 38

Imperial muffins 73
Indian bread 62
Insulated bakeware 244

Jelly roll pan 255

Lemon meringue pie 135
Liquid measuring cups 182
Low milling process 38

Maillard reaction 1
Malt cereal syrup 48
Measuring cups 179
Measuring spoons 178
Metal individual cups 180
Microwave ovens 264
Muffin tin 256

Non-stick bakeware 245

Parker house rolls 63
Pastry blender 119
Patent flour 42
Pie plate 257
Pies 1
Poppy seeds 49

Rolled oats bread 60
Roller milling 38
Rope 52
Rusks 67
Rye biscuit 60
Rye flour 43

Salmonellosis 10
Scoops 183
Seame seeds 49
Sensor cooking 288
Shortenings 47
Silicone bakeware 246
Silver anodized bakeware 242
Single measure cups 180
Springform pan 259
Squash biscuits 72
Stainless steel bakeware 247
Stale bread 50
Staling of crumb 49
Staphylococci 10
Starch retrogradation 49
Sticks 64
Stone bakeware 247

Tart pan 259
Third bread 59
Tinned steel bakeware 248
Tube pan 260
Two-crust cherry pie 137

Water bread 57
Water toast 74
White flour bread 59
Wood fired ovens 266